Surface Preparation and Cleaning Conference 2016 (SPCC 2016)

Santa Clara, California, USA
19 - 20 April 2016

ISBN: 978-1-5108-2572-7

Printed from e-media with permission by:

Curran Associates, Inc.
57 Morehouse Lane
Red Hook, NY 12571

Some format issues inherent in the e-media version may also appear in this print version.

Printed by Curran Associates, Inc. (2016)

For permission requests, please contact Surface Preparation and Cleaning Conference (SPCC) at the address below.

Surface Preparation and Cleaning Conference (SPCC)
c/o Mark Thirsk
Linx Consulting-PO Box 384
Mendon, MA 01756-0384
USA

Phone: 617-273-8837

mthirsk@linx-consulting.com

Additional copies of this publication are available from:

Curran Associates, Inc.
57 Morehouse Lane
Red Hook, NY 12571 USA
Phone: 845-758-0400
Fax: 845-758-2633
Email: curran@proceedings.com
Web: www.proceedings.com

TABLE OF CONTENTS

SESSION 5

SESSION 6: CONTAMINATION FREE MANUFACTURING

SESSION 6: PANEL DISCUSSION – THE FUTURE OF FORMULATED CHEMISTRIES

POSTERS

Emerging Interconnect Technologies

Prof. Krishna Saraswat

Department of Electrical Engineering
Stanford University, Stanford, CA, USA

Outline

- Scaling limits of interconnects
  - ➢ Alternatives to Cu
- Performance simulations
  - Cu, CNT, optical interconnects
- Technology for novel optical interconnects
- 3-D integration
- Summary

Effect of Scaling on interconnect performance

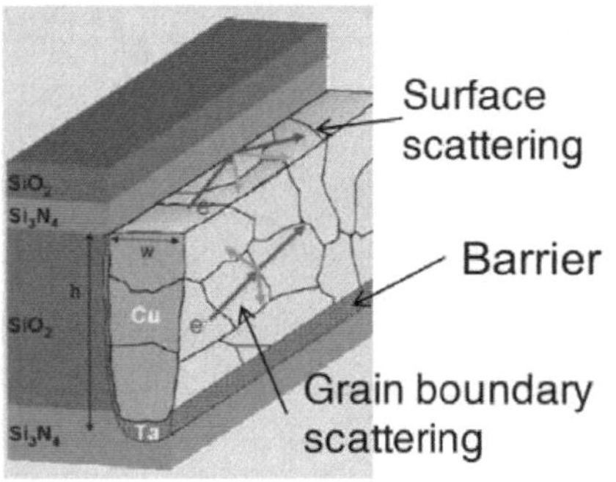

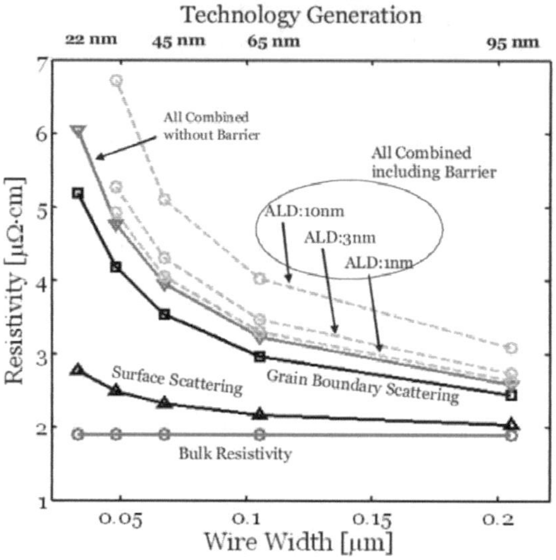

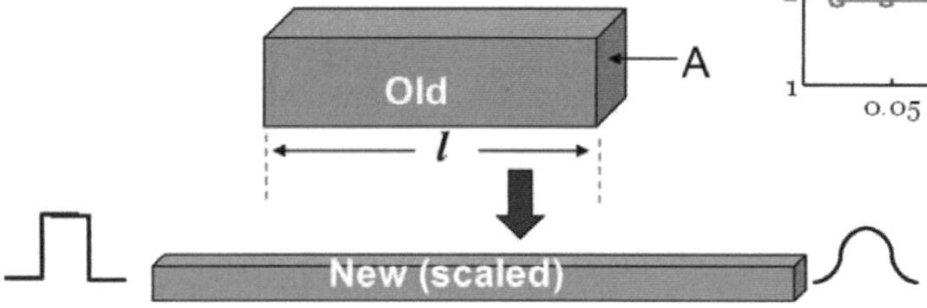

Bit rate $\propto A/l^2$

Delay $\propto l^2/A$

Power dissipation $\sim CV^2f$

K.-H. Koo and K. C. Saraswat, in Nanoelectronic Circuit Design, Springer, 2011, pp. 377-408

Graphene vs. TaN Barrier for Cu

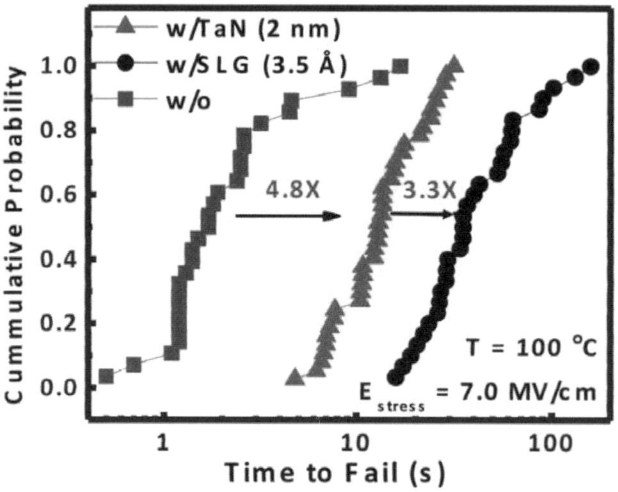

L. Li … H.-S. P. Wong, *Symp. VLSI Tech.* 2015
L. Li … H.-S. P. Wong, *ACS Nano* 2015

Thinner barrier: 3.5 Å single layer graphene is better than 2 nm TaN

How can we improve interconnect performance?

– **Carbon nanotubes/ Graphene**

– **Optical interconnects**

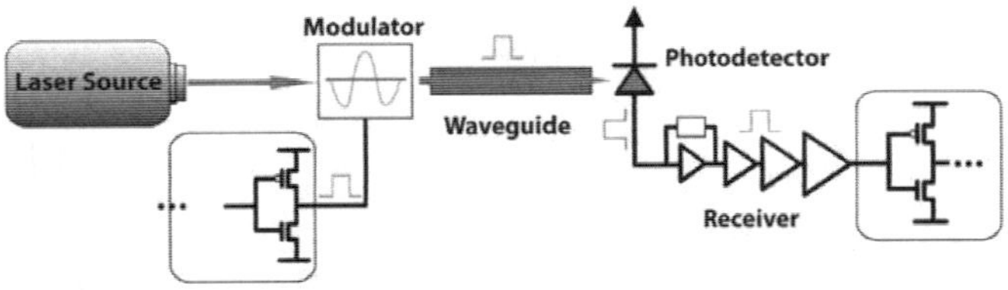

– **3D**

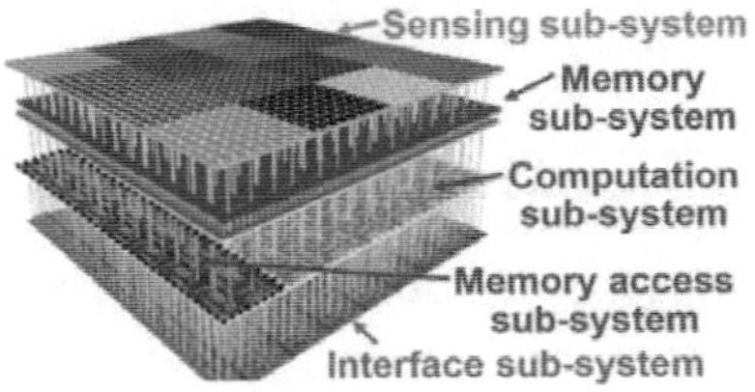

Optical & Electrical Wires: Schematic

On-chip Electrical Interconnect (Cu or CNT)

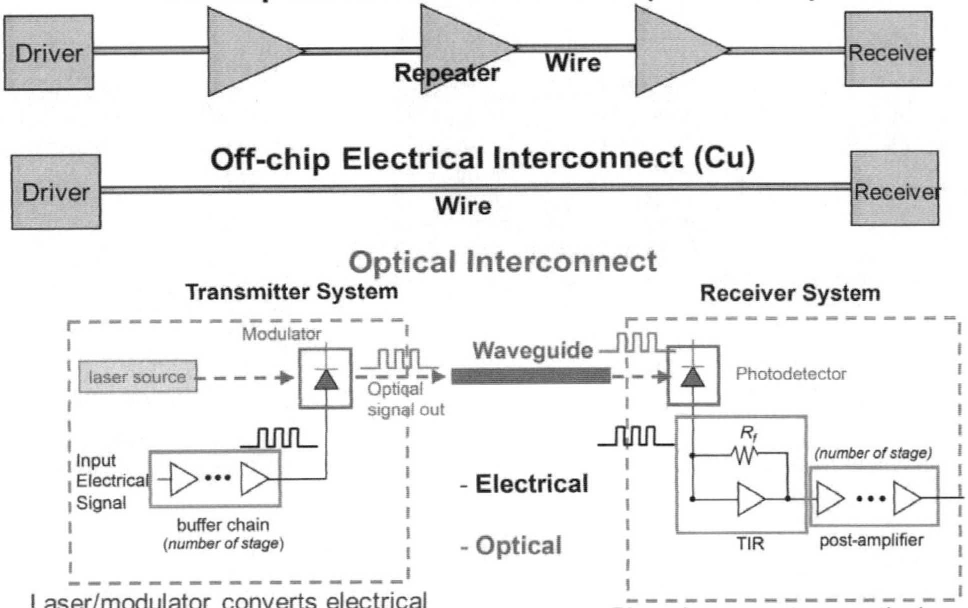

Laser/modulator converts electrical signal into optical signal

Photodector restores optical signal into electrical signal

Carbon Nanotubes

1-D conductors:

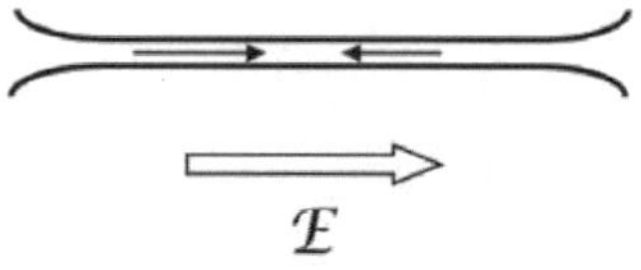

Quantum Wires:

- 1D system with limited density of states. Hence quantum effects play an important role in determining the values of R, L and C
- Mean free paths as large as 1.6μm.

3-D conductors:

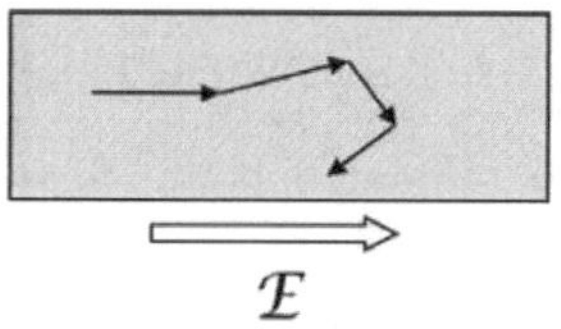

Conventional wires :

- Backscattering through a series of small angle scatterings.
- Mean free paths ~ 30nm.

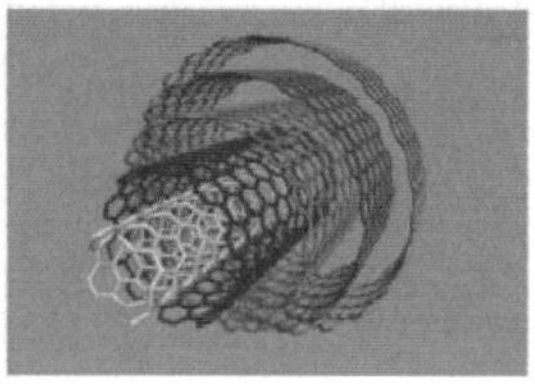

Potential Candidates for GSI Interconnects.

Formidable Task: Dense Bundles of SWNTs

Horizontal Growth for Interconnects

Y. Nishi and H.-S. Philip Wong (Stanford)

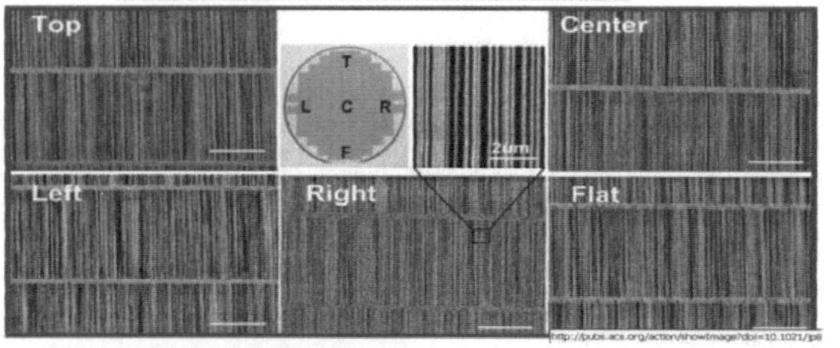

Vertical Growth

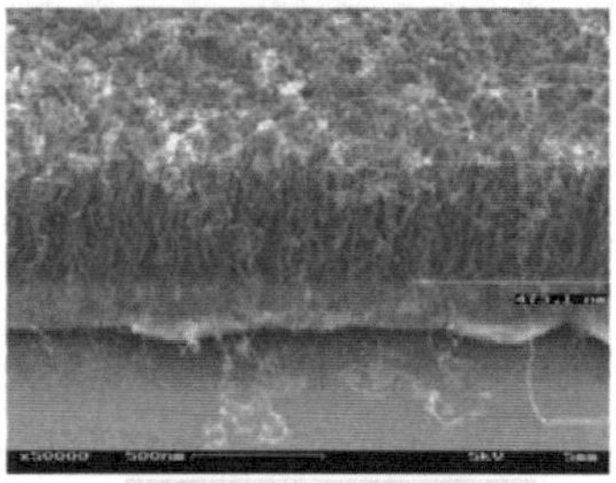

C.V. Thompson, MIT

256-Element CNT Ring Oscillator

H.-S. Philip Wong (Stanford)

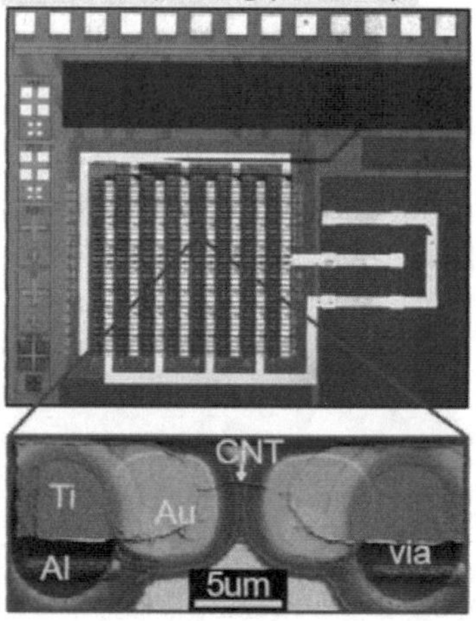

Promising progress in creating aligned isolated SWNTs by transferring SWNTs grown on sapphire to other substrates

Potential reliability performance comparison

- **Good thermal conductivity**
  - Graphene: $4.84\times10^3 \sim 5.30\times10^3 W/mK$
  - CNT: $1.75\times10^3 \sim 5.80\times10^3 W/mK$
  - Copper: $385\ W/mK$

- **High breakdown current**
  - Graphene: $\sim 10^8\ A/cm^2$
  - CNT: $\sim 10^9\ A/cm^2$
  - Copper (EM threshold): $\sim 10^7\ A/cm^2$

Latency and Energy/bit vs. Wire Length

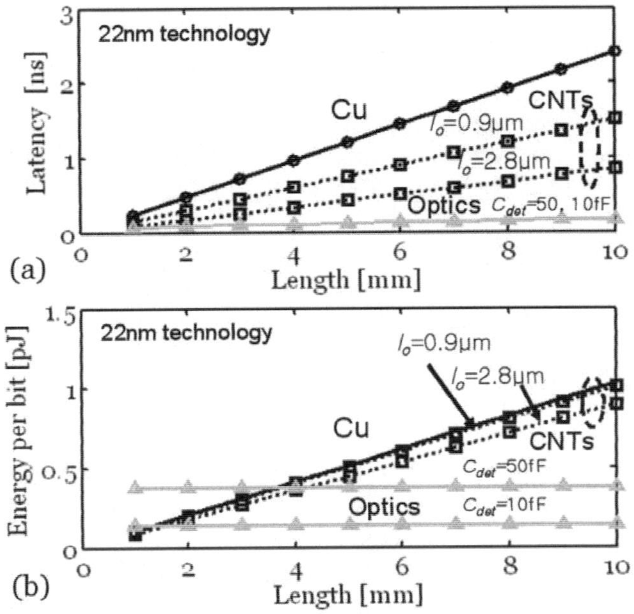

- Cu, CNT: small wire width → more repeaters, wire capacitance → latency ↑
- CNTs are favorable for shorter wires
- Optics favorable for longer wires

- Cu, CNT: small wire width → Energy per bit decreases as wire pitch is scaling (CV^2)
- CNTs is favorable for shorter global wire
- Optics: transmitter receiver power ↓
- Optics favorable for longer wires

Koo, Kapur and Saraswat, IEEE Trans. Electron Dev., Sept. 2009

Comparison Study: Global Interconnect CNTs, Cu, Optics

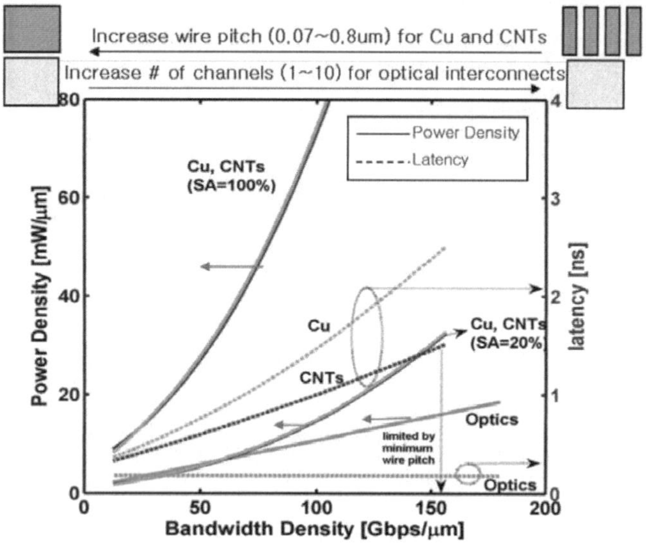

- **BW density**

  Cu and CNTs: f_{clk} / $pitch_{wire}$

  Optics: no. of wavelength of WDM

- **Power density**

  Cu & CNTs (non-linear):

  Cap and wire pitch

  Optics (linear):

  no. of wavelength channel

- **Latency**

  Optics < CNTs < Cu

Koo, Kapur and Saraswat, IEEE Trans. Electron Dev., Sept. 2009

Off-Chip Interconnect Performance: Electrical vs. Optical

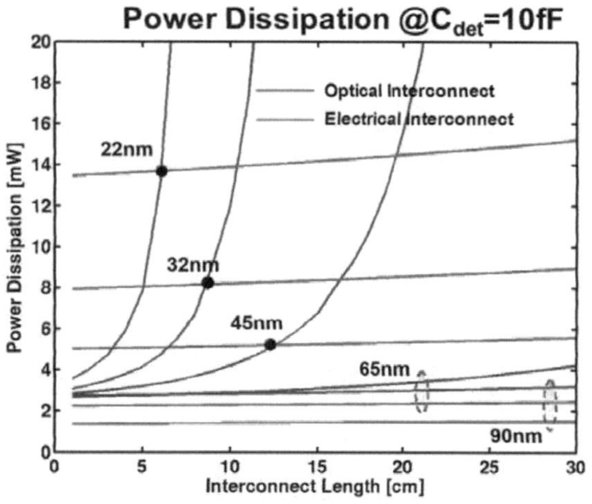

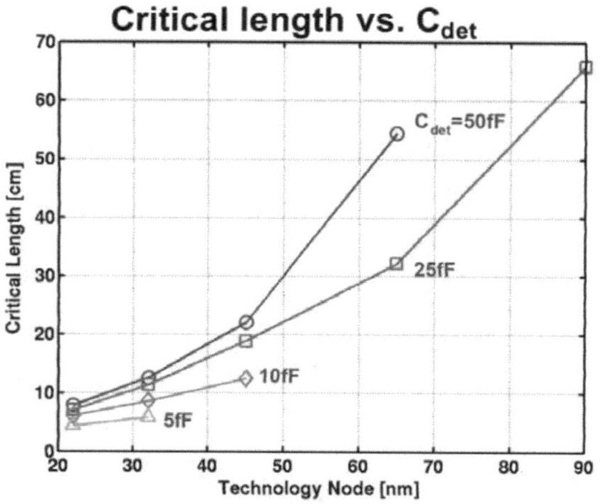

- **Beyond certain length optical I/O is more power efficient**
- **Critical length decreases at higher bit rate & lower detector capacitance**
- **Beyond 32nm Technology node critical length < 10cm**

Cho, Kapur and Saraswat, IEEE IITC, June 2005

Why Really Photonics?

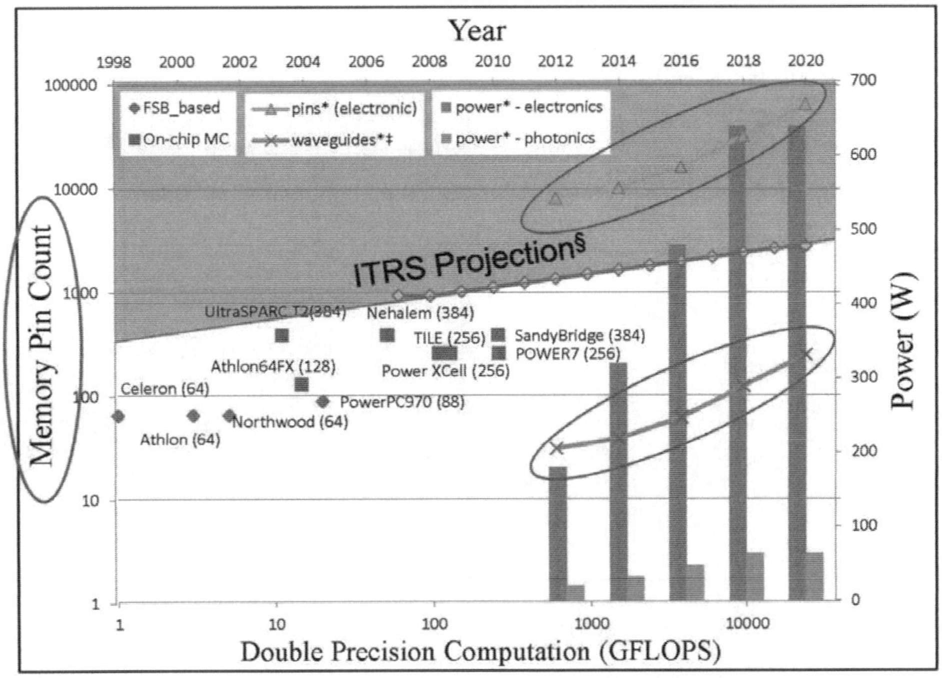

Source: Keren Bergman, Columbia University

Communication Dominates Power

On-Chip Power Breakdown 50nm node

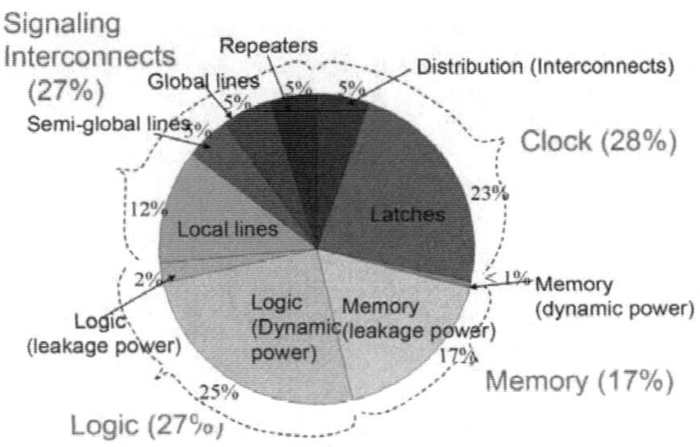

More than half the power can be attributed to interconnects

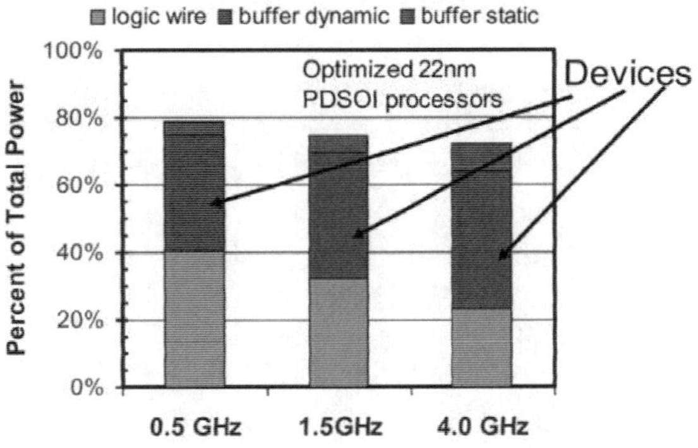

70-80% of total logic power is for communication

– Need proper consideration of wires!!

Chandra, Kapur and Saraswat, IEEE IITC, June 2002

Wilfried Haensch, (IBM) Data Abundant Systems Workshop, Stanford Univ., April 2014

Why Off-chip Photonic Interconnects

Copper wires

Optical Interconnects

- Copper wires are reaching physical limits
- Photonic interconnects offer the solution for the future

The Interconnect Problem

"For the long term, material innovation with traditional scaling will no longer satisfy performance requirements. Interconnect innovation with optical, RF, or vertical integration ... will deliver the solution" (ITRS)

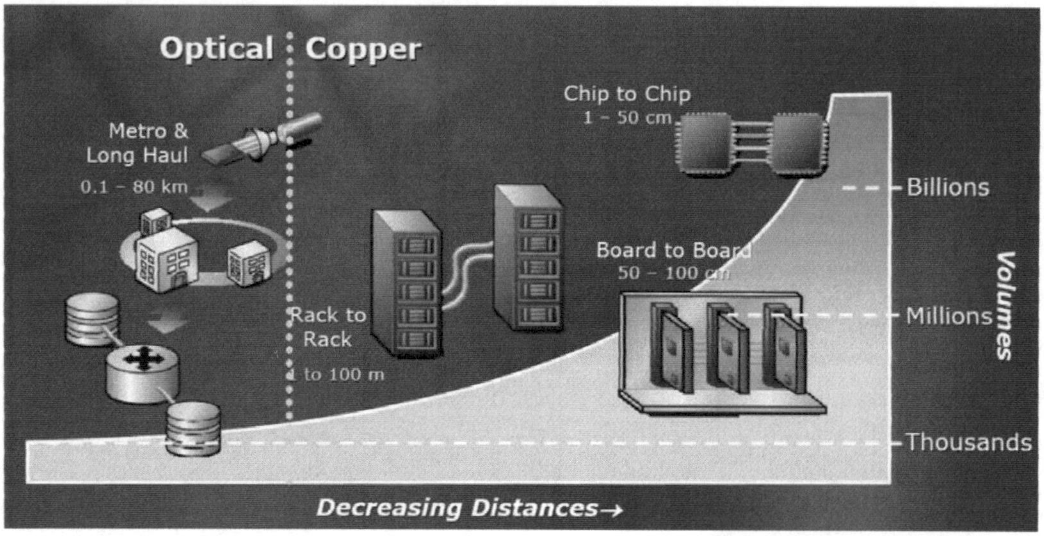

Source: Intel

Material Options for Optical Interconnects

- What are the right optical devices to use?
  - Need to be cheap, available in large numbers
  - Compatible with CMOS
- Silicon devices are a long shot
  - Need 3D heterogeneous integration
- Flip bond III-V to Si CMOS
  - Current process
  - Cost, resources, yield?
- How about germanium?
  - Bandgap ideal for λ = 1.5 μm
  - Can be monolithically integrated on Si
  - Becomes direct bandgap material by straining or adding tin

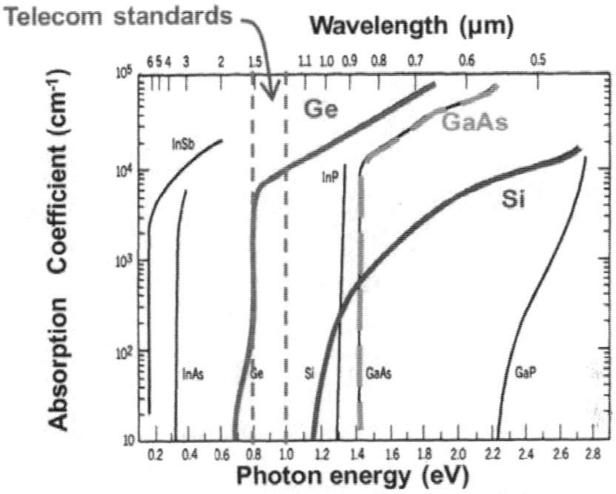

Selective and Lateral Overgrowth of Ge on SiO_2

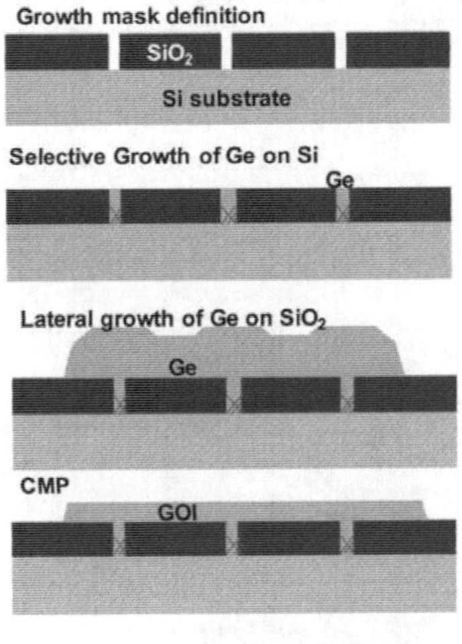

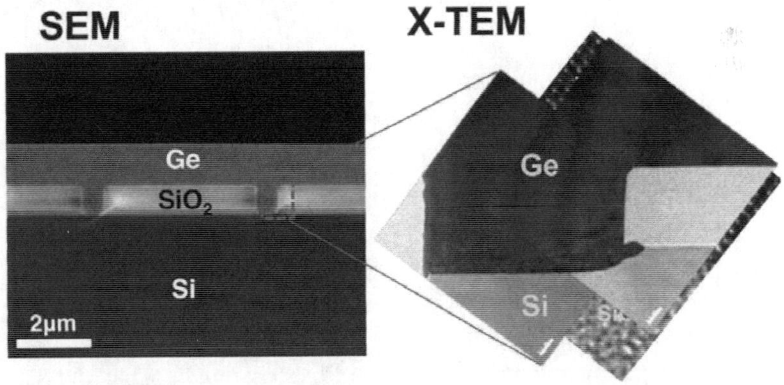

400°C deposition + 800°C H_2 anneal + 600°C deposition

- Lateral Ge growth on SiO_2 window achieved
- Dislocation density of Ge on SiO_2 < 10^6 cm^{-2} (same thickness Ge on Si: 1×10^8 /cm^2)
- Surface RMS roughness ~0.4 nm after CMP
- Ge is ~0.2% tensile strained due to thermal mismatch with Si

Ju Hyung Nam...........K. Saraswat, J. Crystal Growth, April 2015

Si Compatible Photonic Interconnect

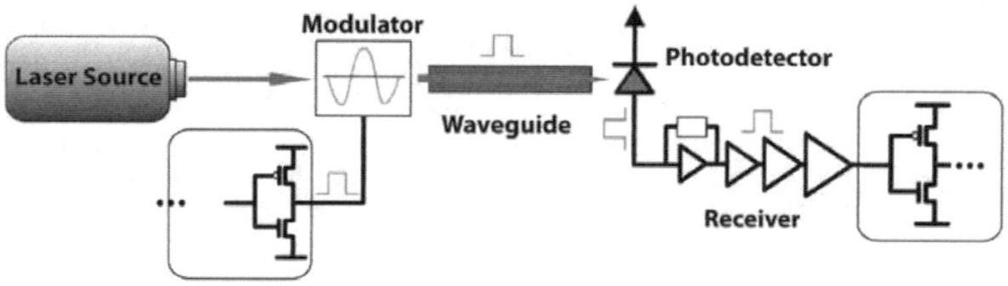

Laser/modulator converts electrical signal into optical signal

Photodector restores optical signal into electrical signal

Laser

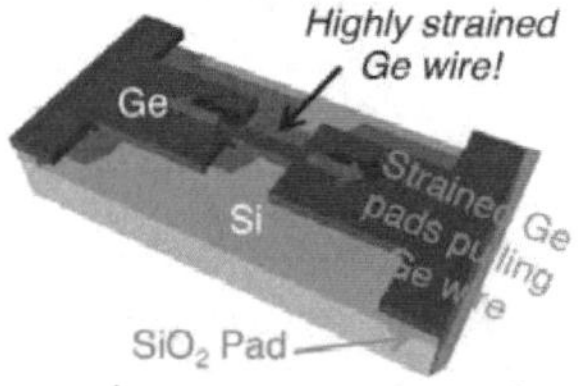

Nam............Saraswat, Nano Letters, June 11, 2013.

Modulator

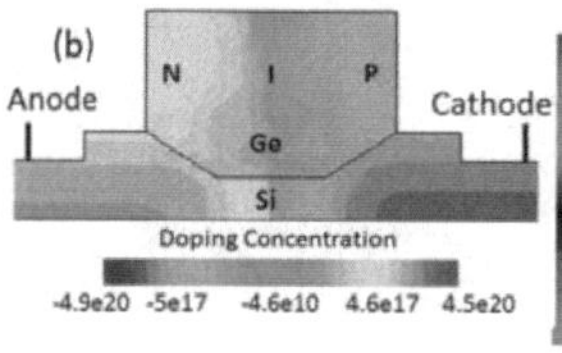

Gupta..........Saraswat, OFC, March 2015

Waveguide

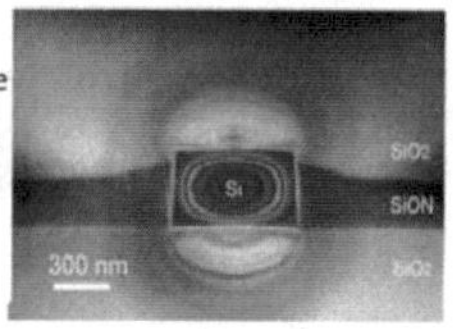

Detector

Okyay......... Saraswat, Optics Lett. 2006

- Germanium devices can be monolithically integrated on silicon
- Laser is the only missing component

Technology for Optical Interconnects on Silicon: Optical Transmission Media

Waveguides

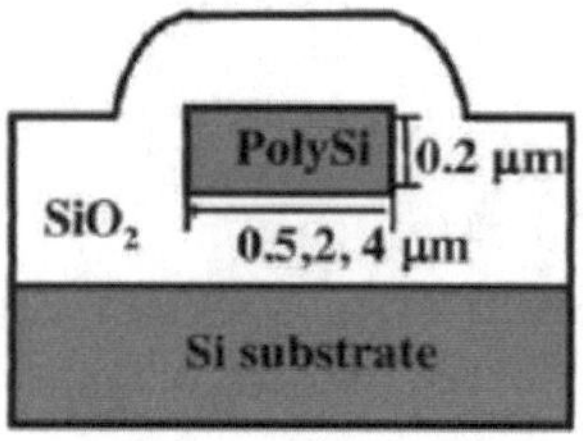

Bending Waveguides

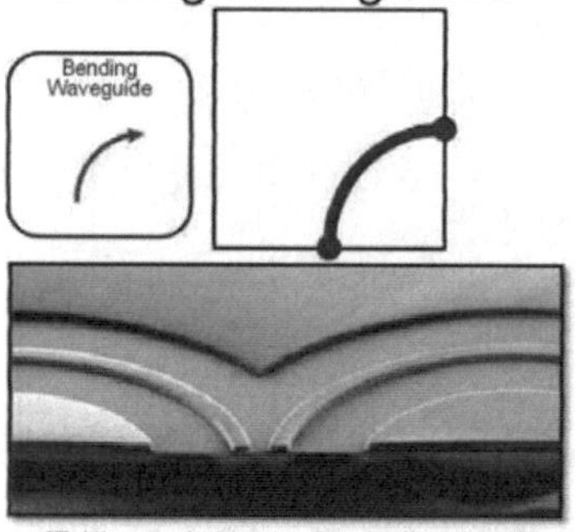

[F. Xia *et al.*, Nature Photonics, 2006]

Couplers

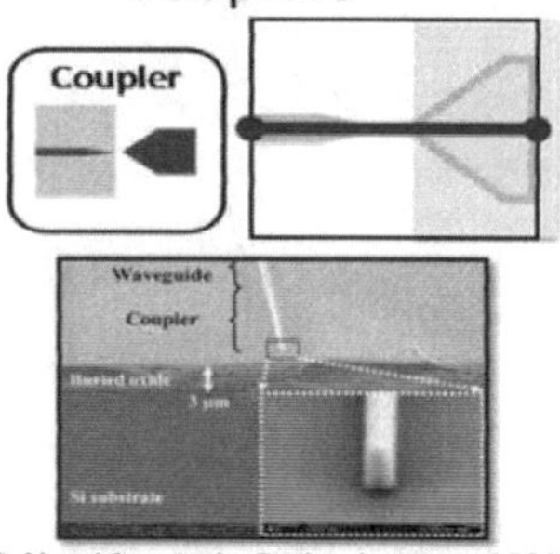

[V. Almeida *et al.*, Optics Letters, 2003]

Splitter

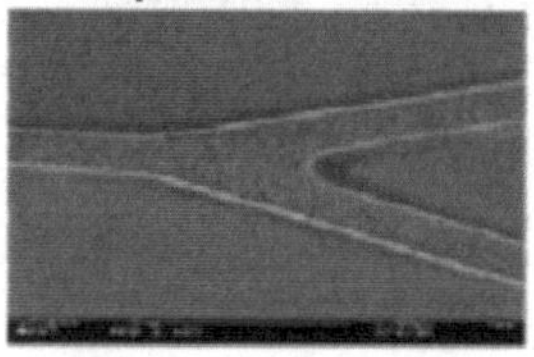

(Kimmerlingr, MIT)

Waveguide Crossings

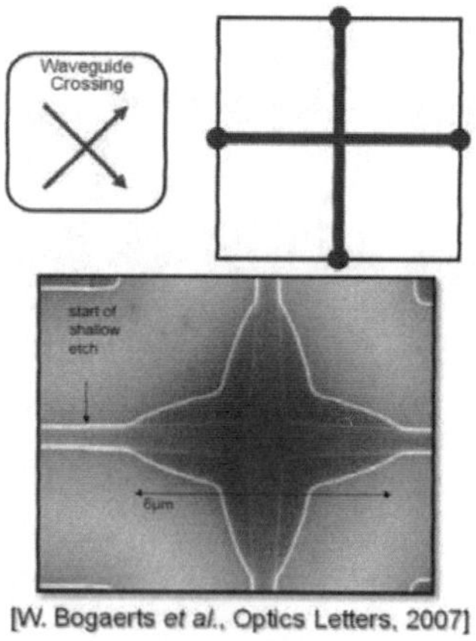

[W. Bogaerts *et al.*, Optics Letters, 2007]

High Efficiency Ge p-i-n Photodetectors on Si

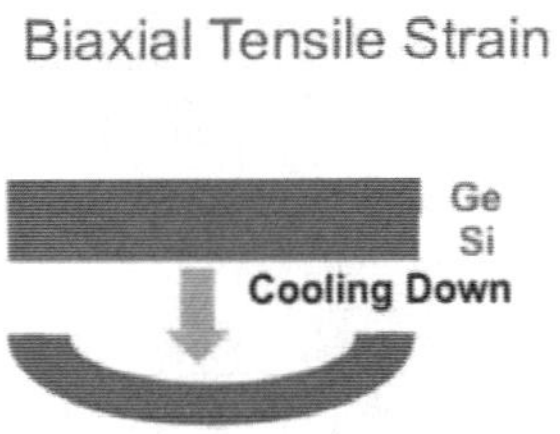

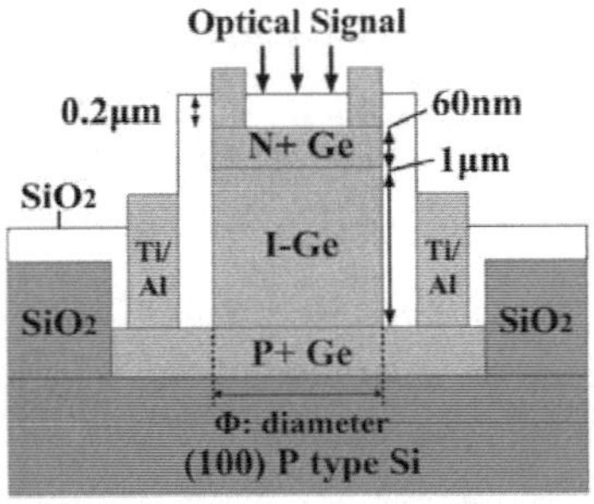

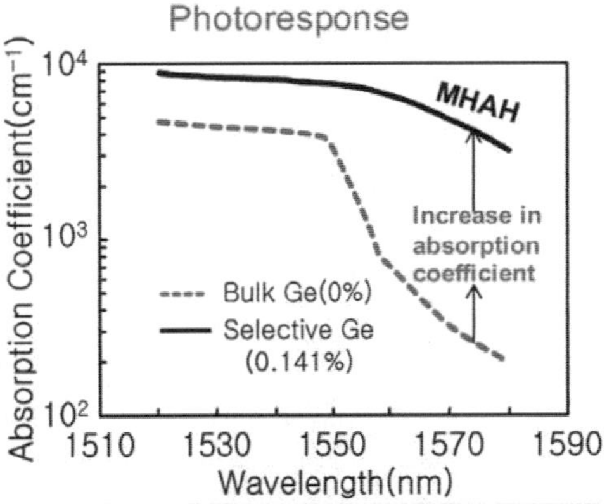

- Ge grown on Si by **M**ultiple **H**ydrogen **A**nneal and **H**eteroepitaxy (MHAH) Technique
- Ge film complies with Si substrate on cooling down resulting in tensile strain => bandgap reduces
- Detector efficiency improves at 1550nm due to tensile stress
- Dark Current high

Hyun-Yong Yu,........... K. Saraswat, IEEE Electron Dev. Lett., Nov. 2009.

Optical Modulator

Electro-optic Modulators

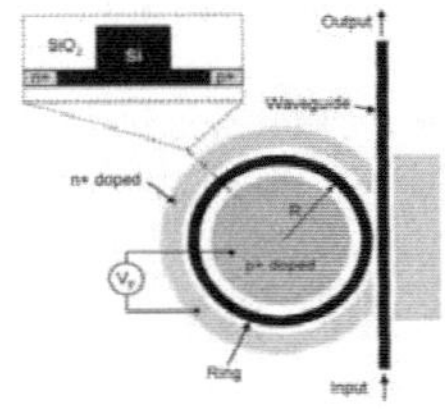

- Index change
- Weak mechanism
- High Q
- Temperature tuning

Lipson (Cornell)

Mach-Zehnder Modulators

- Phase shift effect in waveguides
- Large size and power consumption

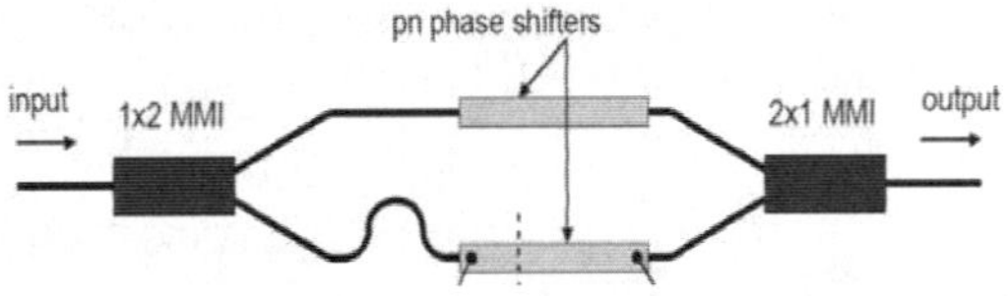

Electro-absorption Modulators

Saraswat (Stanford)

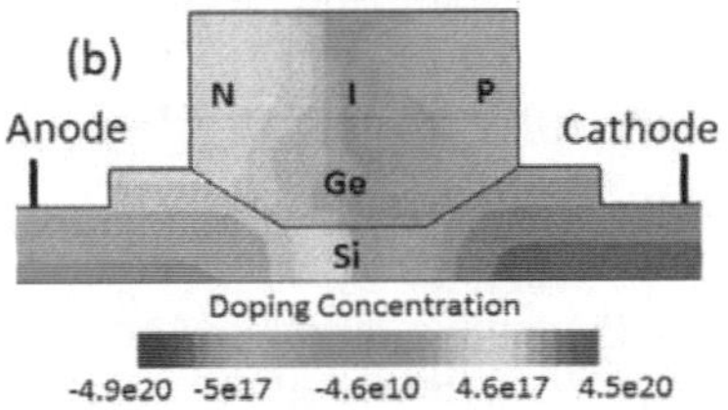

- Franz-Keldysh effect in bulk material

QCSE Ge/SiGe Modulator

Harris and Miller (Stanford)

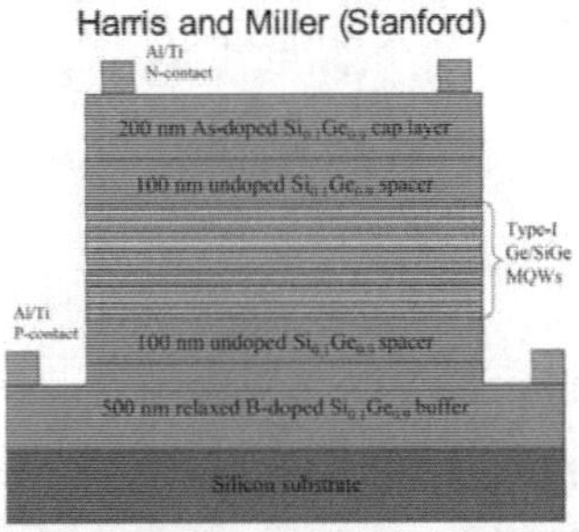

Ge/Si Electro-Absorption Franz-Keldysh Effect Modulator

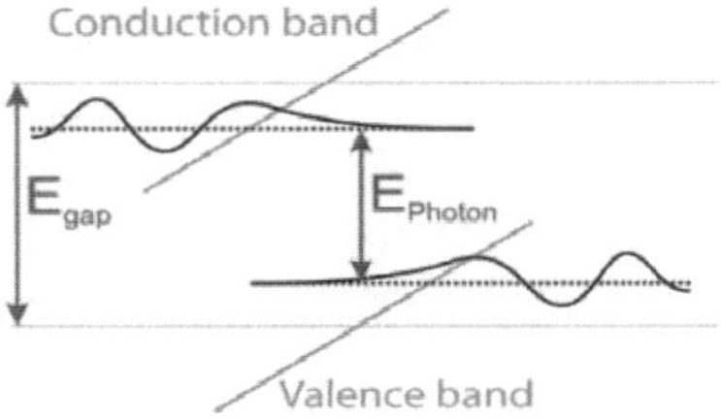

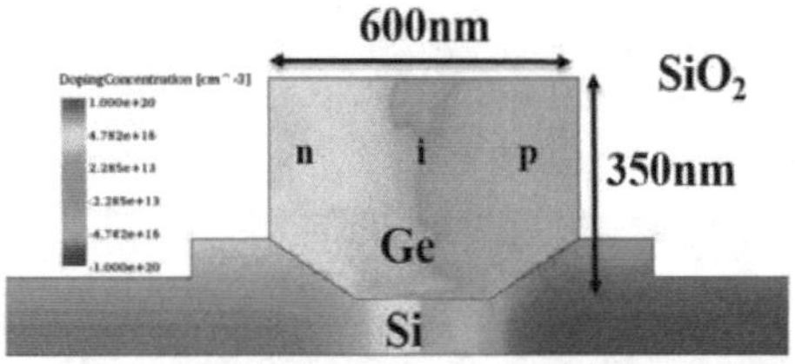

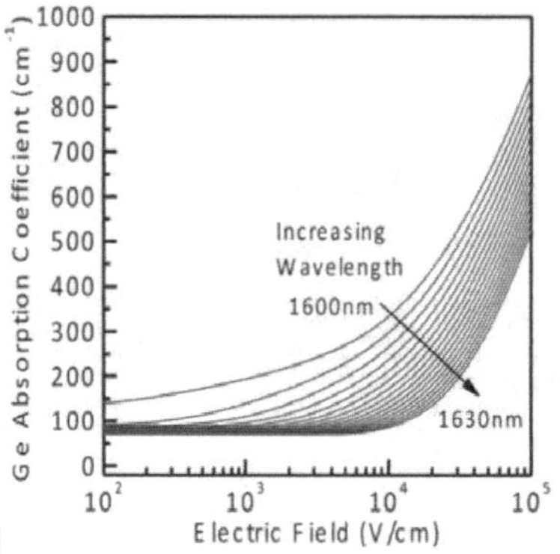

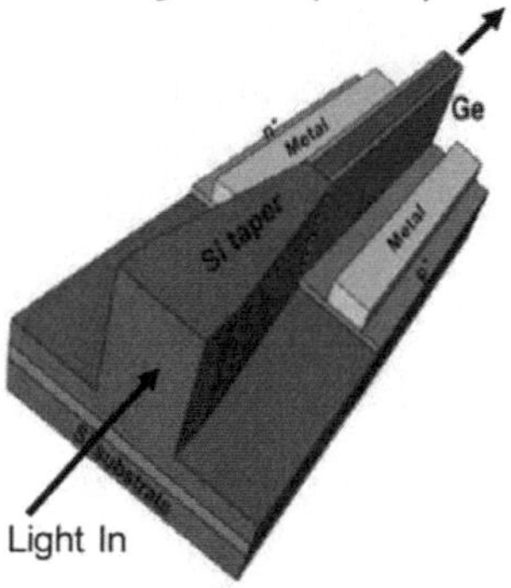

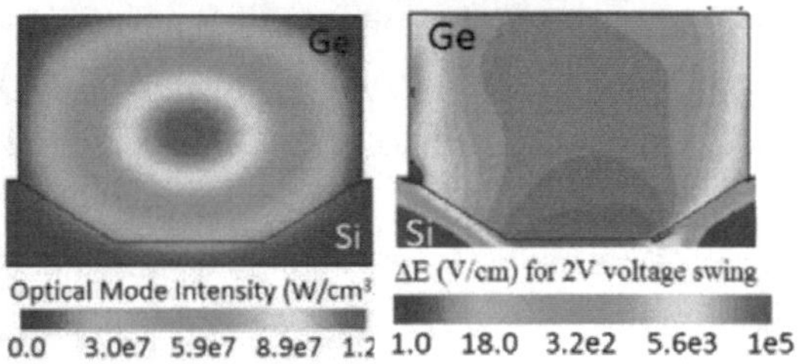

In the presence of an electric field, the conduction and valence bands of a semiconductor tilt. Application of an electric field leads to overlap in valence and conduction band wave functions, and hence optical absorption, at energies below the semiconductor bandgap.

S. Gupta.................K. Saraswat, OFC, Los Angeles, Paper No. Tu2A.4, March 2015

Structure Needs for Efficient Lasing

Heterojunction

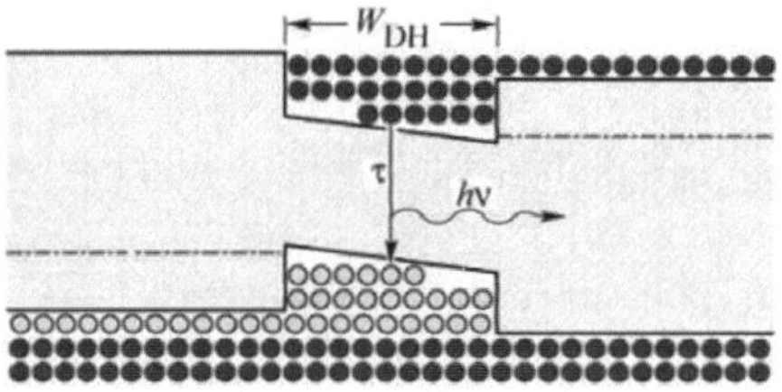

For efficient laser

- Direct bandgap cavity
- Hetrojunction quantum well for carrier confinement

E. F. Schubert, Light Emitting Diodes (Cambridge Univ. Press)

Engineering Ge for light emission- Doping

- N-type doping can be used to fill electrons into the L valley upto the level of Γ valley
- But it is difficult to heavily dope Ge n-type
- Increases free carrier absorption and auger recombination
- Inefficient light emission

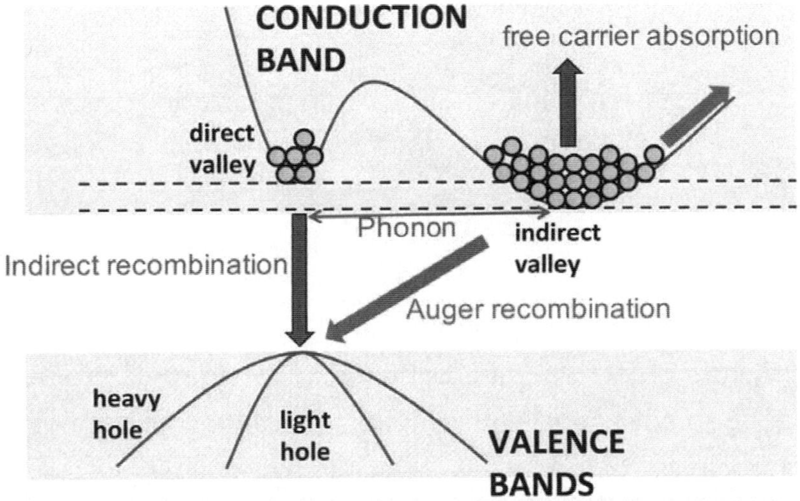

Engineering the Ge band structure for light emission: Strain

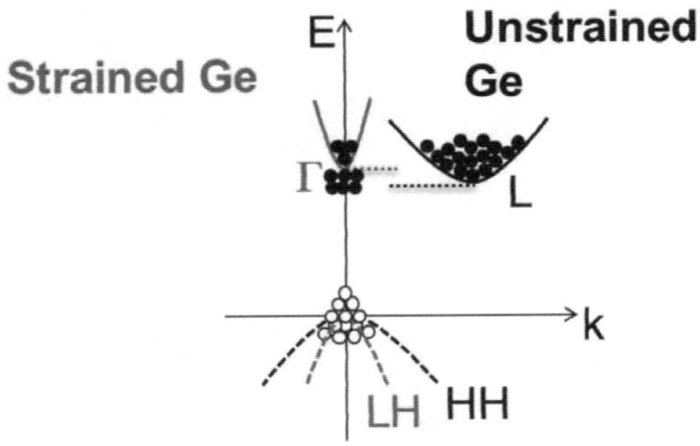

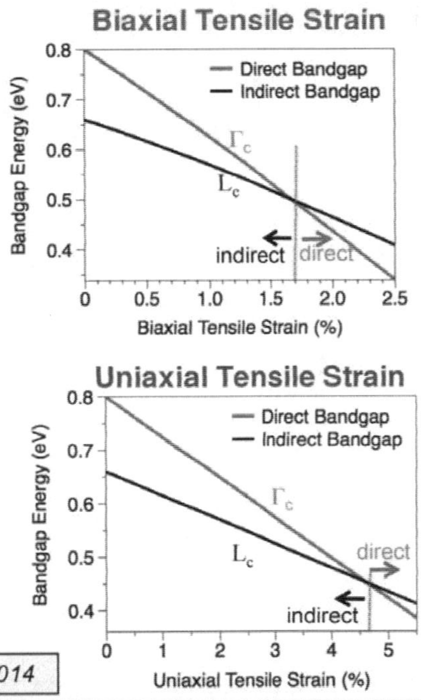

>1.7% biaxial tensile strain or > 5% uniaxial tensile strain turns Ge into a direct bandgap material, making light emission possible

Sukhdeo, ... Saraswat, Photonics Research, 2014

Heterostructure in a Single Material: Strained Ge

SEM

2D Strain Distribution

Band Diagram

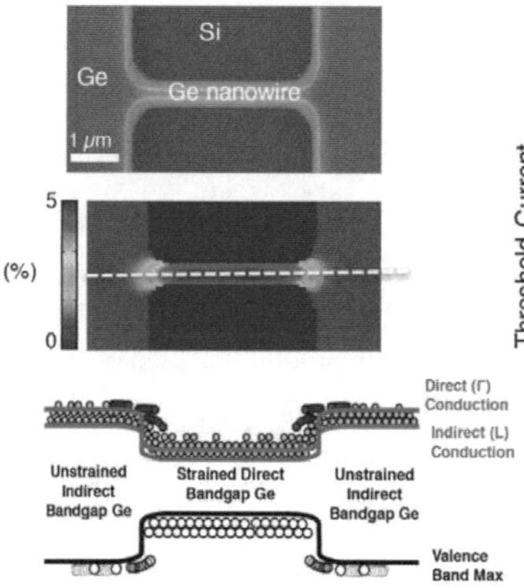

Threshold Current Simulations

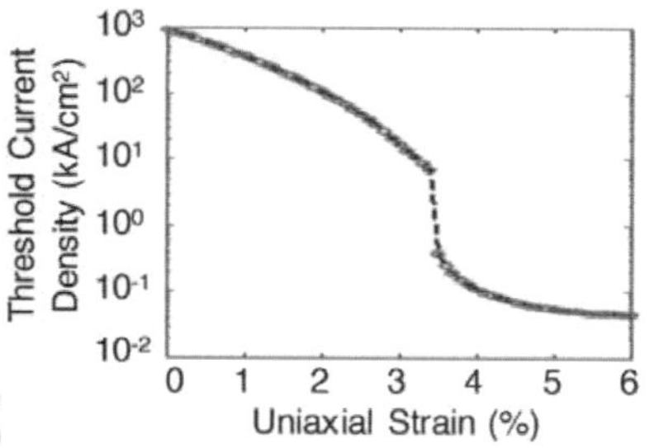

- Strain can be tunable with geometry
- Heterostructure created due to reduction in bandgap of strained Ge
- Direct bandgap cavity and hetrojunction quantum well in single material

Nam, Saraswat, et al., Nano Letters, June 11, 2013.

Saraswat
Stanford University

27

Engineering the Ge band structure by alloying with tin for GeSn CMOS and photonics

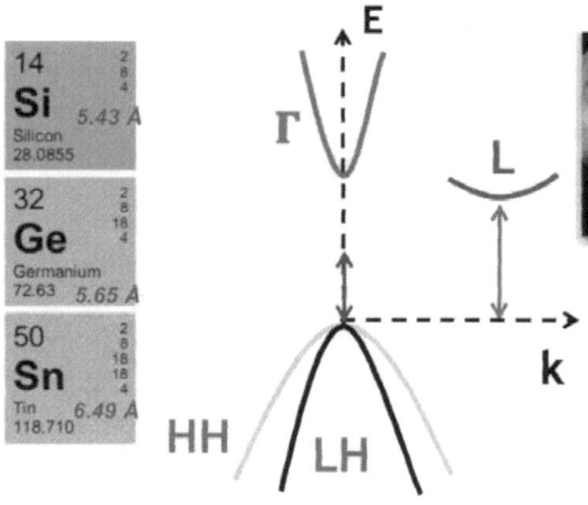

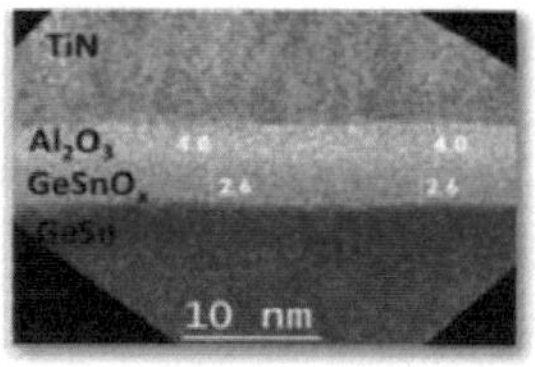

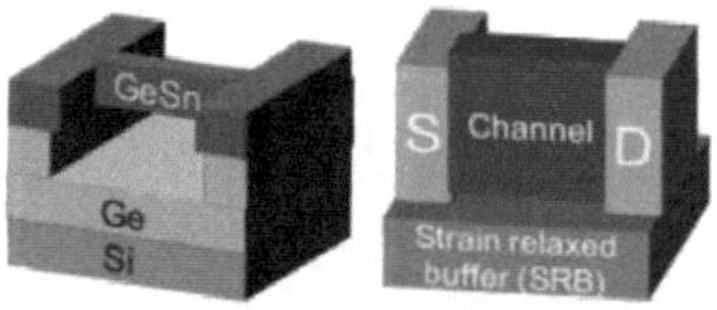

Strain engineering in FinFETs

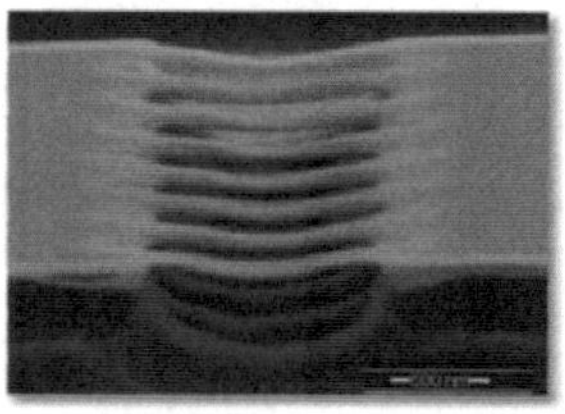

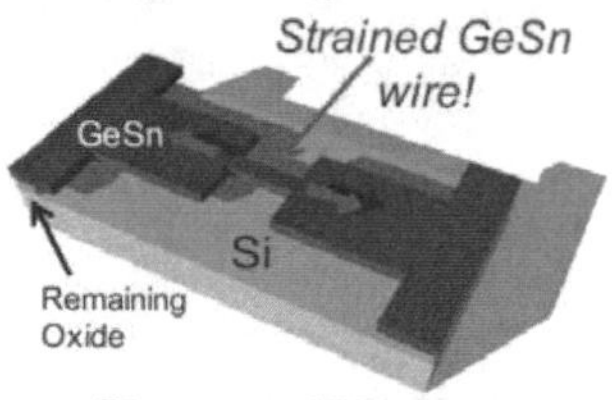

Si-compatible Laser

At ~ 7 Sn% $Ge_{1-x}Sn_x$ becomes direct band gap!

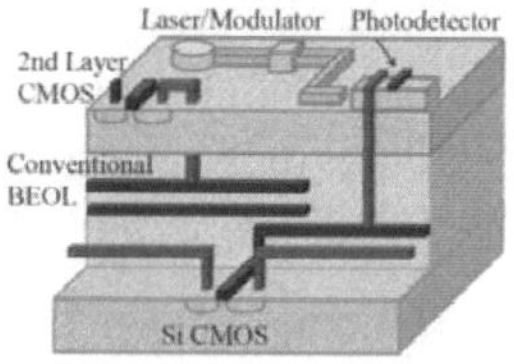

Gupta, Yeo, Takagi, Saraswat, et al., MRS Bulletin, Aug 2014

3-D IC: CMOS, Photonics co-integration

Saraswat
Stanford University

28

Demonstration of Optically Pumped GeSn Laser

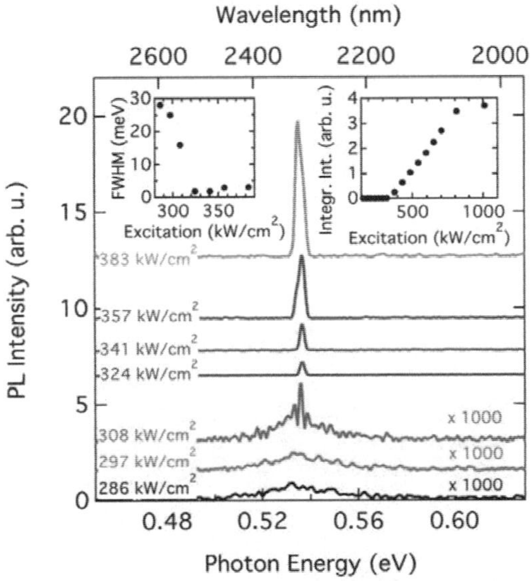

- Lasing emission spectra measured from the facet of a 5 µm x1 mm long Fabry-Perot waveguide $Ge_{0.87}Sn_{0.13}$ cavity under optical pumping at 20 K.
- More work needed for room temperature laser

R. Geiger, T. Zabel and H. Sigg, Frontiers in Materials, July 2015

Direct Bandgap GeSn Microdisk Laser on a Si-Platform

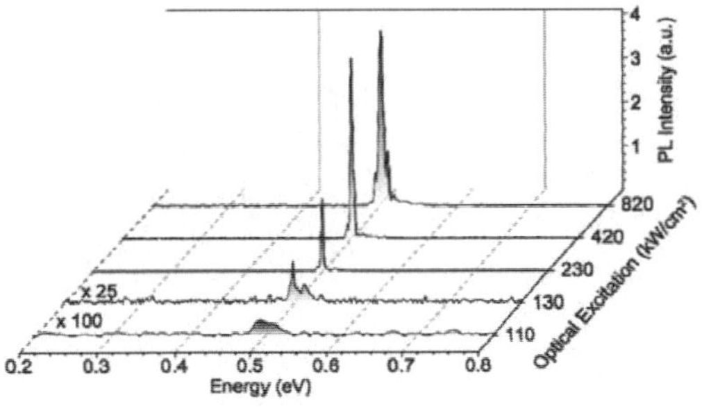

- Power-dependent PL spectra of an 8 µm diameter $Ge_{0.875}$ $Sn_{0.125}$ microdisk at 50 K.
- More work needed for room temperature laser

S. Wirth, et al., IEEE IEDM, Dec. 2015

Multiple Knobs to Turn for Direct-Gap: Strained GeSn

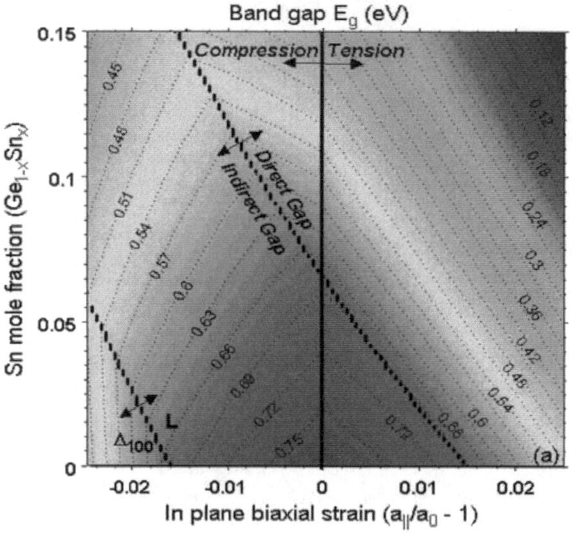

- A combination of alloying Ge with Sn and strain can also give us a direct bandgap material
  - Efficiency would be comparable to present III-V lasers

Gupta, Magyari-Köpe, Nishi, Saraswat, J. Appl. Phys. **113**, 073707 (2013)

3-D Integration: Motivation

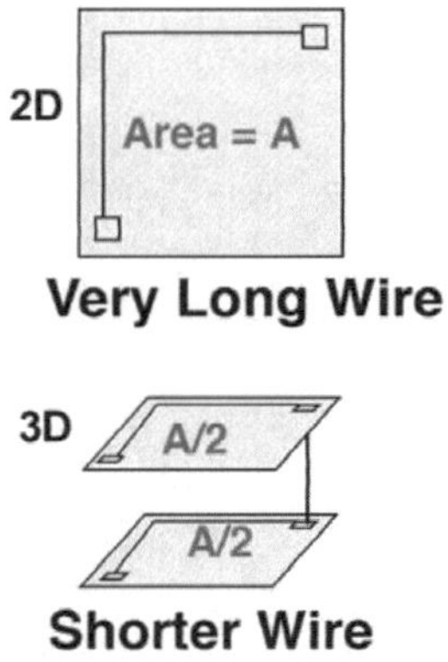

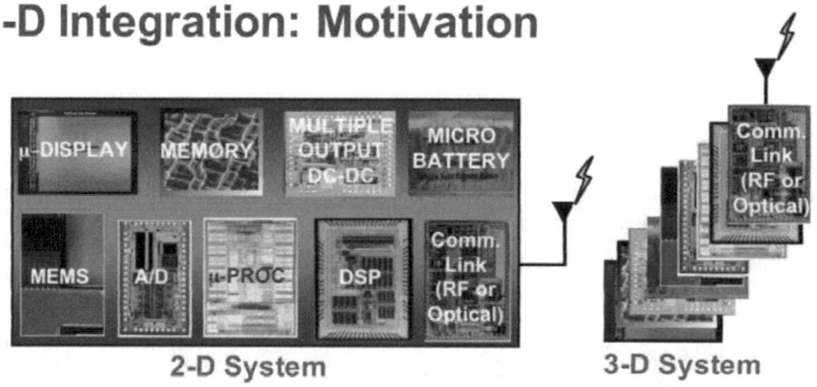

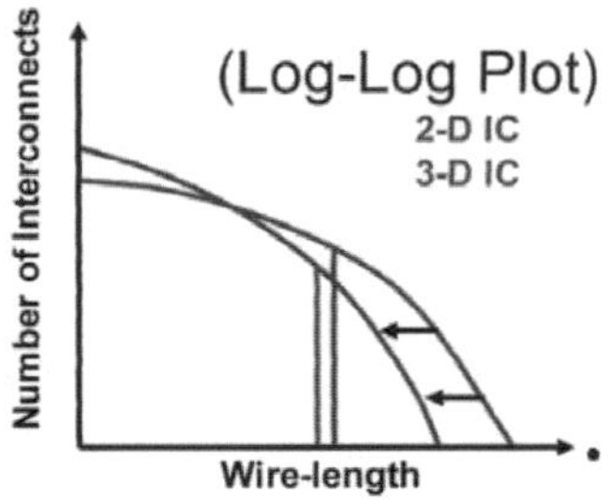

- **Reduce Chip footprint**
  - **Improved form factor**
  - **Interconnect length ↓ and therefore R, L, C ↓**
    - **Delay reduction**
    - **Power reduction**
    - **Higher bandwidth**
- **Integration of heterogeneous technologies possible, e.g., memory & logic, sensors, optical I/O**

Technology to Fabricate 2.5D/3D ICs

- **2.5D packaging** (mature technology)
  - Wire bonded
  - Bump
  - vertical interconnect density < 20/mm or 400/mm^2
- **3D bonding/TSV** (emerging technology)
  - Die stacking
  - wafer stacking
  - vertical interconnect density < 40,000/mm^2
- **3D crystallization** (near future technology)
  - Epitaxial growth
  - Laser melting and crystallization
  - Seeded crystallization
  - Liquid phase crystallization
  - vertical interconnect density < 25M/mm^2
- **3D self assembled devices** (future technology)
  - Si and Ge nanowires
  - Carbon nanotubes
  - Organic semiconductors

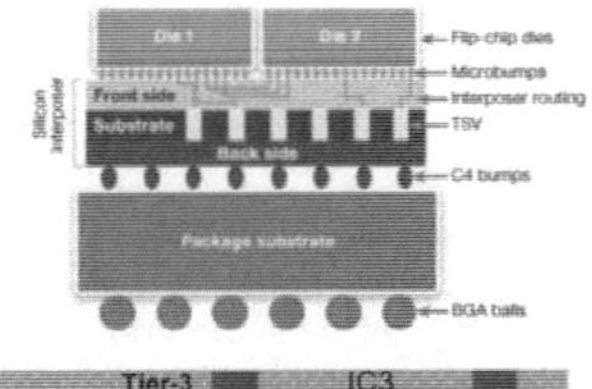

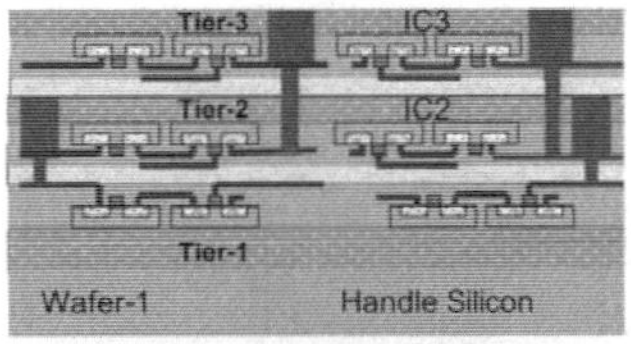

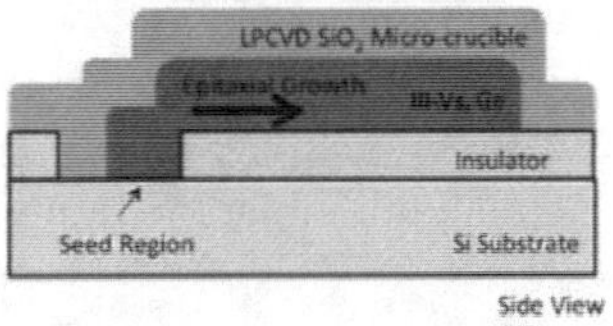

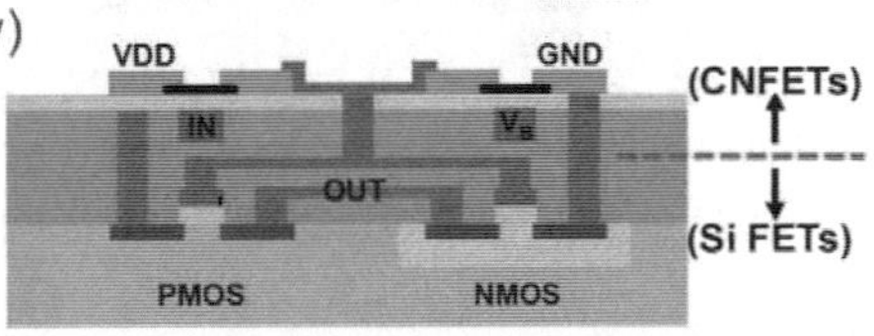

Monolithic 3D Integration of Si MOSFETs with RRAMs and CNTFETs

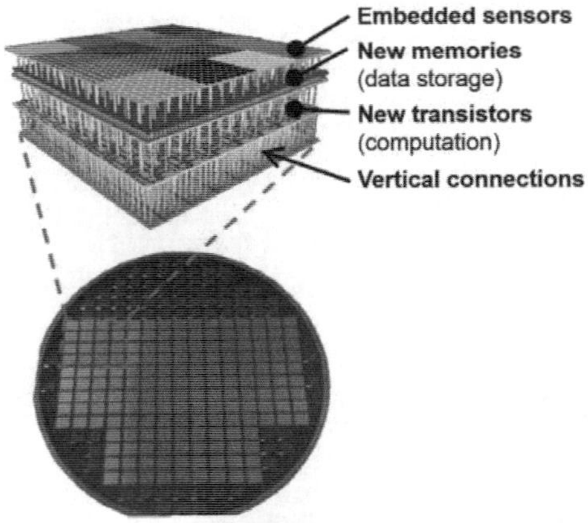

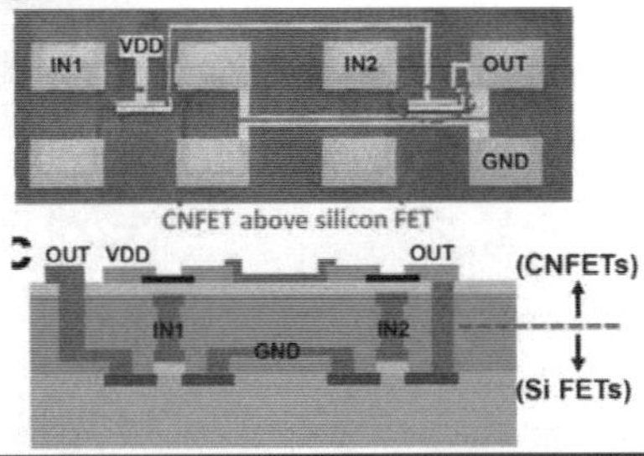

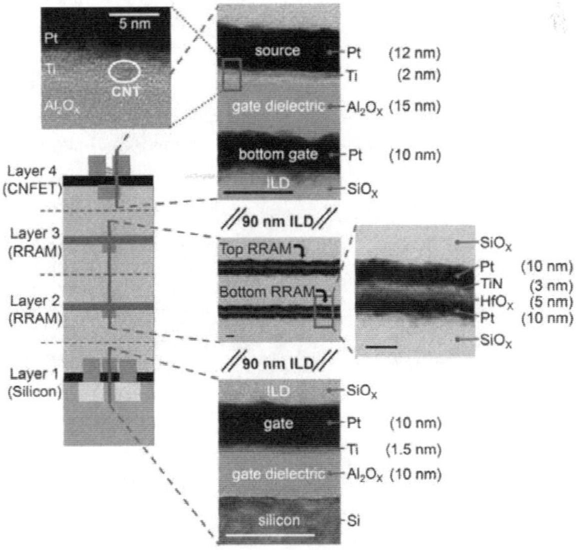

Max Shulaker, et al., IEDM, December 2014, VLSI Symp. 2014

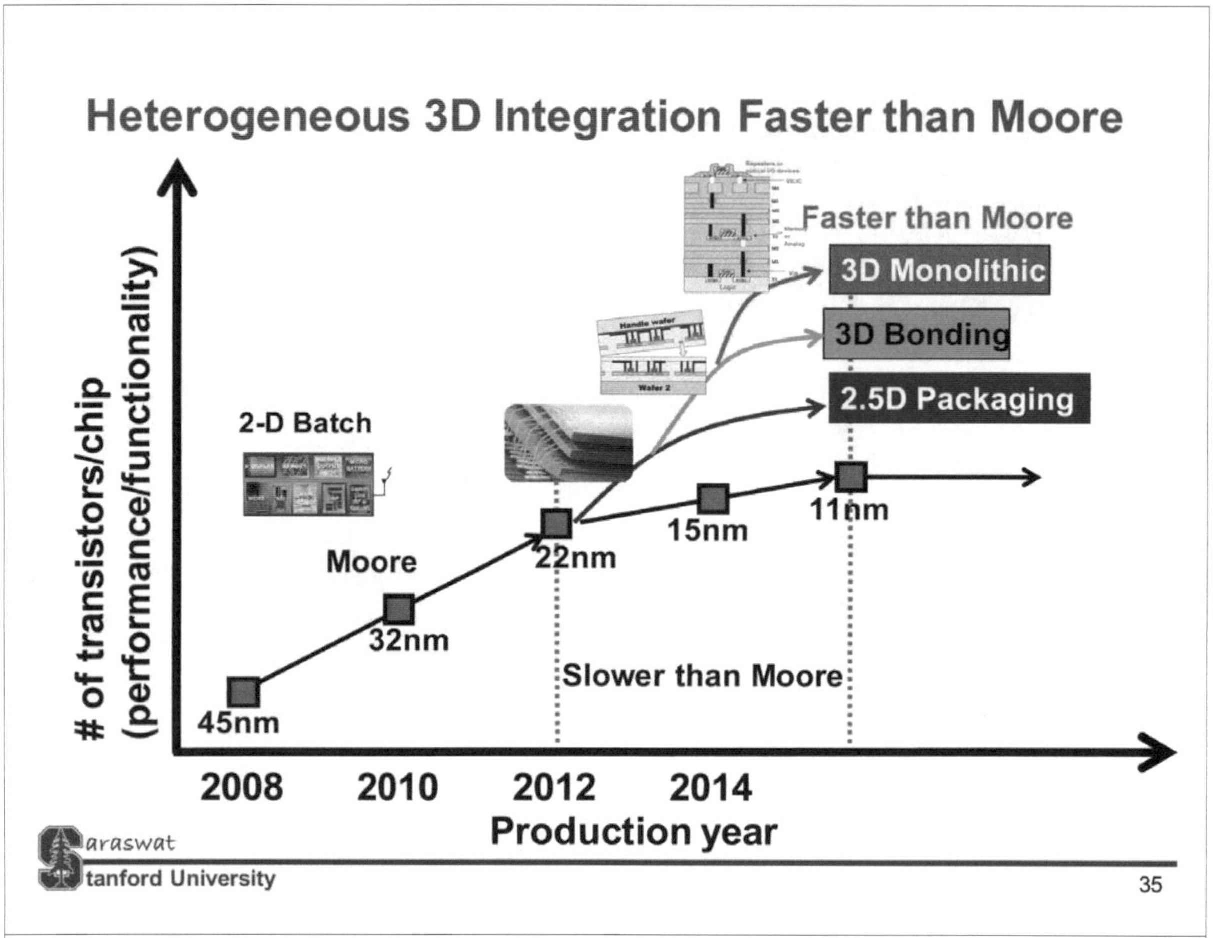

Heterogeneous 3D Integration Faster than Moore
# of transistors/chip (performance/functionality)
Faster than Moore
3D Monolithic
3D Bonding
2.5D Packaging
2-D Batch
Moore
45nm
32nm
22nm
15nm
11nm
Slower than Moore
2008
2010
2012
2014
Production year
Saraswat
Stanford University
35

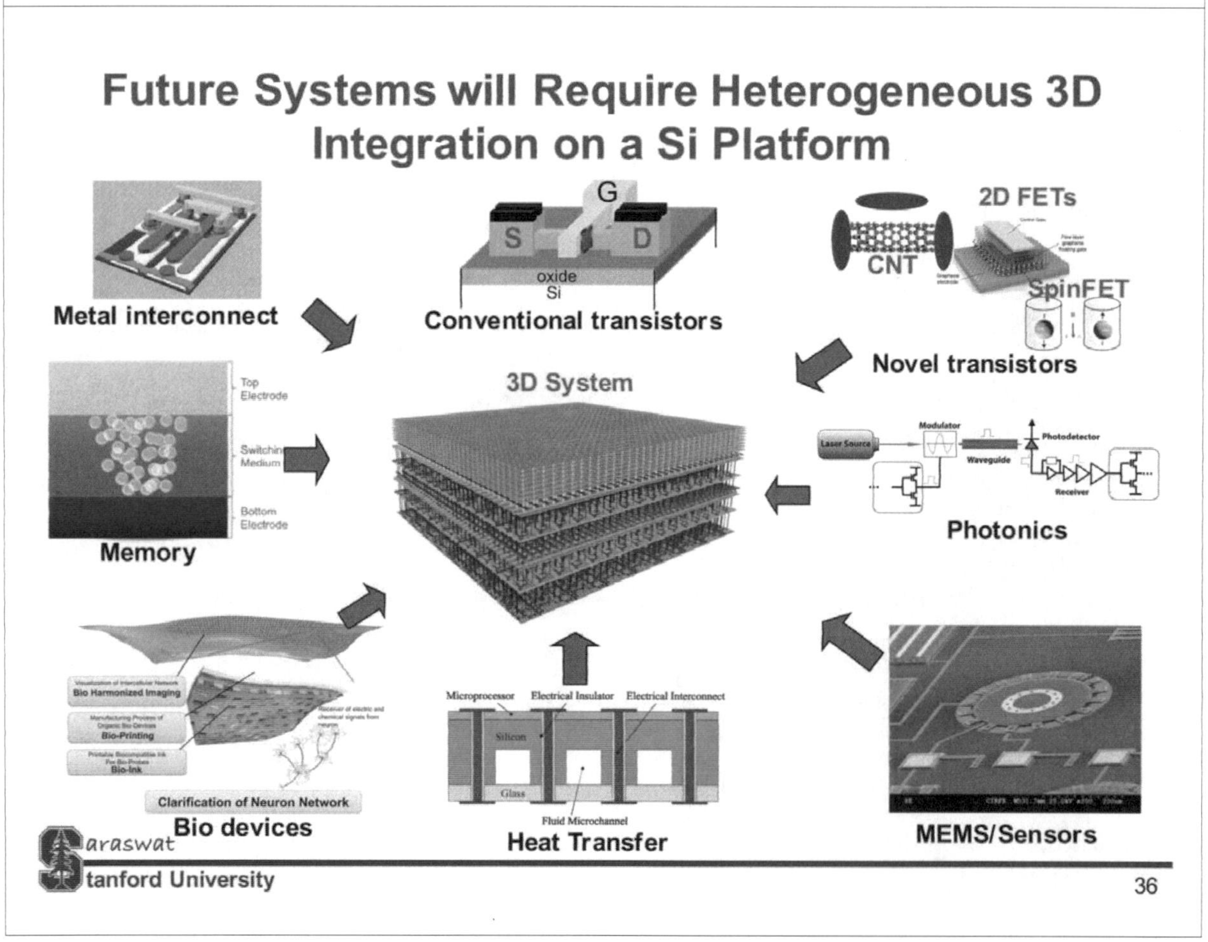

Future Systems will Require Heterogeneous 3D Integration on a Si Platform
Metal interconnect
G
S
D
oxide
Si
Conventional transistors
2D FETs
CNT
SpinFET
Novel transistors
3D System
Top Electrode
Switching Medium
Bottom Electrode
Memory
Photonics
Bio devices
Microprocessor
Electrical Insulator
Electrical Interconnect
Silicon
Glass
Fluid Microchannel
Heat Transfer
MEMS/Sensors
Saraswat
Stanford University
36

Conclusion

- ☹ Cu resistivity increases as technology scales down. This will be a bottleneck of future high-performance chip.
- ☺ CNTs have a significant advantage over Cu wires especially for local interconnects
- ☺ Optical links have smallest latency and energy per bit for longer global interconnects requiring higher band width
- ☺ 3D heterogeneous integration will keep the Moore's law going for awhile.

Saraswat
Stanford University 37

Contributors / Collaborators:

Hoyeol Cho, Raj Dutt, Shashank Gupta, Suyog Gupta, Jim Harris, Pawan Kapur, Kyung-Ho Koo, Donguk Nam, Ju Hyung Nam, Ammar Nayfeh, Ali Okyay, Jan Petykiewicz, Dave Sukhdeo, Jelena Vuckovic and Hyun-Yong Yu.

Saraswat
Stanford University 38

2016 Surface Preparation and Cleaning Conference

Please contact the author for a copy of this presentation:

peethala@us.ibm.com

Tungsten Post-CMP Cleaning Formulations for Advanced Nodes: 10 nm and 7 nm

Daniela White*[a], Thomas Parson[a], Shining Jenq[b], Don Frye[a], Steve Medd[a], Ruben Lieten[c] and Michael White[a]

a. Entegris Inc., R&D Surface Preparation and Integration, 7 Commerce Drive, Danbury CT 06810, Ph: 203-739-1470, daniela.white@entegris.com

b. TTC, Entegris, Inc. 1F, No 669, Sec. 4, ZhongXing Road | Zhudong Town | Hsin-Chu County 310 Taiwan

c. Entegris GmbH | Hugo-Junkers-Ring 5 | Gebäude 107/W 01109 Dresden, Germany

Entegris
creating a material advantage

Outline

- Development of an efficient W post-CMP cleaner for 10 nm and 7 nm nodes, compatible with barrier liners (TiN) and dielectrics (TEOS, Si_3N_4, SiC, polysilicon);
- Mechanistic considerations and AG-W100 formulation design;
- Particle defect count results on W and dielectric surfaces using AG-W100 formulation;
- W/TiN galvanic corrosion data and knobs to control it;
- SEM on W/TiN 45 nm pattern wafers cleaned with AG-W100;
- Understanding and cleaning organic residue post-CMP on dielectric substrates: contact angle and XPS metrology;
- New AG-W100 CIP formulations with tunable and improved performance.

Entegris

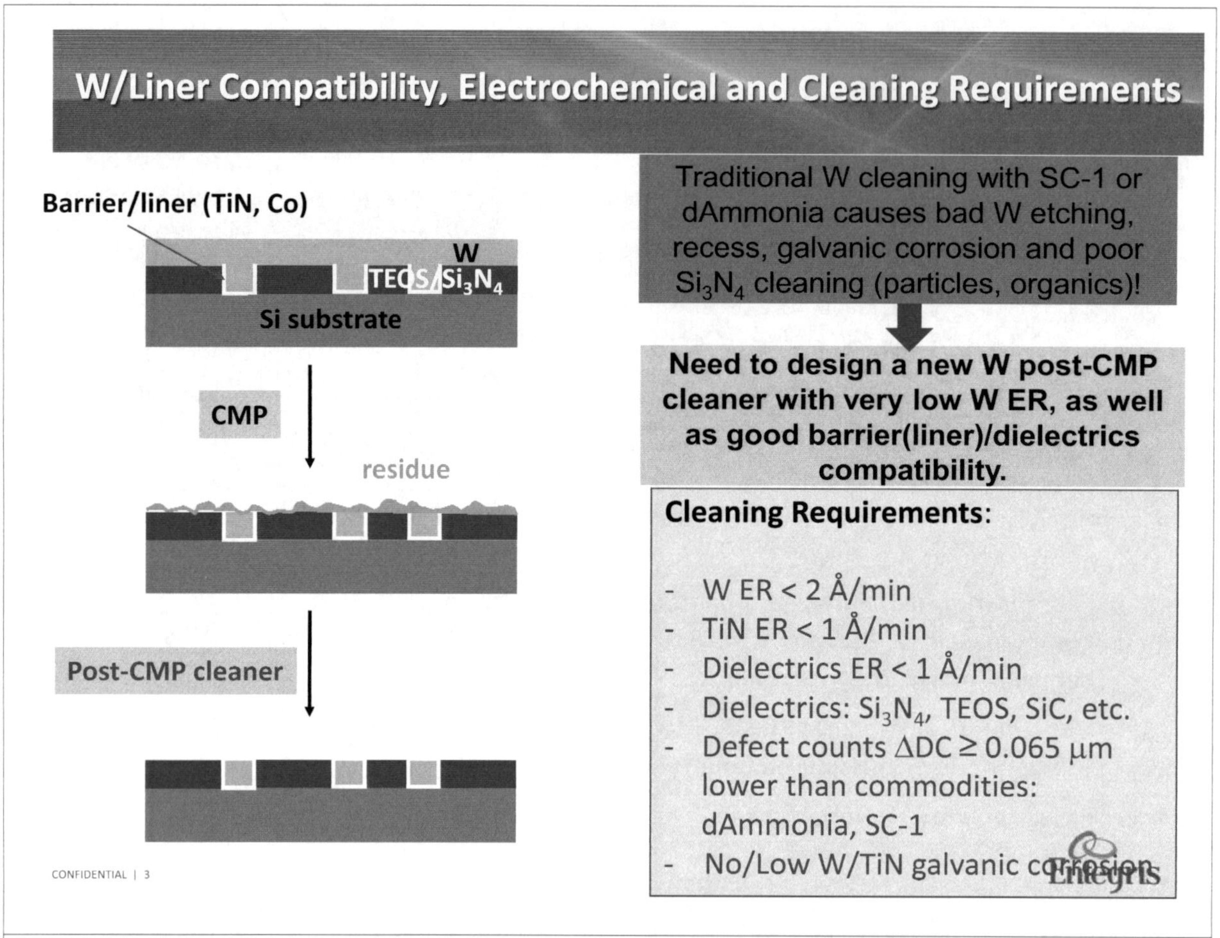
W/Liner Compatibility, Electrochemical and Cleaning Requirements
Barrier/liner (TiN, Co)
W
TEOS/Si_3N_4
Si substrate
CMP
residue
Post-CMP cleaner
CONFIDENTIAL | 3
Traditional W cleaning with SC-1 or dAmmonia causes bad W etching, recess, galvanic corrosion and poor Si_3N_4 cleaning (particles, organics)!
Need to design a new W post-CMP cleaner with very low W ER, as well as good barrier(liner)/dielectrics compatibility.
Cleaning Requirements:
- W ER < 2 Å/min
- TiN ER < 1 Å/min
- Dielectrics ER < 1 Å/min
- Dielectrics: Si_3N_4, TEOS, SiC, etc.
- Defect counts ΔDC ≥ 0.065 μm lower than commodities: dAmmonia, SC-1
- No/Low W/TiN galvanic corrosion
Entegris

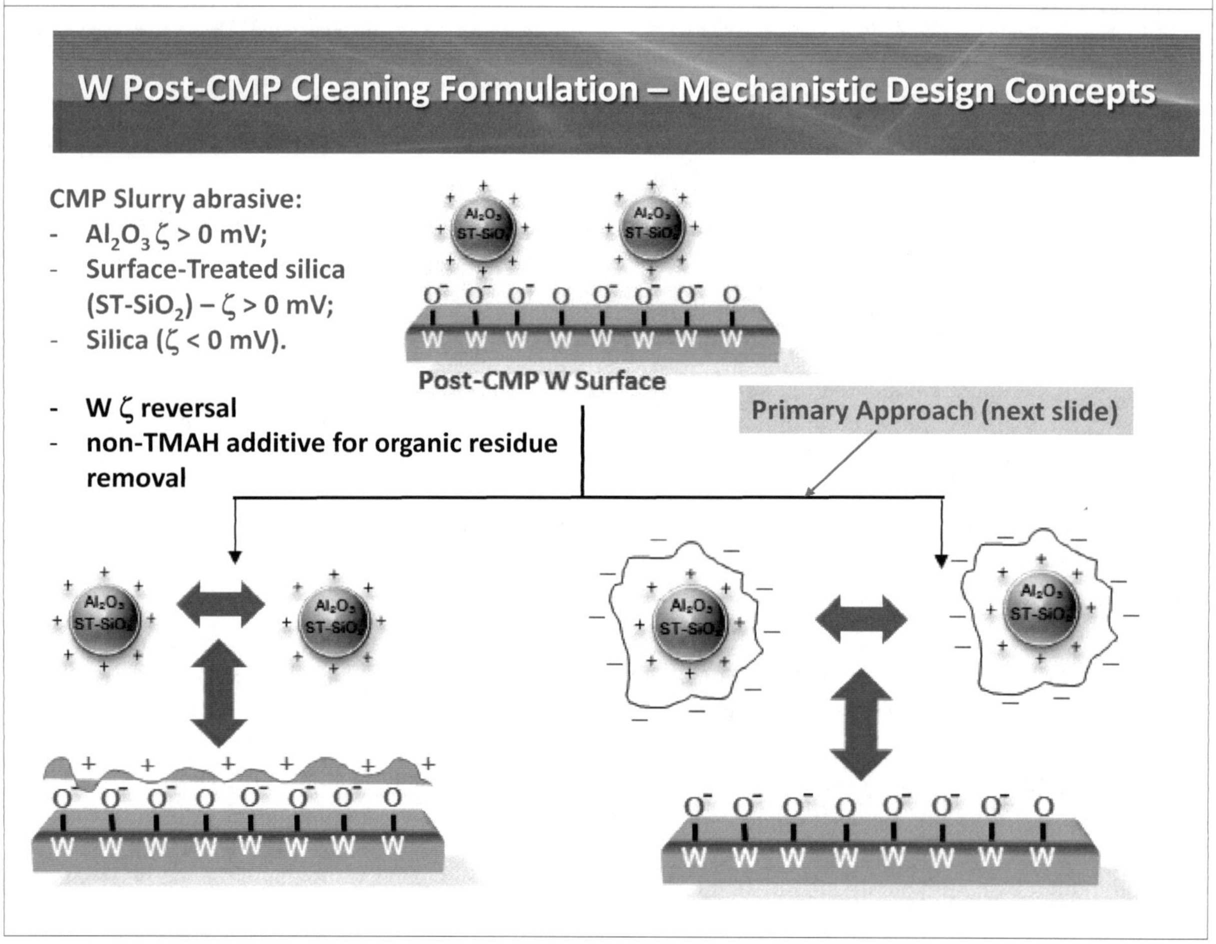
W Post-CMP Cleaning Formulation – Mechanistic Design Concepts
CMP Slurry abrasive:
- Al_2O_3 $\zeta > 0$ mV;
- Surface-Treated silica ($ST\text{-}SiO_2$) – $\zeta > 0$ mV;
- Silica ($\zeta < 0$ mV).
- W ζ reversal
- non-TMAH additive for organic residue removal
Al_2O_3
$ST\text{-}SiO_2$
Post-CMP W Surface
Primary Approach (next slide)

PlanarClean® AG-W100 Formulation Additives List – Function and Mechanism

Component	Function	Mechanism
A	non-TMAH pH Adjustor	• Provides the hydroxyl anions and adjust pH needed for W surface hydroxylation and good wetting • Ensures negative surface charge on both wafer and contamination, by being adsorbed on inorganic and organic residues.
B & C	Complexing Agents	• Surface Modification (ST-SiO_2 and Al_2O_3 complexants) • Stabilization of particle with electrostatic repulsion (prevent agglomeration and re-precipitation)
D	Dispersing Agent	• Interacts with particles and wafer surfaces to prevent aggregation and control etch rate.

CONFIDENTIAL | 5

Entegris

Zeta Potential for Slurry NP (Alumina, Silica), W and Dielectrics Surfaces in Contact with AG-W100 Formulation

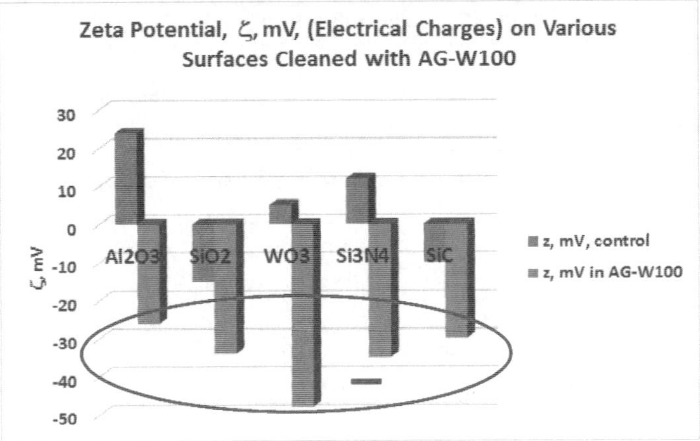

- Control dispersions, pH ~ 3-4;
- **All surfaces are strongly negatively charged in AG-W100;**
- Charge reversal for alumina, silicon nitride and silicon carbide;
- pH_{IEP}TiN ~ 4-5, TiN surfaces negatively charged in AG-W100;
- **Strong repulsion-type interactions** between all surfaces to be cleaned (W/WO_3, TiN, TEOS, Si_3N_4, SiC) and CMP slurry particles (alumina, silica).

CONFIDENTIAL | 6

Entegris

Etch Rates of CVD W, TiN, TEOS, Si_3N_4 and SiC with PlanarClean® AG-W100

Formulation	Avg. ER(Å/min)	Std. Dev.
DHF 100:1	0.3889	0.0038
$NH_4OH:H_2O$ = 1:2850	1.1844	0.0234
PC AG-W100 60:1	0.5822	0.0139

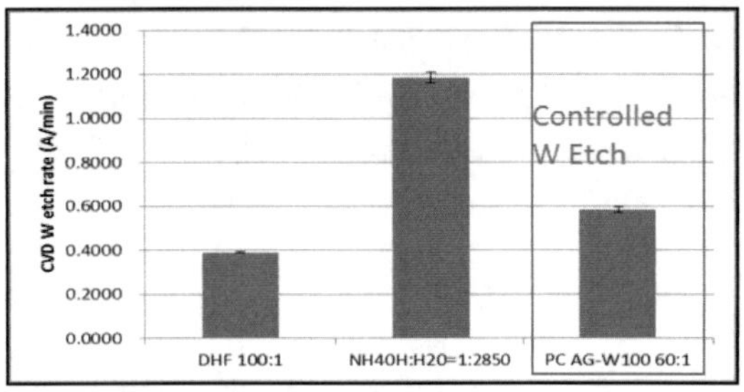

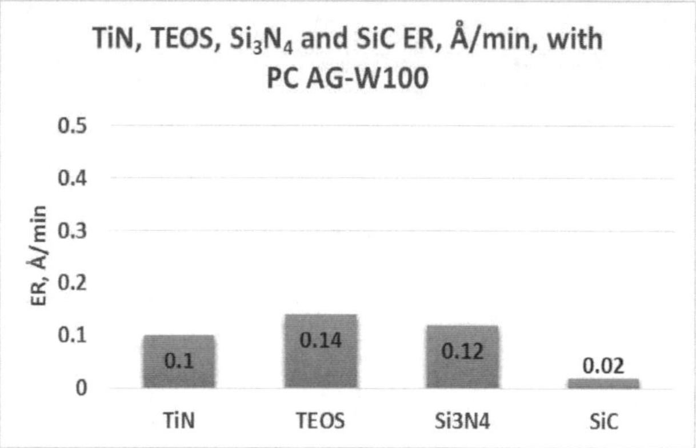

- **PlanarClean® AG-W100 has W etch rate < 0.6 Å/min.**
- **PlanarClean® AG-W100 has lower (controlled) W etch rate than $NH_4OH:H_2O$ = 1:2850.**
- **TiN, TEOS, Si_3N_4 and SiC have ERs < 0.2Å/min**

Entegris

PC AG-W100 vs. dAmmonia Cleaning Data

imec

Experiment done in collaboration with IMEC - Leuven, Belgium

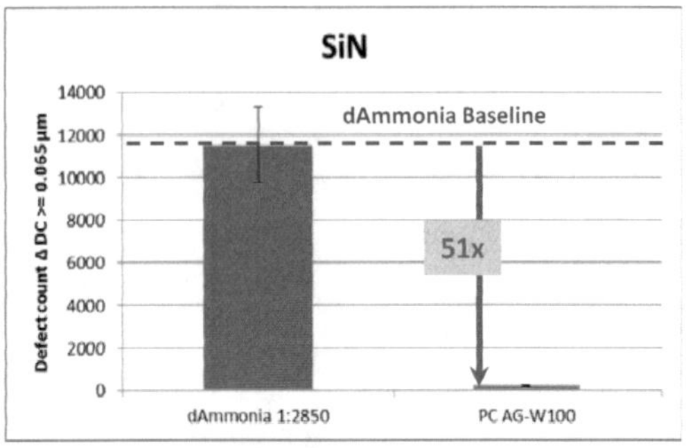

- AMAT Reflexion LK polishing tool
- KLA SP-3 defect inspection tool

- **Si_3N_4 – 51X less defects than dAmmonia;**
- **SiO_2 defects –** comparable to dAmmonia;
- **W defects – very low;**
- **Metal impurities on W wafers –** much lower.

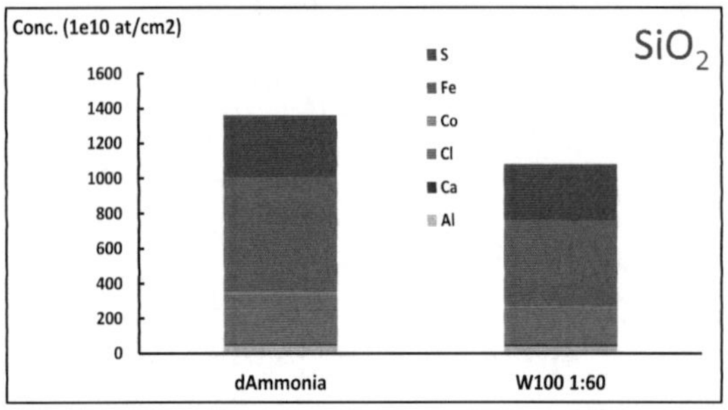

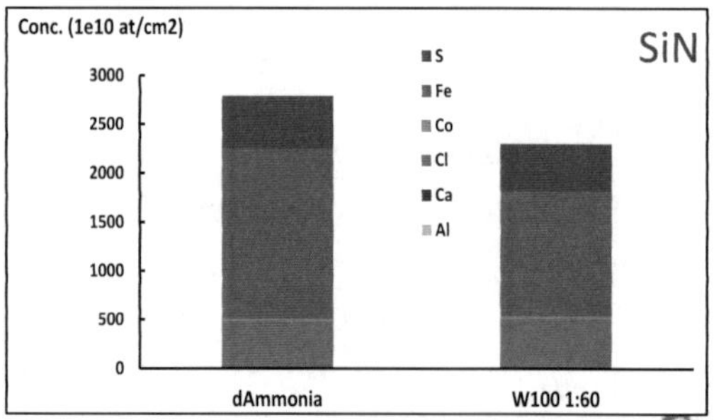

Entegris

Defect Counts on 8" TEOS & Si_3N_4 Wafers After Cleaning : PlanarClean® AG-W100 vs. SC-1 – 3X-5X Less Defects

Experiment done in collaboration with AXUS - AZ

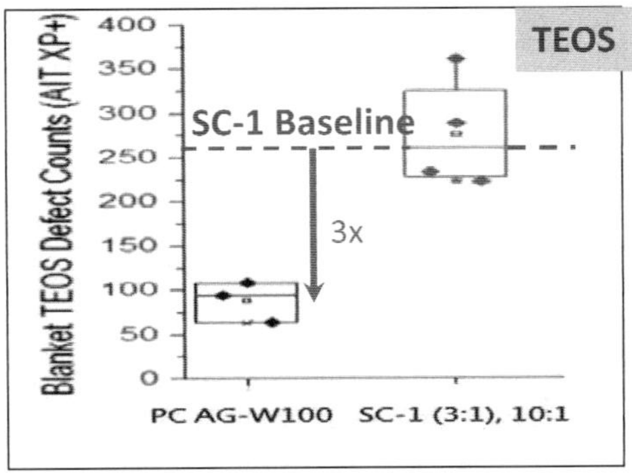

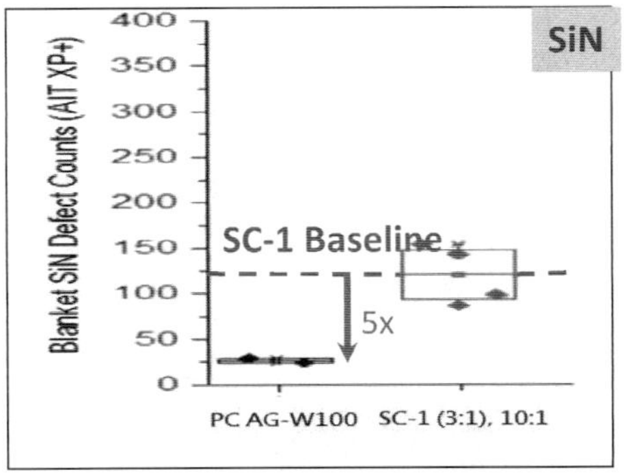

The defect counts on TEOS with PC AG-W100 are 3X better than SC-1

The defect on SiN with PC AG-W100 is 5X better than SC-1

- Detection limit: 150 nm

Entegris

PC AG-W100 W/TiN Galvanic Corrosion Tafel Plots

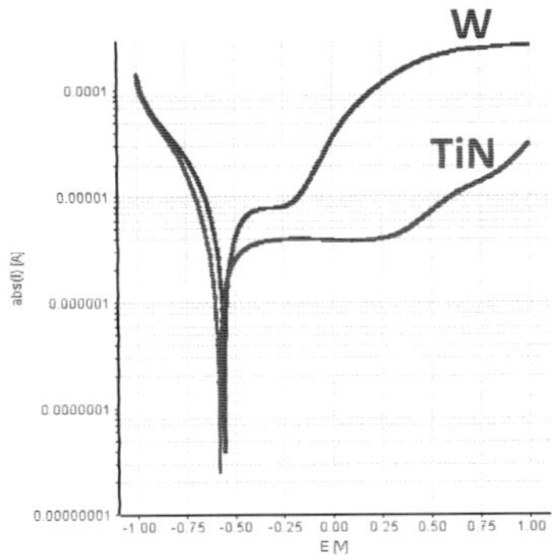

PC AGW-100 baseline

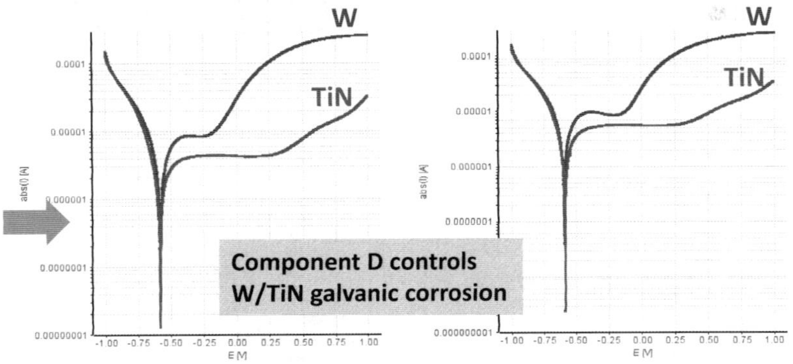

PC AGW-100 baseline + different % [Component D]

	TiN	Galvanic Corrosion	W
Galvanic Corrosion of TiN (Å/min)	**0.069**		
Material	Anode: TiN		Cathode: W
Corrosion Potential (V)	-0.584	-0.566	-0.557
Corrosion Current (A)	9.85E-07	7.24E-07	1.78E-06
Corrosion Rate (Å/min)	0.093	0.069	0.101
Slope	4.48E-05		8.09E-05

W (cathode) protected
TiN (anode) low corrosion

	TiN	Galvanic Corrosion	W
Galvanic Corrosion of TiN (Å/min)	**0.013**		
Material	Anode: TiN		Cathode: W
Corrosion Potential (V)	-0.584	-0.581	-0.579
Corrosion Current (A)	1.08E-06	1.37E-07	1.90E-06
Corrosion Rate (Å/min)	0.102	0.013	0.108
Slope	4.89E-05		8.64E-05

Good overlay
W (cathode) protected
Minimal galvanic corrosion

	TiN	Galvanic Corrosion	W
Galvanic Corrosion of W (Å/min)			**0.013**
Material	Cathode: TiN		Anode: W
Corrosion Potential (V)	-0.59	-0.594	-0.596
Corrosion Current (A)	1.26E-06	2.26E-07	2.10E-06
Corrosion Rate (Å/min)	0.119	0.013	0.119
Slope	5.72E-05		9.54E-05

TiN (cathode) protected
Minimal galvanic corrosion

PC AGW-300 - 45nm W/TiN Pattern SEM Characterization

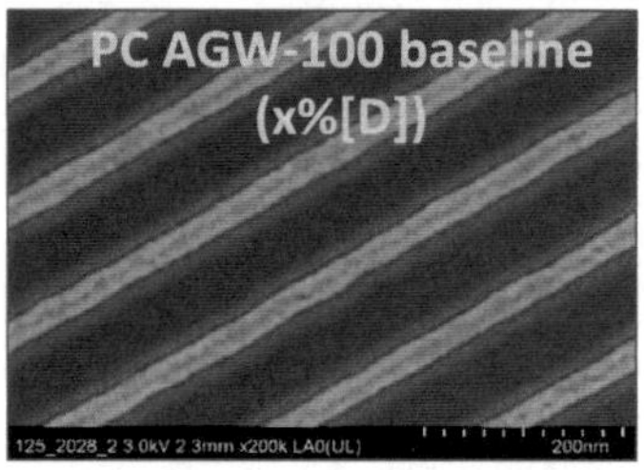

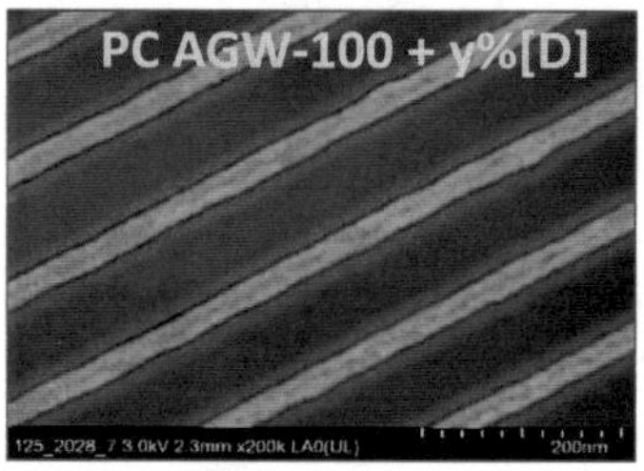

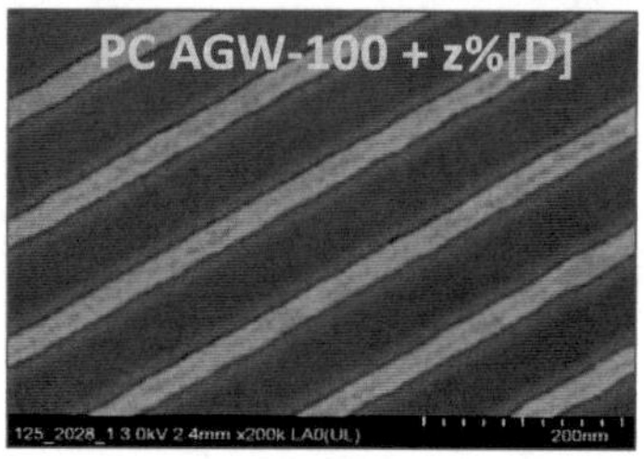

- W lines (control) already etched in the center by the CMP process
- Dirty control generates fuzzy SEM images;
- AG-W100 cleaned controls have **sharper lines @ W/TiN interfaces and generate clear SEM pictures**

Entegris

Silicon Nitride Contamination Study with W CMP Slurry Additives: Organic Contaminants

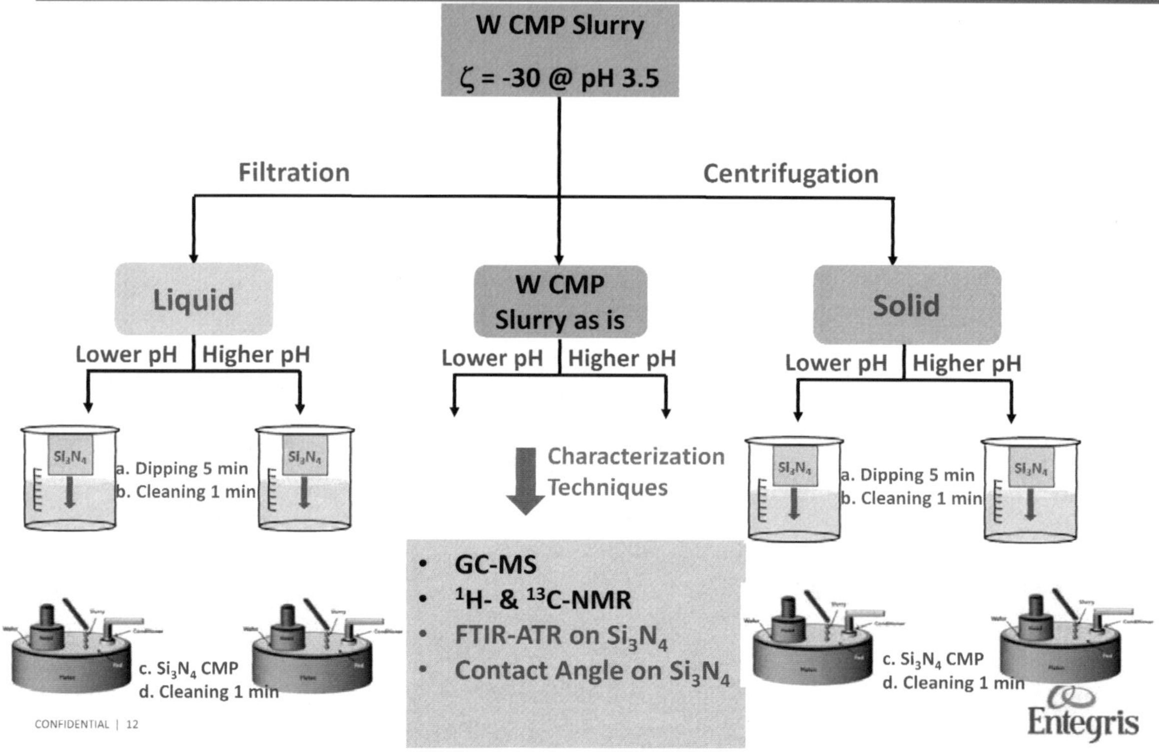

Entegris

Organic Contaminants Precipitated During Post-CMP DIW Rinse

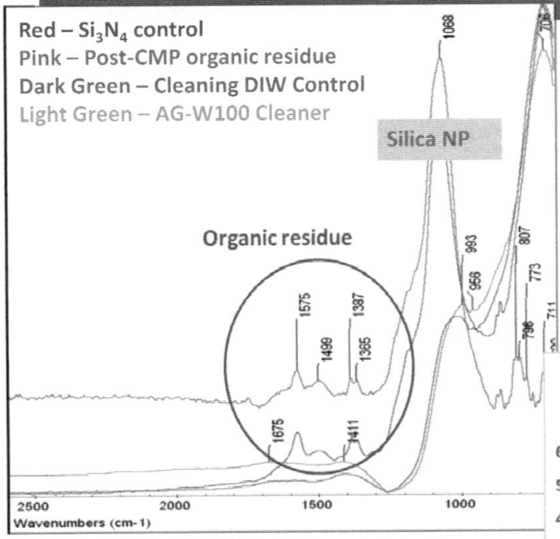

FTIR-ATR spectrum

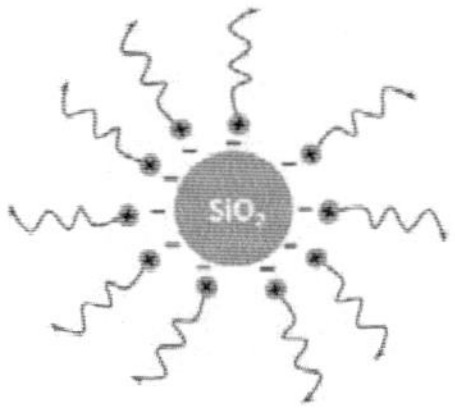

Organic contaminant (W inhibitor) adsorbed on silica particles @ pH 3

- No/minimal changes for slurry filtrate with pH
- CMP slurry releases organic residue during post-CMP rinse;
- Most sensitive: Formamide values variation

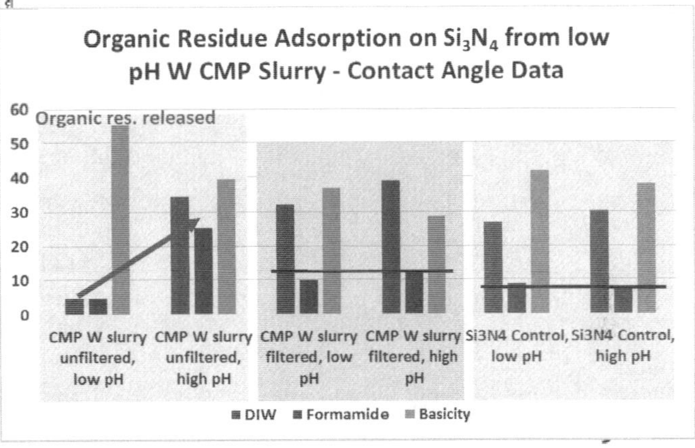

Cleaning Organic Residue on Si_3N_4 with PC AGW-100 - Metrology

1. Contact Angle – Acid-Base Van Oss Theory

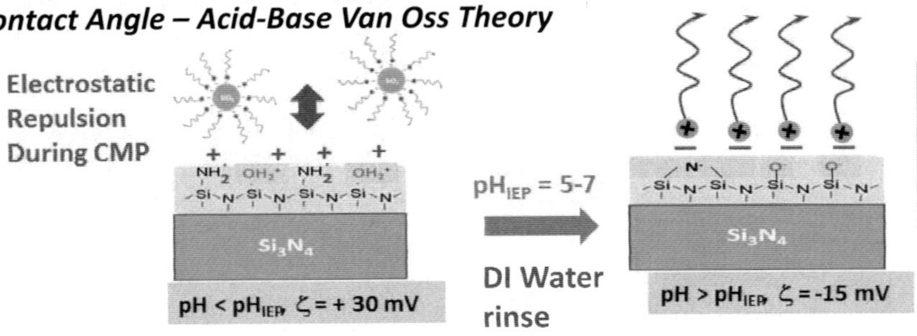

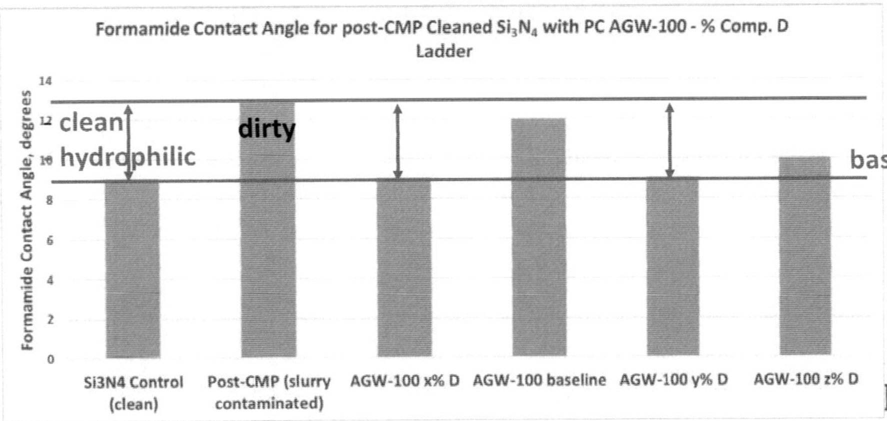

CONFIDENTIAL | 14

Entegris

Cleaning Organic Residue on Si_3N_4 with PC AG-W100 - Metrology

2. X-Ray Photoelectron Spectroscopy (XPS)

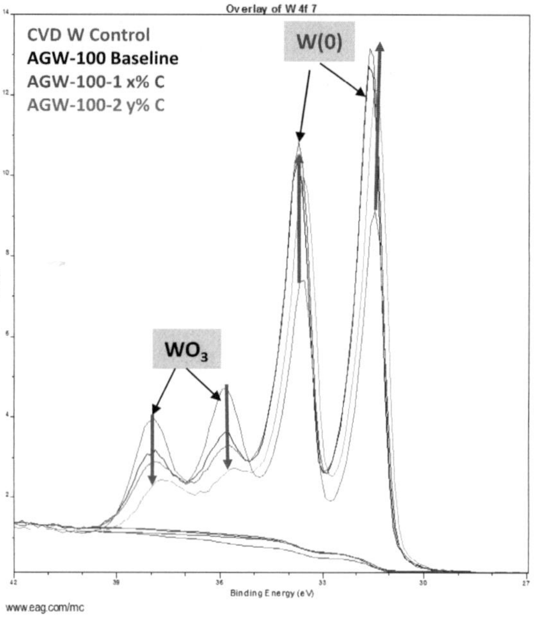

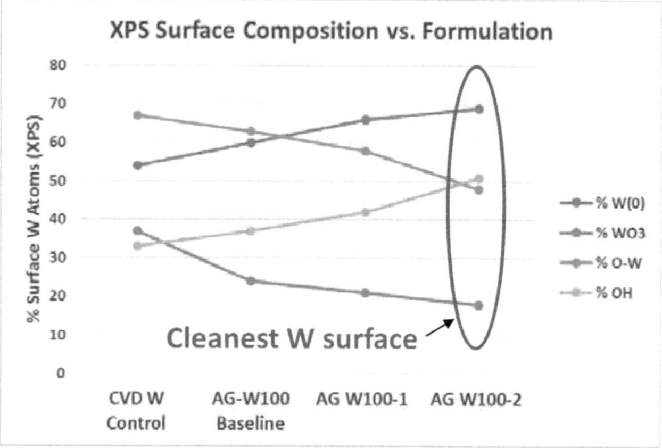

Best cleaner generates:

- **the thinnest WO_3 surface layer;**
- **lowest % surface atoms: C, N, O, etc.**

PC AG-W100 CIP Formulations

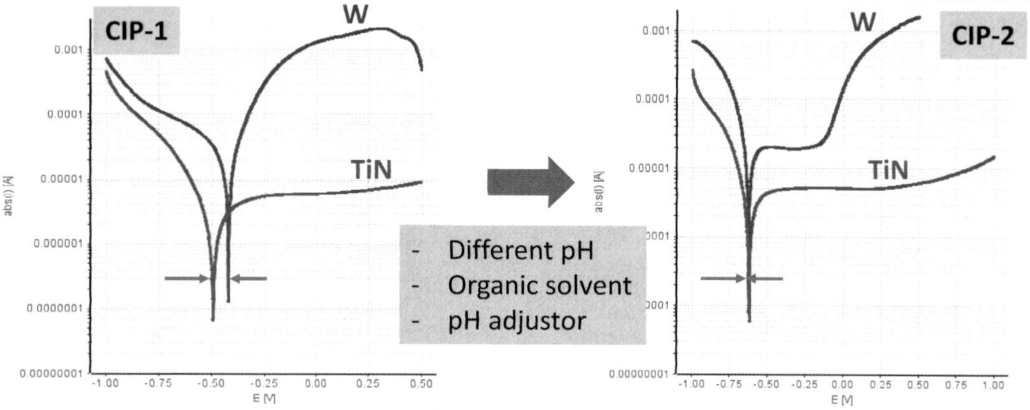

	TiN	Galvanic Corrosion	W
Galvanic Corrosion of TiN (Å/min)	0.268		
Material	Anode: TiN		Cathode: W
Corrosion Potential (V)	-0.494	-0.426	-0.422
Corrosion Current (A)	1.22E-06	2.83E-06	1.31E-05
Corrosion Rate (Å/min)	0.115	0.268	0.744
Slope	5.53E-05		5.97E-04

	TiN	Galvanic Corrosion	W
Galvanic Corrosion of W (Å/min)	0.013		
Material	Cathode: TiN		Anode: W
Corrosion Potential (V)	-0.622	-0.626	-0.625
Corrosion Current (A)	1.76E-06	2.35E-07	1.13E-05
Corrosion Rate (Å/min)	0.167	0.013	0.642
Slope	8.02E-05		5.16E-04

Cleaning Performance for PC AG-W100 CIP 1 and CIP 2 Formulations

Experiment done in collaboration with IMEC, Leuven, Belgium

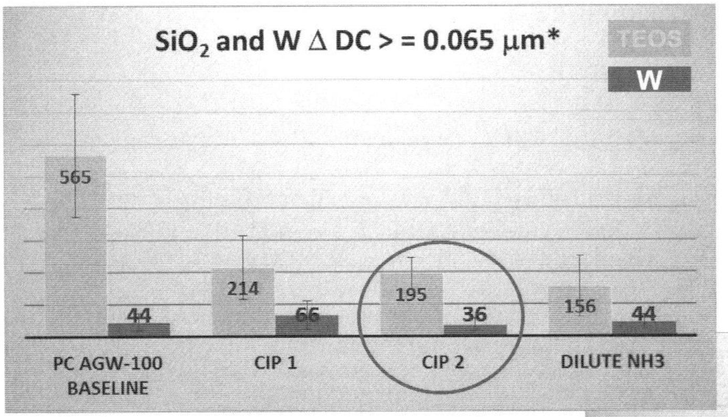

- AMAT Reflexion LK polishing tool
- KLA SP-3 defect inspection tool

- Si_3N_4 – ~125x less defects than dNH_3
- W - ~ 3.5x less defects than dNH_3
- SiO_2 – comparable with dNH_3

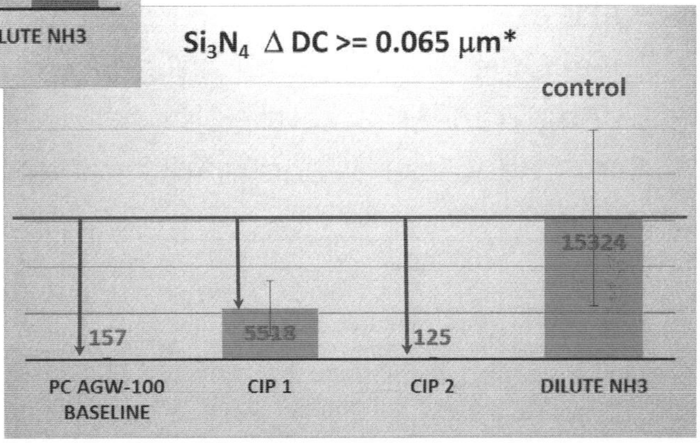

Conclusions

- W/TiN/Dielectrics post-CMP cleaning formulation PlanarClean® AG-W100, was successfully developed and tested for performance:
- Low particle count defects (on SP3) – big improvement over the commodities like dAmmonia and SC-1;
- Low/no W/TiN galvanic corrosion (Tafel plots) – tunable by varying components;
- Low/no W/TiN interfacial corrosion on 45 nm pattern wafers (SEM);
- Low/no post-CMP organic residue on dielectric substrates (FTIR-ATR, contact angle, XPS);
- CIP versions with tunable and improved performance in development.

Entegris

Acknowledgements

- Robin Van Den Nieuwenhuizen – financial support
- Emanuel Cooper, Jun Liu, Liz Thomas, Volley Wang, Spencer Tu – brainstorming and consulting
- Atanu Das – electrochemistry
- Mike Deangelo, Asa Frye – SEM characterization
- Mike Owens – lab formulations
- Cuong Tran – marketing advice
- Fadi Coder – sales and marketing advice

CONFIDENTIAL | 19

Entegris

Entegris

creating a material advantage

2016 Surface Preparation and Cleaning Conference

Please contact the author for a copy of this presentation:

Er-Xuan_Ping@amat.com

2016 Surface Preparation and Cleaning Conference

Please contact the author for a copy of this presentation:

peethala@us.ibm.com

Advanced Metal Nitride Select Etch for 7 nm FEOL and BEOL Applications

Chien-Pin Sherman Hsu, Ph.D.

Avantor Performance Materials
Hsinchu, Taiwan
sherman.hsu@avantormaterials.com

J.T.Baker® | AVANTOR™ Electronic Materials

Agenda

- **Metal Nitride (MNx) Select Etch Requirements and Challenges**
- Effect of Nitridation, Temperature and Surface Film Properties on TiN Etch Rates
- Development of Select Etch Chemistries for FEOL and BEOL Applications
- Results
- Conclusion

J.T.Baker® | AVANTOR™ Electronic Materials

2

Metal Nitride (MNx) Select Etch Requirements and Challenges

- Single-wafer processing; milder processing conditions
- Partially remove ("pull back" 10-30 Å removal) or totally remove hard mask (HM, 200-300 Å removal), profile trimming (< 30 Å)
- Wide range of selectivity/compatibility requirements (e.g., high k, low k, TEOS, BARC, metal gate, W, Al, Cu, SiGe, Ge, and III-V)
- Management/compatibility of:
  - Thin ALD films
  - Work function metal/metal alloys
  - BARC in FEOL & BEOL patterning/profiling

J.T.Baker® | AVANTOR Electronic Materials

3

Metal Nitride (MNx) Select Etch Challenges

Performance of Standard/Conventional Etch Chemistries

Description	Temp.	TiN	TaN	SiN	TEOS	BARC	W	Al	AlN	CU
100:1 DHF	25° C	3	1	7	> 60	< 0.4	2	> 500	> 35	< 5
SC-1 (5:1:1 H_2O-28%NH4OH-30%H_2O_2)	65° C	> 100	0	1	1	> 100 (lift-off)	> 500	> 800	> 100	> 30

Standard HF-based etchants
- Ineffective MNx etch; TiN etch rates are too low to be useful
- Poor compatibility with TEOS, Al and AlN

Standard alkaline peroxide-based etchants, such as SC-1
- Good TiN etch rates
- Poor compatibility with W, Al, AlN, Cu and BARC (lift-off)

Conventional etch chemistries
- Cannot meet MNx etch needs for 7 nm nodes

J.T.Baker® | AVANTOR Electronic Materials

4

Agenda

- Metal Nitride (MNx) Select Etch Requirements and Challenges
- **Effect of Nitridation, Temperature and Surface Film Properties on TiN Etch Rates**
- Development of Select Etch Chemistries for FEOL and BEOL Applications
- Results
- Conclusion

J.T.Baker | AVANTOR Electronic Materials

5

Effect of Nitridation on TiN Etch Rate

Sample Description	Atomic % Nitrogen	Atomic % Titanium	Etch Rate (Å/min) by 100:1 DHF at 25 °C	Etch Rate (Å/min) by XM-7 at 35 °C
PVD TiN #1	1X (standard)	1Y	> 100	200
PVD TiN #2	1.63X	0.31Y	0	31
CVD TiN #3	1.67X	0.27Y	1	25

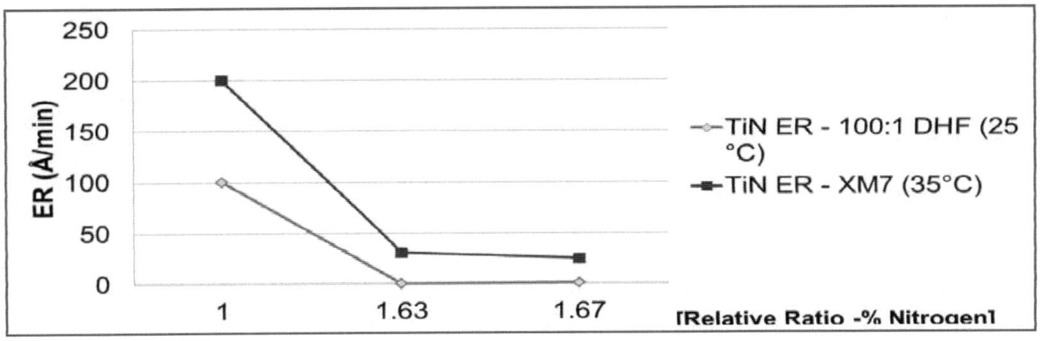

- Relative % Nitrogen of TiN films increase from 1X to 1.6-1.7X
- TiN etch rates by 100:1 DHF decrease to < 1%
- TiN etch rates by XM-7 decrease to < 16%
- ***For ≤ 14 nodes (including 7 nm), most TiN films are higher % Nitrogen and more etch resistant***

J.T.Baker | AVANTOR Electronic Materials

6

Effect of Temperature on Etch Rate: Example XSE-1

Material Description	Temp.	Etchant XSE-1
TiN	65 °C	88
	75 °C	290
	85 °C	520
W	65 °C	0.4
	75 °C	2.2
	85 °C	3
Low-k	65 °C	0
	75 °C	0.1
	85 °C	0.6

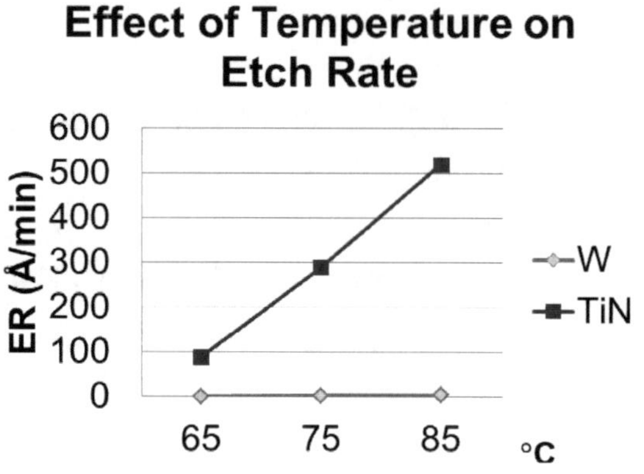

Observations for XSE-1:

- TiN etch rate is impacted significantly by temperature
- XSE-1 has superior W and low-k compatibility

J.T.Baker | AVANTOR Electronic Materials

7

Effect of Surface Film Properties on Etch Rates

Co Etch Rate/Compatibility Study
3:1 XSE-7 with H_2O_2

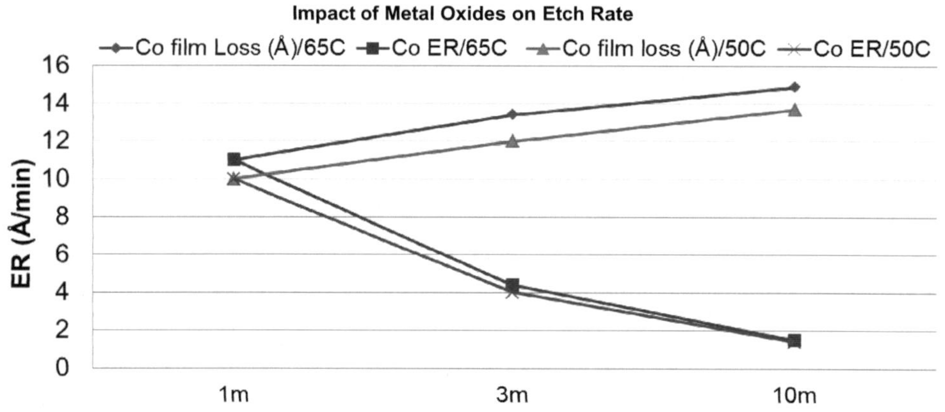

Results:

- **"Co film" loss: large initial loss within first minute; minimal loss after**
- Likely ~ 10 Å CoO (cobalt oxide) on 15-20 Å Co film readily removed by this chemistry
- Etch rate: 1.5-11Å/min for CoO/Co film at 65 °C (depends on dipping time)
- **Literature: ~ 25 Å CoO for > 50 Å Co film. Metal oxides much less on < 25 Å Co film.** [L. Smardz et. Al. J. Appl. Phys, 5199 (1992)]

J.T.Baker | AVANTOR Electronic Materials

8

Metal Nitride – A Number of Functional Constituents

Constituent Type	Titanium	Tantalum
Metal-Nitride	TiN	TaN
Metal Oxynitride	TiOxNy	TaOxNy
Metal Oxide	TiO_2, Ti_2O_3	Ta_2O_5
Metal-Metal	Ti-Ti	Ta-Ta

- Multiple types of bonding in metal nitride films (aside from M-N bonds)
- Other bonding types include metal-metal (m-m), metal oxides and metal oxynitrides
- Need to consider all of the bondings referenced above in evaluating metal nitride etch behaviors

J.T.Baker® | AVANTOR Electronic Materials

9

Agenda

- Metal Nitride (MNx) Select Etch Requirements and Challenges
- Effect of Nitridation, Temperature and Surface Film Properties on TiN Etch Rates
- **Development of Select Etch Chemistries for FEOL and BEOL Applications**
- Results
- Conclusion

J.T.Baker® | AVANTOR Electronic Materials

10

Development of Metal Nitride Select Etch Chemistries

J.T.Baker® chemistries are designed to:

- Include solubility factors at various pH and redox states (i.e., the full dissolution of all components and products/by-products from etches or cleans in solvent matrices)
- Eliminate particle-generating reactions: no self-internal reactions or reactions, especially those with the atmosphere/environment (e.g., air, oxygen, moisture)
- Manage and remove all film constituents: metal nitrides, metal oxynitrides and metal oxides – including pre-existing or generating in-situ surface oxides
- Address broad, expanding substrate/metallization compatibility requirements: ultrathin ALD films, BARC use in FEOL processing, work function alloys, etc.

J.T.Baker® | AVANTOR Electronic Materials

11

Agenda

- Metal Nitride (MNx) Select Etch Requirements and Challenges
- Effect of Nitridation, Temperature and Surface Film Properties on TiN Etch Rates
- Development of Select Etch Chemistries for FEOL and BEOL Applications
- **Results**
- Conclusion

J.T.Baker® | AVANTOR Electronic Materials

12

Metal Nitride Select Etch Chemistries

Summary Table: FEOL, MOL and BEOL Applications

Description	Temp.	TiN	TaN	SiN	Low-K	BARC	HfOx	W	Al	AlN	Cu
XSE-1	70 °C	180	12	5	< 2	> 100	14	0.4	4	-	> 800
XSE-2	70 °C	125	6	1.5	< 2	> 100	7	0.4	9	< 3	> 800
5:1 (XSE-3) – 30% H_2O_2	65 °C	180	0.7	0.4	< 3	0	0.1	> 800	> 800	-	12
5:1 (XSE-4) – 30% H_2O_2	65 °C	180	0	0.1	< 3	0	0.2	> 800	0	-	11
3:1 (XSE-6) – 30% H_2O_2	65 °C	350	3	0.2	< 3	< 5	0.1	> 400	< 10	1.5	< 3
3:1 (XSE-7) – 30% H_2O_2	50 °C	103	-	-	0	-	-	-	-	1	1
XSE-5	45 °C	20	< 0.9	0.7	< 3	< 0.4	5.4	3	> 800	> 35	> 800
	55 °C	73	< 0.9	2.4	-	< 0.4	>16	15	> 800	-	-
	65 °C	150	0.9	6	-	0.4	>16	40	> 800	> 35	-
100:1 DHF	25 °C	3	1	7	< 3	< 0.4	2	2	> 500	> 35	< 5
SC-1 (5:1:1)	65 °C	> 100	0	1	< 2	> 100 (lift-off)	0.1	> 500	> 800	> 100	> 30

Avantor has developed novel solutions that enable MNx select etch

J.T.Baker AVANTOR Electronic Materials

13

TiN Etch/Removal with Broad Capability

Compatible with Cu, Co, Al, AlN, High-k, Low-k, BARC

Description	Temp.	TiN	TaN	SiN	Low-k	HfOx	BARC
3:1 XSE7-30% H_2O_2	50 °C	100	-	-	0	-	-
3:1 XSE6-30% H_2O_2	65 °C	350	3	0.2	< 3	0.1	< 5
5:1 XSE4-30% H_2O_2	65 °C	180	0	0.1	< 3	0.2	0

Description	Temp.	Cu	Co	Al	AlN	W	
3:1 XSE7-30% H_2O_2	50 °C	1	1.5	-	1	-	
3:1 XSE6-30% H_2O_2	65 °C	< 3	< 3	< 10	1.5	> 400	
5:1 XSE4-30% H_2O_2	65 °C	11	1	0	-	> 800	

- **XSE-7 with peroxides**: TiN pull back/removal, Cu/Co/AlN/low-k compatibility
- **XSE-6 with peroxides:** TiN pull back/total HM removal; Cu/Co/AlN/low-k and HfOx/BARC compatibility
- **XSE-4 with peroxides:** TiN pull back/HM removal, ≈ XSE-6, + Al compatibility

Multiple options for various, customized needs

J.T.Baker AVANTOR Electronic Materials

14

TiN Etch/Removal with Broad Capability

Compatible with W plus Low-k, BARC, Al, AlN (suitable for M1 and other applications)

Description	Temp.	TiN	TaN	SiN	Low-k	BARC	HfOx	W	Al	AlN
XSE-1	70 °C	180	12	5	< 2	> 100	14	0.4	4	-
XSE-2	70 °C	125	6	1.5	< 2	> 100	7	0.4	9	< 3
XSE-5	45 °C	20	<0.9	0.7	< 3	< 0.4	5.4	3	> 800	> 35
	55 °C	73	< 0.9	2.4	-	< 0.4	> 16	15	> 800	-
	65 °C	150	0.9	6	-	0.4	> 16	40	> 800	> 35

- **XSE-1**: TiN pull back/total removal; W/low-k/Al compatibility
- **XSE-2**: TiN pull back/total removal; ≈ XSE-1, + SiN/AlN compatibility
- **XSE-5**: Select TiN etch vs. TaN, plus BARC compatibility

Multiple options for various, customized needs

J.T.Baker | AVANTOR Electronic Materials

15

Agenda

- Metal Nitride (MNx) Select Etch Requirements and Challenges
- Effect of Nitridation, Temperature and Surface Film Properties on TiN Etch Rates
- Development of Select Etch Chemistries for FEOL and BEOL Applications
- Results
- **Conclusion**

J.T.Baker | AVANTOR Electronic Materials

16

Conclusion

- **Surface film property studies provide a better understanding of metal nitride etch mechanisms**
  - Higher nitridation levels increase etch resistance
  - Certain nitrides show large temperature effects
  - Surface metal oxides and oxynitrides impact MNx etch behaviors
- **J.T.Baker® chemistries enable metal nitride etch**
  - Multiple chemistries for a wide range of compatibility needs: W, TEOS, low-k, high-k, BARC, Cu, Al and Al alloys
  - Excellent etch selectivity: 100x or greater
  - Tunable metal nitride etch rates: 0 to > 600 Å/min
  - Suitable for FEOL, MOL and BEOL applications on single wafer tools

J.T.Baker | AVANTOR Electronic Materials

17

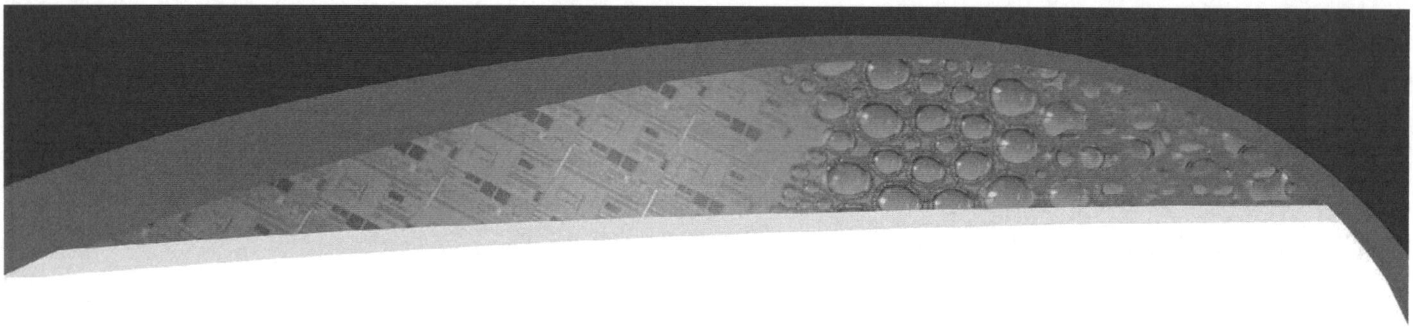

2016 Surface Preparation and Cleaning Conference

Please contact the author for a copy of this presentation:

shariq@us.ibm.com

BALAZS™ nanoanalysis

AIR LIQUIDE™

Analytical Toolbox for Technology Enabling and Troubleshooting

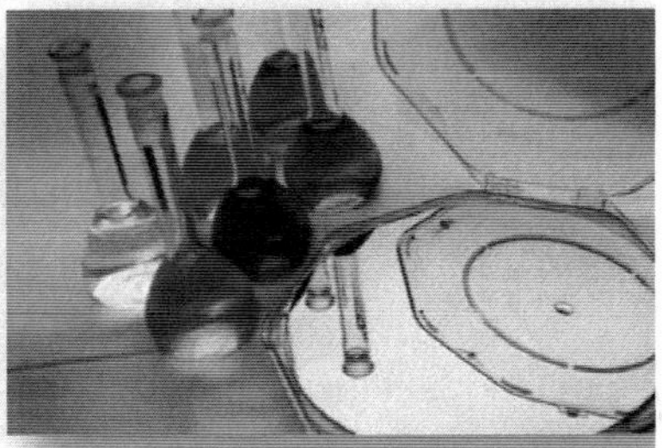

SEMATECH SPCC2016

Agenda

- The Problem Statement
- Airborne molecular contamination becoming Surface molecular contaminants
- Industrial standard methods as tools
- The Toolbox

- Trouble shooting guides

Semiconductor Integrated Circuit (IC) Chip

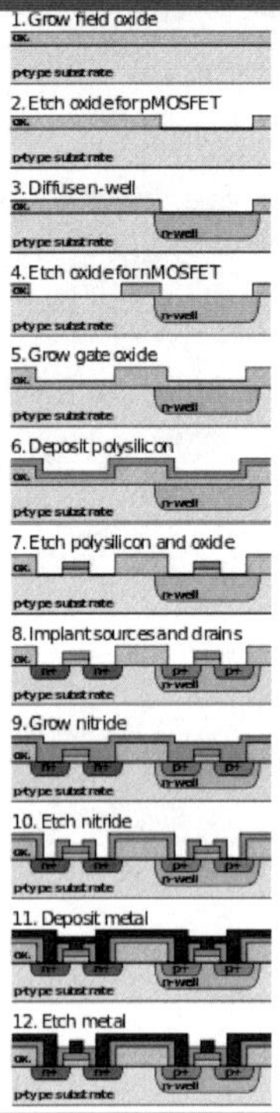

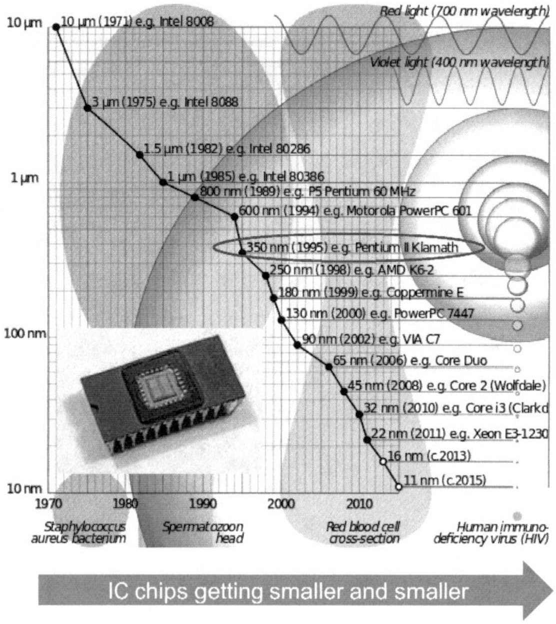

Semiconductor manufacturing processes (gate length/line width)

10 µm — 1971
3 µm — 1975
1.5 µm — 1982
1 µm — 1985
800 nm (.80 µm) — 1989
600 nm (.60 µm) — 1994
350 nm (.35 µm) — 1995
250 nm (.25 µm) — 1998
180 nm (.18 µm) — 1999
130 nm (.13 µm) — 2000
90 nm — 2002
65 nm — 2006
45 nm — 2008
32 nm — 2010
22 nm — 2012
14 nm — approx. 2013
10 nm — approx. 2015
7 nm — approx. 2020
5 nm — approx. 2022

Cleanroom Surface Cleanliness

IEST-RP-CC043: SURFACE MOLECULAR CONTAMINATION

- This Recommended Practice (RP), IEST-RP-CC043 describes the types of surface molecular contamination (SMC) that may affect critical surfaces and processes
- Types of SMC include organic and inorganic contaminants that may be adsorbed or reacted onto the surface
- Effects are manifested in the physical, electrical, chemical, or optical properties of the surface
- The document also describes sources as well as measurement and control methods

Cleanroom Surface Cleanliness

- SMC (surface molecular contamination)
  - AMC can form particles leading to SMC
  - If > monolayer, SMC can make films, homogeneous or islands
  - SMC is often < ML (~5Å) or approximately 10^{15} atoms or ions/cm^2

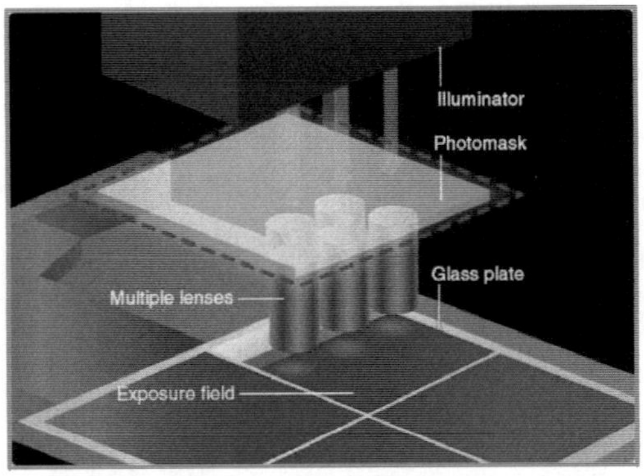

Flat panel display (FPD) lithography system and photomask

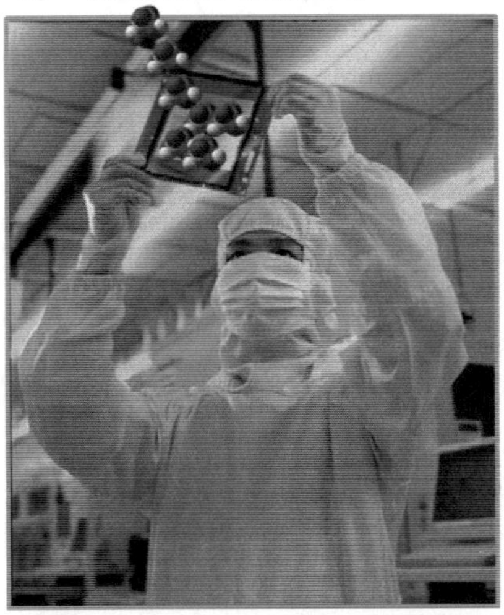

AMC Sources

- Outside air: autos, power plants, smog, industry, roofing, paving, fertilizers, pesticides, farming, sewers, fab exhaust, ocean/saline water
- Process chemicals (esp. hot), reaction by-products, reactor exhaust
- Wet cleaning, wet- and dry-etching, electroplating baths
- Solvents: lithography, cleaning solutions
- People: ammonia, sulfides, organics
- Equipment outgassing: robots, motors, pumps, fans, electronics, computers, heaters
- Materials outgassing into air or onto sealed products
- Disasters, internal or external:
- Spills, leaks (coolants), accidents, fires, power outages
- Failures of air handlers and scrubbers
- Recirculating air between areas
- FOUPs, Pods, shippers, carriers, minienvironments

SMC Effects

- DUV photoresist T-topping
- Uncontrolled boron or phosphorus doping
- Surface issues: adhesion, wafer bonding, delamination, electrical conductivity, high contact resistance, shorts, leakage currents, wetting, cleaning, etch rate shifts, spotting, particle removal, electroplating defects
- Wafer hazing: time dependent haze
- Optics hazing: hazing by adsorption, reactions, etching or photochemistry on lenses, lasers, steppers, masks, reticles, pellicles - especially for 157 and 193 nm lithography
- Corrosion: process wafers (Al, Cu), flat panel displays, equipment, instruments, wiring and facility (over many years)
- SiC/Si3N4 formation following pre-oxidation clean
- Threshold voltage shifts
- Nucleation irregularities

Airborne Molecular Contamination

Molecular Condensable (AMC-MC) Leading to SMC

Plasticizers

TXIB, bp 280 °C
(Texanol isobutyrate)

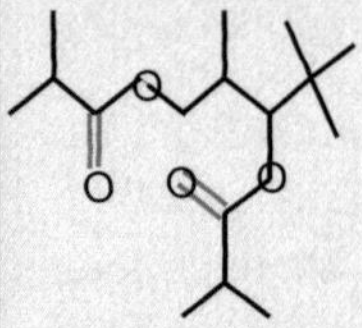

DOP, bp 384 °C
(Dioctyl phthalate)

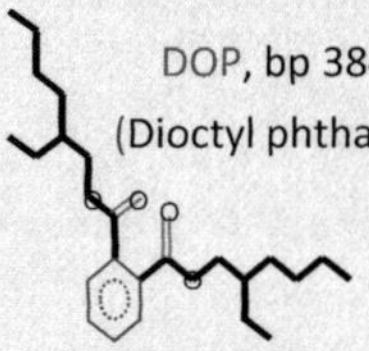

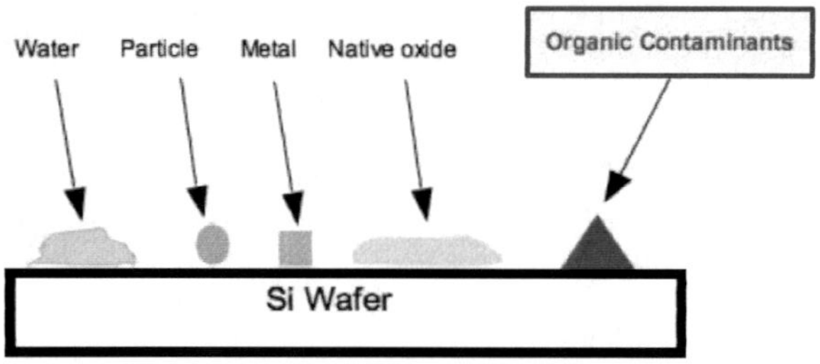

Antioxidants

BHT, bp 233 °C
(Butylated hydroxytoluene)

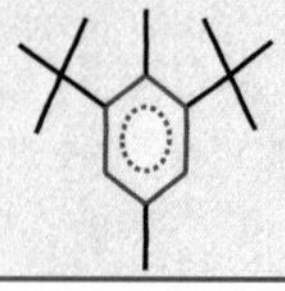

Silicones / Siloxanes

PDMS
Poly(dimethyl silicone)

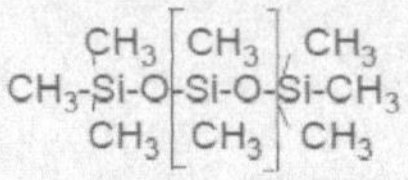

Decamethylpentasiloxane
bp 211°C

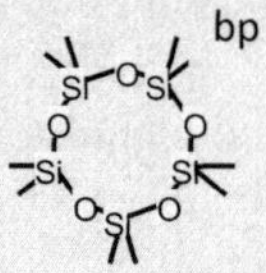

Phosphates

TEP, bp 215°C
(Triethyl phosphate)

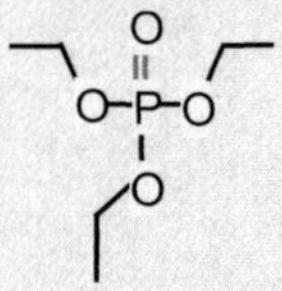

Sources of Molecular Condensables

Classification	Compound	Comments
ITRS wafer recommendation for >C7 is targeting 2 ng/cm2 (~ 0.1 ML) for 24 h exposure.		
Plasticizers	Phthalates	Phthalates are plasticizers and present in organic contruction materials.
	TXIB	TXIB (texanol isobutyrate) is commonly observed on wafers and is from airborne deposition from outgassing vinyl materials in the cleanroom.
	Diethyl phthalate Diisobutyl phthalate Butylbenzyl phthalate	Common source may be from the wafer carrier or from outgassing of copolymers used in flexible cable housing, ties, etc. Problem is phthalates affects gate oxides and can decompose to make SiC.
	DOP (dioctyl phthalate)	DOP is from cleanroom flooring material outgassing and from vinyl and plastic type materials such as wafer carriers and flexible duct connectors. Presence not acceptable in most fabs.
Antioxidants	Butylated hydroxytoluene	BHT is common on wafers from wafer carrier outgassing. BHT found in urethane foam sealants for HVAC, vinyl curtain, and floor sheet.
Silicones	Cyclo-dimethylsiloxane	Compound used in silicon HEPA gel seal.
Organometals	Organophosphates	Its presence can be an issue as phosphorus could be a counter n-dopant resulting in 10-15% yield loss. Common phosphates in the cleanroom such as TEP (triethyl phosphate) used in urethane foam sealants for HVAC and HEPA gel seals, TBP (tributyl phosphate) found in vinyl materials, and TCPP (tris{beta-chloroisipropyl} phosphate or Fyrol PCF used as a fire retardent.
	Organoborates	Would not expect this in a tool unless it was a reactive ion etcher using BBr_3. Boron could be a p-dopant and an issue for ion implantation.
Base Compds	NMP (N-methylpyrrolidinone)	Presence would affect lithography (DUV) and also cause hazing. Found in paints and strippers.
Alkyl Esters		Present in many materials including solvents, lubricants and oil.

Molecular Condensables in Air and Wafers

- Can cause delamination or irregular spin-on thickness
- Vacuum degradation in process tool and inspection tool chambers
- Organics may carbonize producing graphite, SiC and carbides resulting in electrical shorts

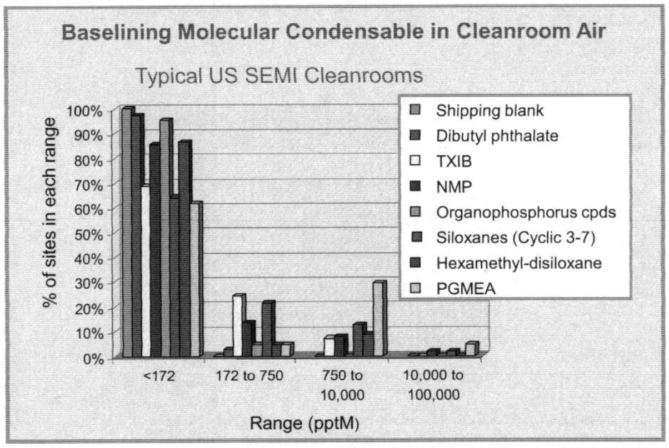

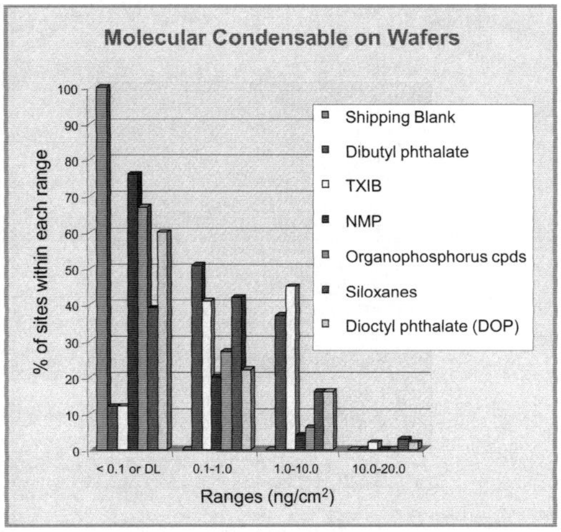

Notes:

- *PGMEA, HMDSO not found on witness wafers*
- *Silicones, phosphates, TXIB, DBP, DOP found on wafers*
- *Both air and wafer tests recommended. Improved detection sensitivity using Si wafers for some compounds.*

Material Outgassing Test Methods

Dynamic Headspace GC-MS (IEST-RP-CC031)

- Method for semi-qualitative analysis of outgassed compounds from cleanroom materials and components

IDEMA M11-99 DHS GC-MS method

- Approved for disk drives and used for cleanrooms
- Good for detecting high boiling compounds outgassed from cleanroom components, disposables

ASTM F1982-99: Analysis for organics on a silicon wafer by TD-GC-MS

Assorted materials for outgasssing characterization

Outgassing onto a substrate of interest

- SEMI E46: Outgassing of pods onto wafers, then IMS analysis
- SEMI E108: Outgassing onto wafer method, GC-MS analysis by ASTM 1982-99

Airborne Molecular Grab Test Methods

- Grab sampling provides excellent sensitivity and specificity
- It is required for determining specificity, i.e. the source of the AMC

MOLECULAR ACIDS (MA)	METHOD	COMMENTS
HF HCl HNO_3 H_2SO_4	Impinger followed by ion chromatography (IC)	Very specific, low DLs
MOLECULAR BASES (MB)		
NH_3 Amines Urea	Impinger followed by ion chromatography (IC)	Very specific, low DLs
NMP	GC-MS	
MOLECULAR CONDESABLES (MC)		
Organic compounds (e.g. silicones and plasticizers)	Adsorbent tubes followed by TD GC-MS	Survey, low DLs
MOLECULAR DOPANTS (MD)		
P, B and As compounds in air	Impinger followed by ICP-MS	Very specific, low DLs
MOLECULAR METALS (MM)		
All metals (e,g, Al, transition metals, alkali)	Witness wafers followed by VPD ICP-MS	Survey, low DLs

Note: MD detection limit best using witness wafer and VPD ICP-MS

AMC and SMC Monitoring Methods

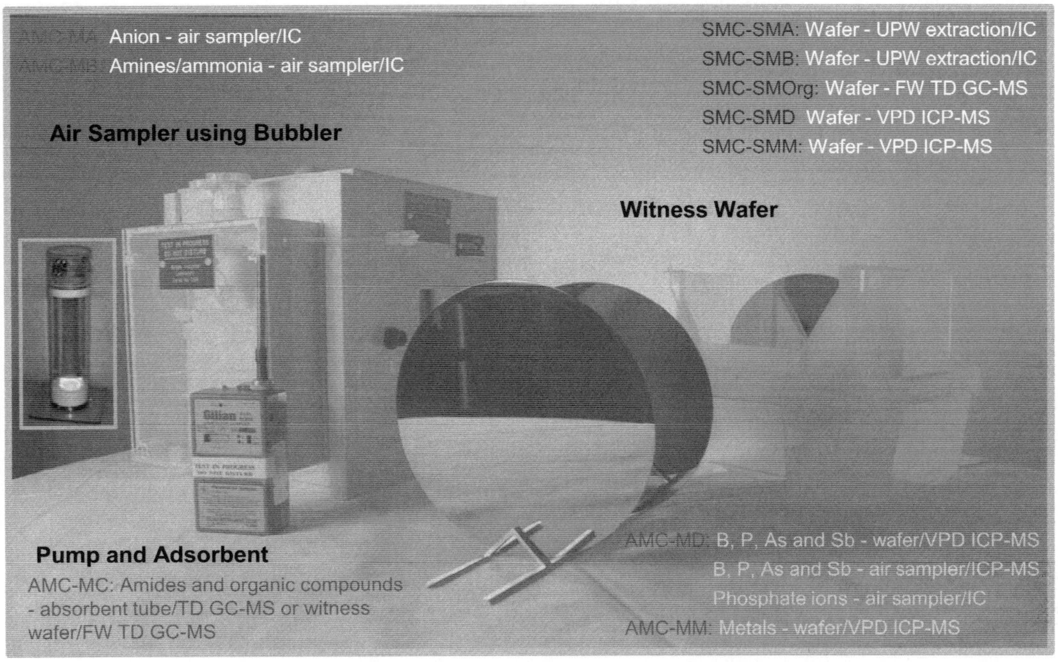

Surface Molecular Test Methods

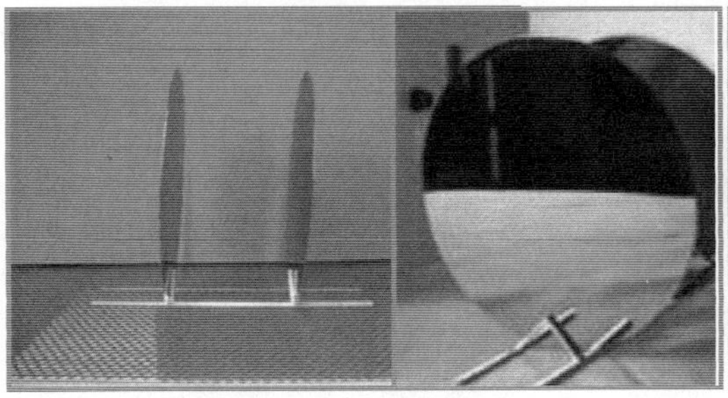

Organic-free wafers on Al rack

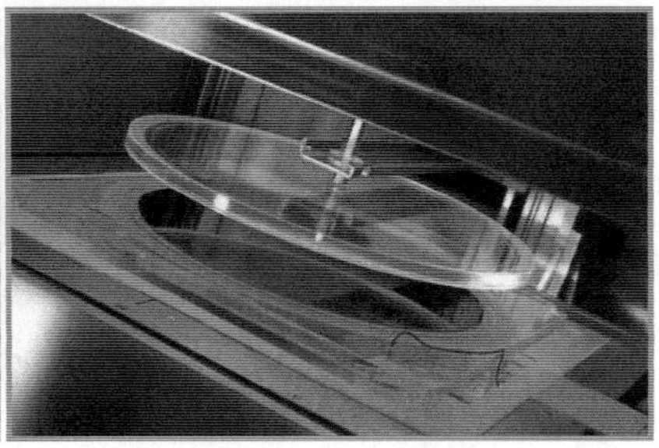

Full wafer outgassing unit

- **SME (metal / particles):** VPD ICP-MS
- **SMD:** Drop scan etch for B, P, As and Sb followed by ICP-MS
- **SMOrg:** Full wafer thermal desorption GC-MS, SEMI MF1982-1103

Trace Metals Analysis by ICP-MS

- There are several ICP instrument configurations available; Quadrupole ICP-MS, Collision Cell-Quadrupole ICP-MS, HMR ICP-MS and Triple Quad ICP-MS

Quadrupole configuration

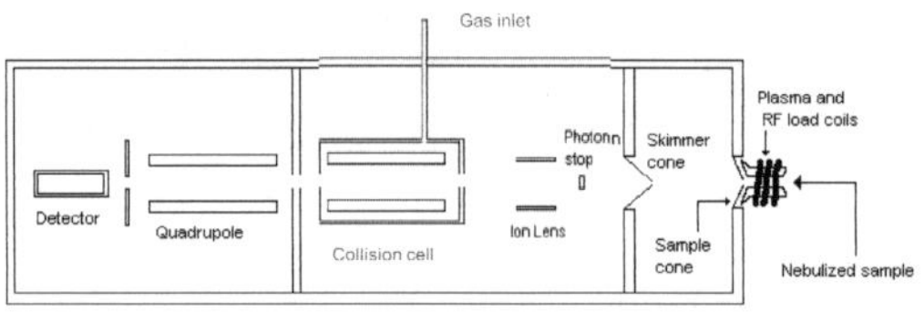

High mass resolution configuration

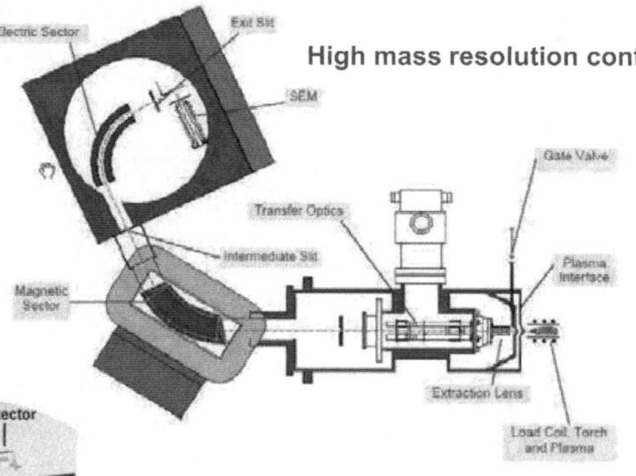

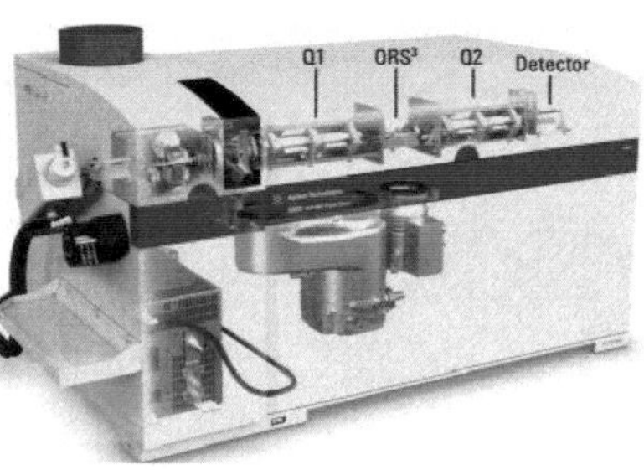

Triple Quadrupole configuration

Molecular Elements (AMC-ME)

- Metals in air may be molecular in some cases
  - $AlCl_3$ (bp 183 °C)
  - WF_6 (bp 18 °C, used for W plugs)
- In the future, more metal problems are likely
  - ALD Organometallic Precursors (organo-Cu, Al, Ti, Ga, As, Ge, In, Ba, Sr, Ta, Zr, Hf, Bi, Nb, La) and hydrides for MOCVD are volatile
  - Etch by-products may also be volatile

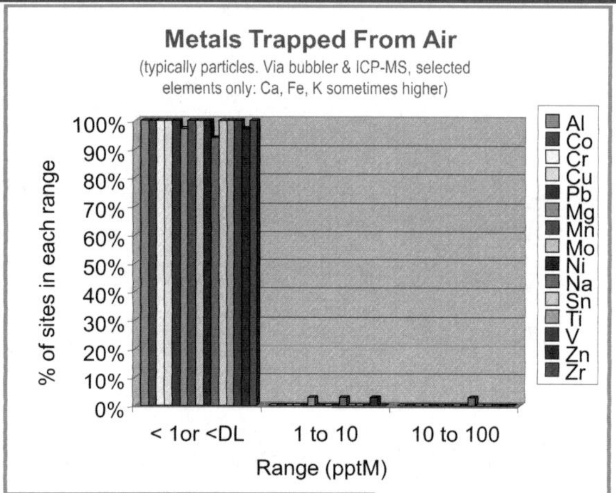

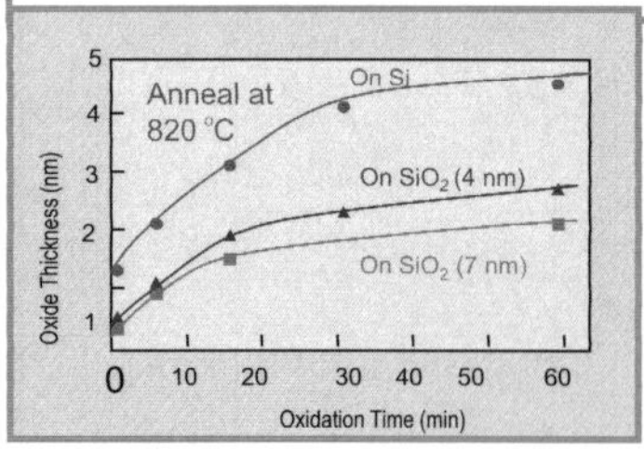

Experiments by Ohmni has shown that 100 ppb Al residue on a wafer surface after a SC1 ($NH_4OH : H_2O_2$: DIW) cleaning process can accelerate oxidation of bare silicon wafers

T. Ohmori, N. Yokoi, and K. Sato, UCPSS, p. 25, 1996

Environmental Reactivity Monitor (ERM)

- Reactivity monitors are capable of estimating AMC concentration levels as low as 1 ppb
- Their main limitation is their inability to provide continuous AMC characterization

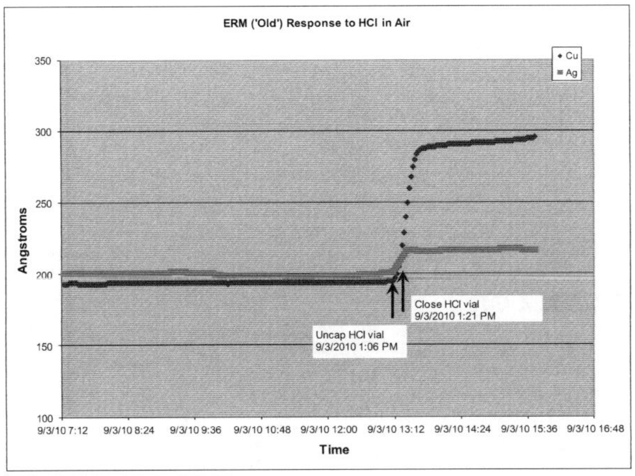

ERM response to HCl in air, 50-ppbV HCl challenge

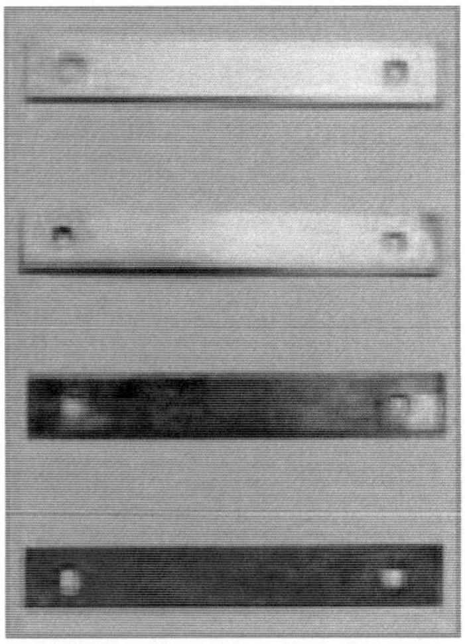

Reactivity monitors are made of Cu or Ag metal strips

Residue Characterization for Organics

Indirect	Solvent extraction / GC-MS
Indirect	Solvent extraction / NVR/FTIR
Direct	TOF-SIMS
Direct	XPS
Direct	TD GC-MS
Direct	FTIR
Direct	Raman
Direct	TGA

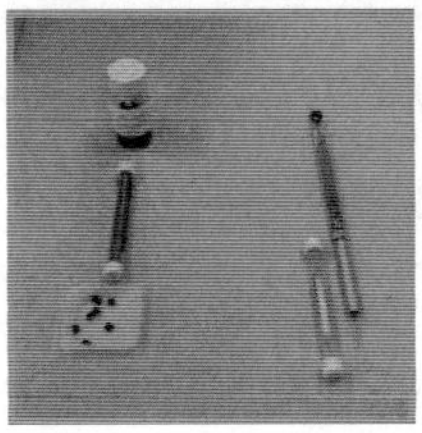

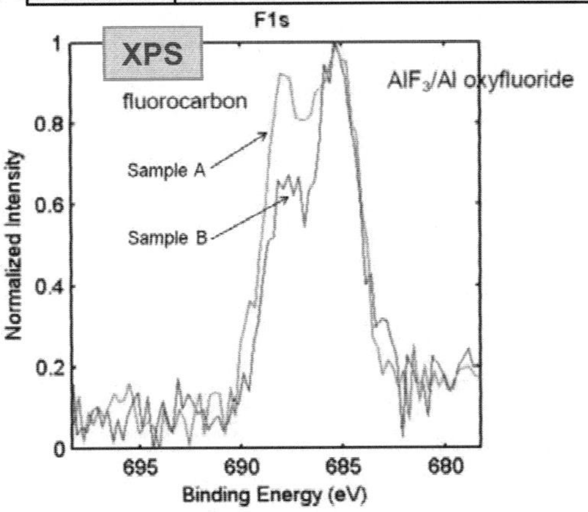

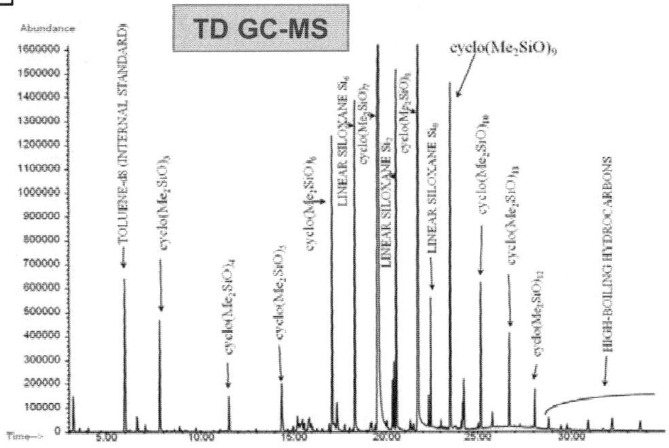

Wafer Exposure Experiments

Experiment

- Organic-free wafers exposed in various environments. Wafers analyzed using TD GC-MS, SEMI MF 1982-1103 Method-B

Results

- Individual shipper provided the best protection – less surface area exposed to the wafer

Conclusion

- Individual shippers provide protection from outside AMC
- FOUP outgassing/carryover issues possible; especially for hot wafers
- Keep wafer exposure in LFH to a minimum

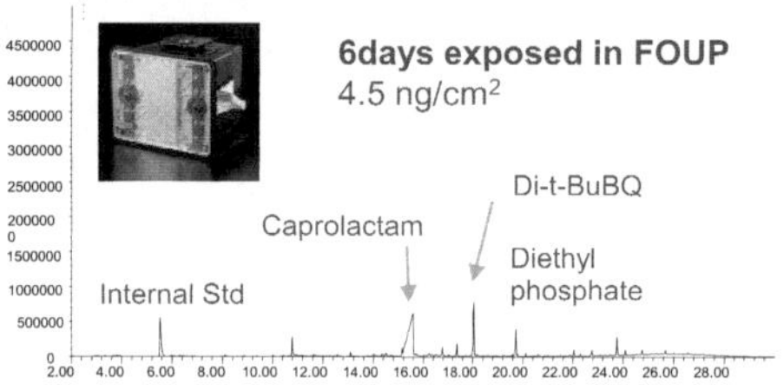

Simulation of Surface Cleaning Procedures

- IPA solvent residue vs. residues from wiping

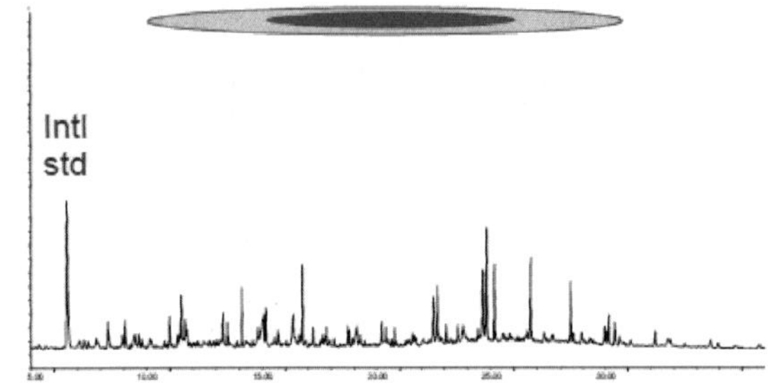

- 8 g IPA (Gigabit)
- **5.3 µg SMOrg residue left**
- 0.67 ppmw SVOC's
- DL <0.05 ppm

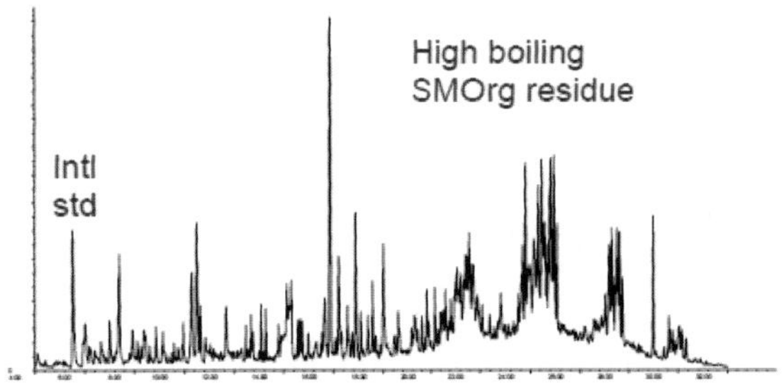

- 6 g polyester wipe + 6 g IPA (Gigabit) + 8 g Nitrile glove
  Wafer wiped 1 min
- **19.3 µg SMOrg residue left**
- 61 ng/cm2 ~ 3 ML
- Most residue from glove/wipe and some IPA

Summa Canister

- A Summa Canister is an evacuated canister that is used to collect an instantaneous air sample. Air sample may be collected over 4 to 24 hrs

- The subsequent analysis by a cryo-focus GC-MS for volatiles identifies the chemical constituents present in the sample often in concentrations to sub-ppbV levels
  - Volatile organics
  - ~165 compounds reported
  - Detection limits in the order of 0.1 to 0.3 ppbV
  - Specific
  - Semi-quantitative

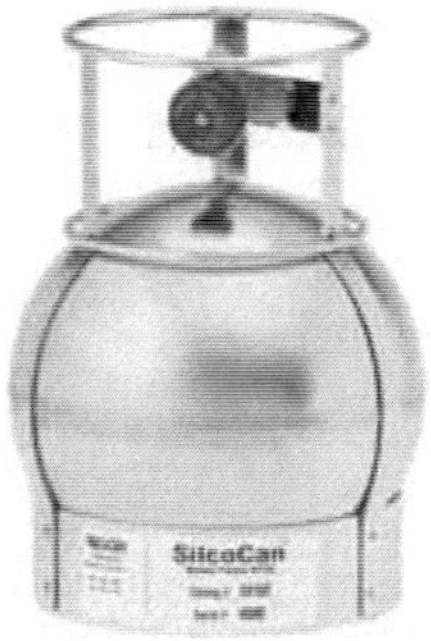

Photo-Ionization Systems

- Portable Photo-ionization Detectors (PID) systems contain built-in correction factors for 100's of compounds.
- This tool is not specific and does not identify the individual volatile material.
- The ppbRAE 3000 measures the presence of VOC's and inorganic gases from 1 ppb-10,000 ppm with widely varying sensitivities.
- System calibrated using standard gas mixture or a specialty custom gas mixture
  - Specifications (Mixture):
    - 10 ppm Isobutylene, Balance Air

ppbRAE 3000

Dräger-Tubes

- The Dräger-Tubes are glass vials filled with a chemical reagent or media that reacts through adsorption or chemisorption to a specific chemical or family of chemicals
- Air is drawn through the tube with a pump
- If the targeted chemical is present the reagent in the tube changes color and the length of the color change typically indicates the measured concentration
- Reagents used in Dräger-tubes enable the removal of potential interfering gases (e.g. aromatic hydrocarbons) to enhance specificity of the targeted chemical
- Measuring range:
  - Fluorine 0.1-2 ppm
  - Chlorine 0.3-10 ppm
  - Ammonia 0.25-3 ppm

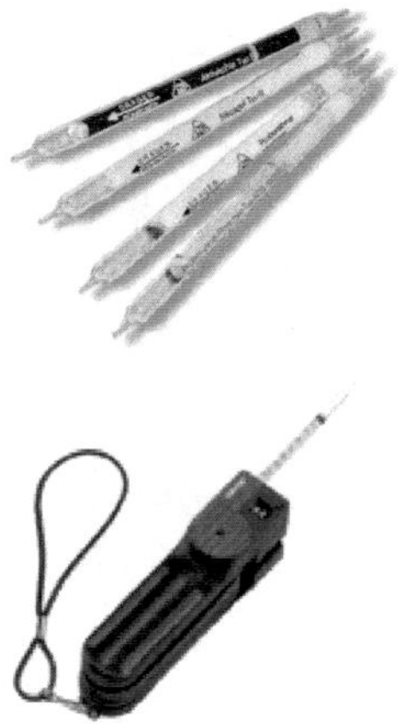

Dräger tubes and pump

Ultra-Sensitive Sulfur Detector

SO_2 and total-sulfur analyzer

- Reporting limit is 200 pptV for SO_2 and/or total sulfur
- Instrument challenge of SO_2 to determine its sensitivity (top right)
- SO_2 concentration in cleanroom make-up air (MUA) over a period of time (lower right)

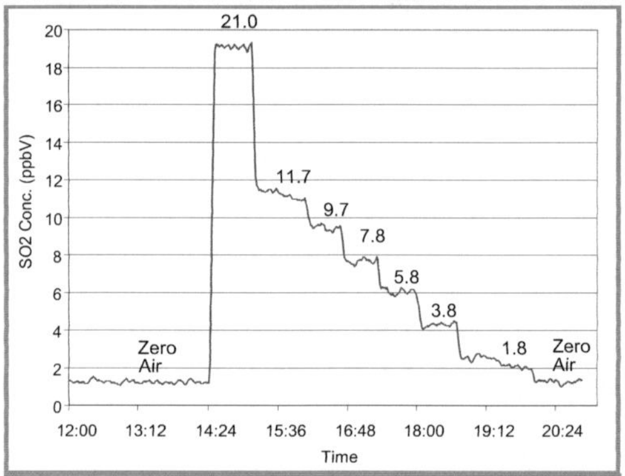

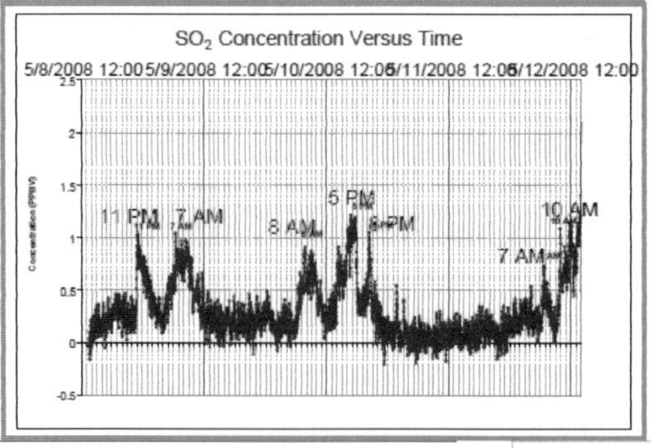

Case Study Using Total Sulfur Monitor

Problem: Experiencing sporadic but confounding problem with substrates hazing

- Analyzed hazed wafers (DIW leach + IC) and found ammonium / sulfate in 2:1 ratio
- Air was sampled using DIW bubblers and identified high concentrations of ammonia and sulfate; suspected wafer TDH due to ammonium sulfate
- TS/SO_2 monitor operated on-site to assess cleanroom air, recirculation air, MUA, that pointed to one MUA unit as the source of contamination

Contamination source: Identified as sodium metabisulfite drums stored outside, but near the building air intake

Remedy: Drums bagged and removed. TS measured in cleanroom air immediately reduced significantly

Lesson learnt: Combination of real-time monitors and highly-sensitive, species-specific methods, are powerful tools for contamination early warning, identification and mitigation

Chemiluminescence (CL)

- Sensitive and mature method for the in-situ analysis of NO_x and many organic molecular bases such as NMP and amines; cannot be applied to most MC and MA compounds
- Can potentially generates ozone, which is undesirable in the cleanroom environment
- The basic chemiluminescence chemistry is:

$$NO + O_3 \rightarrow NO_2^* + O_2$$
$$NO + O_3 \rightarrow NO_2 + O_2$$
$$NO_2^* \rightarrow NO_2 + h\nu \text{ photons } (\sim 600 \text{ to } 3{,}000 \text{ nm})$$

- Light emission with intensity linearly proportional to the concentration of NO

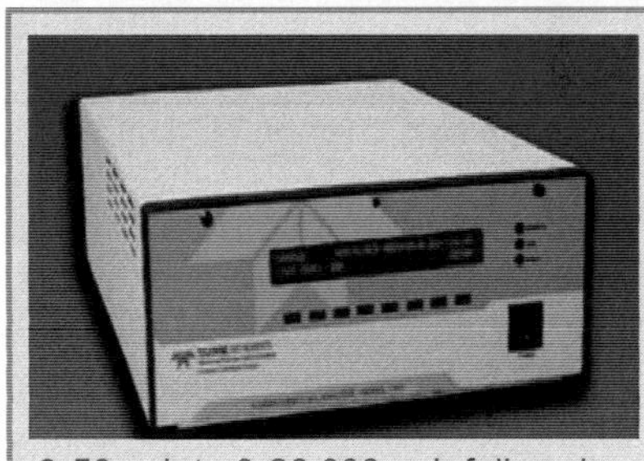

0-50 ppb to 0-20,000 ppb full scale with independent NO, NO_2, NO_x ranges and autoranging EPA's NO_x Reference Test Methods 20 and 7E were based on and written for chemiluminescence NO_x analyzers

AMC Monitoring System

- Tiger Optics' gas phase AMC analyzer is currently available in two platforms; Tiger-i 1000 and the Tiger-i 2000 with parts per trillion (ppt) levels detection
  - The Tiger-i 1000 HF and HCl analyzers employ a split architecture that allows users to place sensor modules up to 50 m away from the central analyzer; employs Tiger Optics' patented Continuous Wave Cavity Ring-Down Spectroscopy (CW-CRDS) technology
  - The Tiger-i 2000 NH_3, HF, HCl, H_2S, CO and CO_2 analyzers provide a compact, integrated architecture for enhanced analysis with a smaller footprint, and is based on HALO line of mini-CRDS analyzers

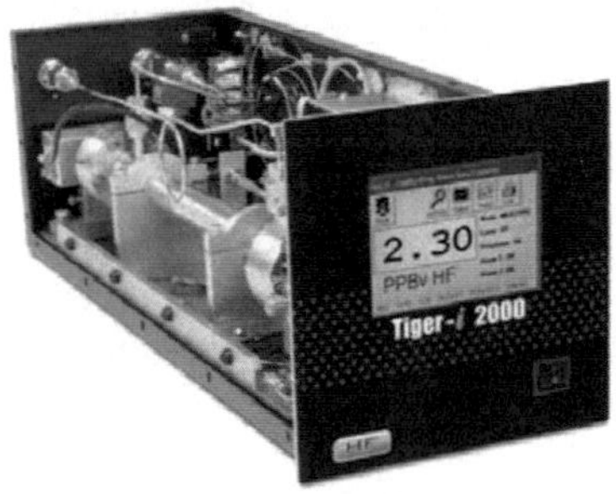

Tiger – i 2000

Ion Mobility Spectrometry (IMS)

- Fundamental papers on ion mobility in electrical fields published around 1905, but IMS is considered to be one of the relatively new analytical methods

$$AB + h\nu \rightarrow AB^* \rightarrow AB^+ + e^-$$
$$AB^* \rightarrow A + B^-$$
$$AB^* + C \rightarrow AB + C$$
$$AB^+ + e^- + C \rightarrow AB + C$$

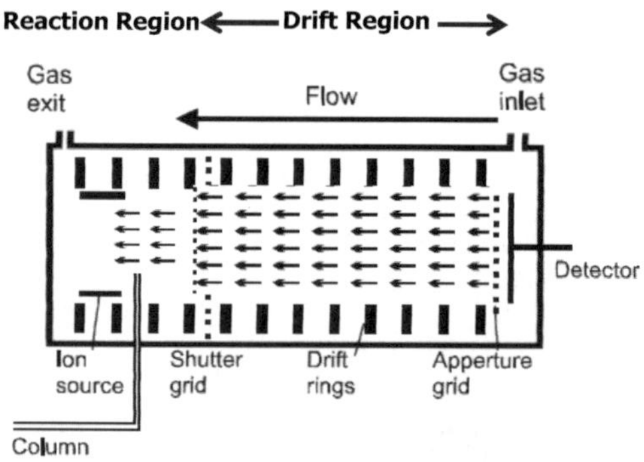

- Advantages of IMS is its capacity to operate at atmospheric pressure with no moving parts
- The IMS cell within the AirSentry® II chloride analyzer provides part-per-trillion (ppt) sensitivity and alerts users to small concentrations or changes in ambient Cl_2 and HCl levels

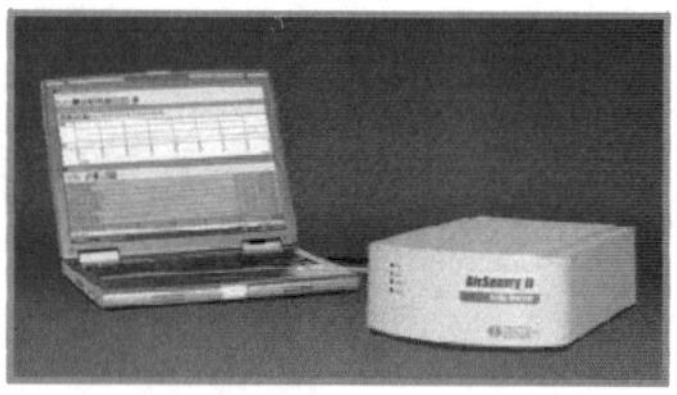

End of presentation

Thank you for your attention

Hugh Gotts, Ph.D.

hugh.gotts@airliquide.com

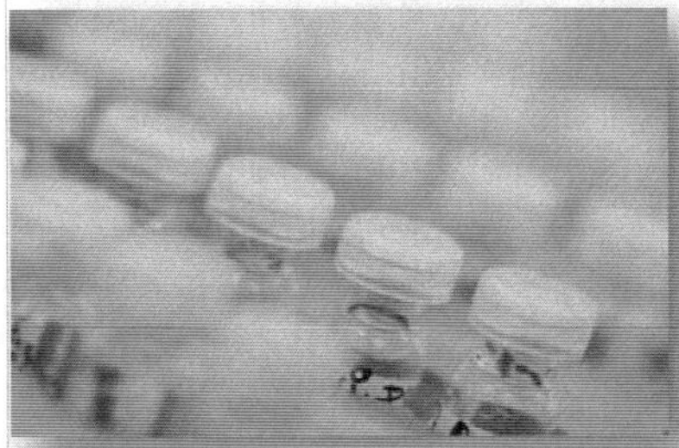

Back-Up Slides

Trouble Shooting Guides

Troubleshooting Guide

- Process impact from building and cleanroom construction materials

Problem or Effect Observed	Target Compound	Categories	Material Component
SiC formation, oxide growth variation	DOP (dioctyl phthalates) TXIB (texanol isobutyrate) BHT (butylated hydroxy toluene) Carbon	Plasticizer Plasticizer Anti-ageing, anti-oxidant Conductive filler	Flooring materials and vinyl tiles
Adhesive failure, surface hydrophobic	Poly dimethyl silicone Hexamethyl tri-silicone Octamethyl tetra-siloxane	Cyclic siloxanes (silicone bi-component)	Fluid seal (ceiling grid), cauking sealant
SiC formation	BHT (butylated hydroxy toluene)	Anti-ageing, anti-oxidant	Cleanroom walls, gaskets
Unintentional doping, surface hazing	Phosphate ester Antimony (Sb) Butyl phthalate	Crossliking agent Plasticizer Anti-oxidant	Fexible connectors and ductwork
CD control, hazing, optics degradation, SiC and SiN parasitic layer formation	Amines Ammonium DBP (dibutyl phthalate) BHT (butylated hydroxy toluene)	Crossliking agent Plasticizer Anti-oxidant	Polyuethane adhesives
CD control, hazing, optics degradation	Amines, polyimides	Crossliking agent Surfactants	Paintings, coatings, concrete fillers
Unintentional doping	TEP (tri-ethyl phosphate) TCEP (tri-chloro ethyl phosphate) TCPP (tri-chloro propyl $BF_3B(O\text{-}R)_3$, boron	Fire retardant Fire retardant Fire retardant Borosilicate fiberglass	ULPA/HEPA filters (media and potting materials)

Troubleshooting Guide

- AMC impact and recommended tests

Problem or Effect Observed	Recommended Test/Source	Other Possible Tests/Comments
High contact resistance, open interconnects	Acids in air Ammonia in air Organics on wafers	Extractable anion tests from wafer surface (e.g. post RIE, post CMP)
Air ionizers (need frequent cleaning >3 times quarterly or "fuzzballs" form)	Acids, bases and organics in air as this is a string indicator of an AMC problem	Identification by EDS on ionizer tip (esp of Si containing) Ammonium salt identification by extraction and analysis by IC
SAW mass build-up	Organics on witness wafers Acids and bases in air for specific contamination identification	Note that reversible binding of water due to humidity changes and binding of alcohols or other
High particle counts in cleanroom or MENV	Acids and bases in air since ammonium aerosols may from particles on witness wafers Metals on witness wafers by VPD ICP-MS	
High particle counts on wafers	VPD ICP-MS for metals	Anions, cations or organics on wafers may sometimes be needed

Troubleshooting Guide

- AMC impact and recommended tests

Problem or Effect Observed	Recommended Test/Source	Other Possible Tests/Comments
Odor complaints	Solvent smell; test for organics in air by GC-MC Irritating smell; test for acids in air by IC Fish-like smell; test for amines in air by IC	If testing needs to be performed for EHS or compliance monitoring appropriate OSHA, NIOSH, or EPA methods may need to be used
Delamination of photoresist,	Organics on wafers at the process	This exposure test is typically 24
Gate oxide thickness measurement variation or Ellipsometry errors	Organics on 24-hwitness wafer by GC-MS	Minimize air exposure time between gate oxidation and Ellipsometry
Metal corrosion on wafers,	Acids and ammonium in air near	Corrosion test via coupon exposure
Electrical tests: poor recombination lifetimes, SPV or leakage currents	Trace metals on witness or process wafers by VPD ICP-MS	May need to test both front and back side of wafers if Cu is used as Cu is fast diffuser
Unwanted doping, resistivity changes, threshold or flatband voltage shifts	Boron and phosphorus in air Boron and phosphorus on air Organics on wafers by GC-MS to assess organophosphorus compounds Metals on wafers by VPD ICP-MS may also be of value	B, P and other dopant profiling by SIMS can be used. Note: if boron is found, recommend testing the air for fluoride and other anions since borosilicate filters may have been attacked and can lead to boron contamination

AMC Comparison Technique Guide

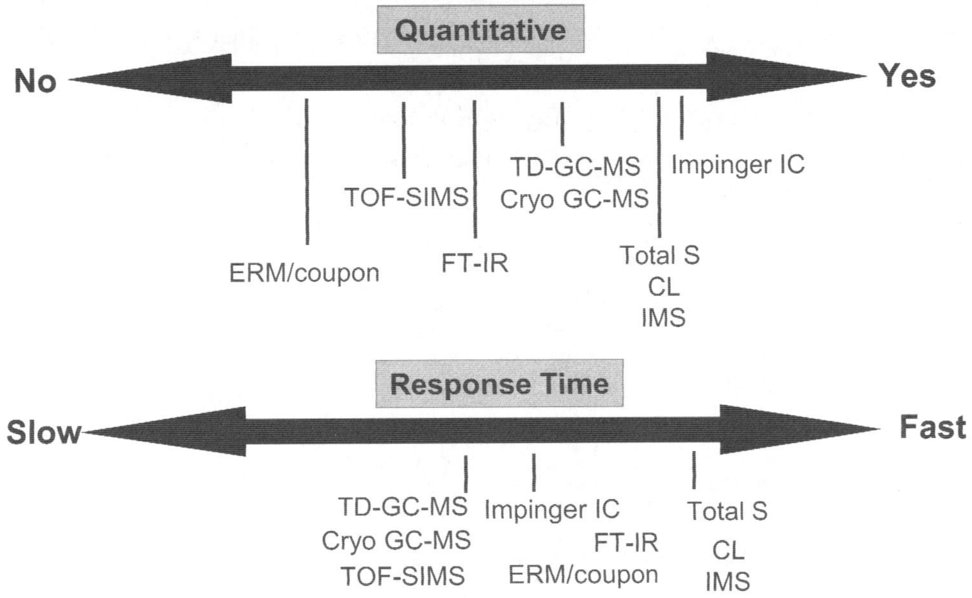

AMC Comparison Technique Guide

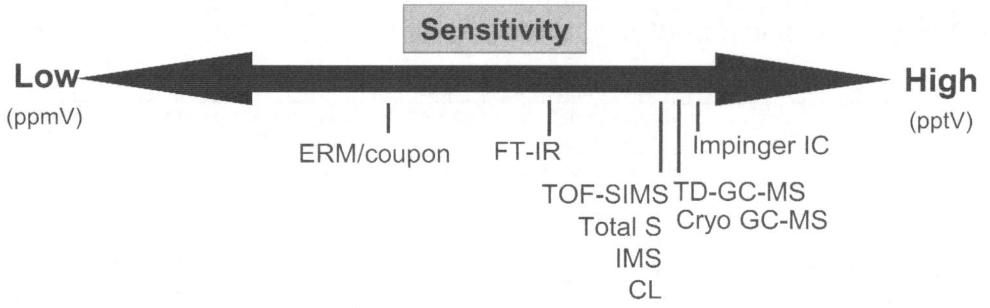

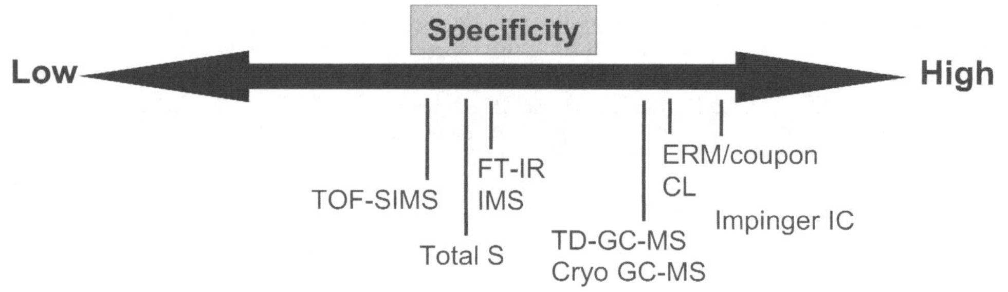

Sources of Molecular Condensables

- **Plasticizers**, such as DOP and TXIB, are used in flooring material, vinyl curtains, flexible duct connectors, wafer carriers
  - TXIB, chemical name is trimethylpentanediol diisobutyrate.
  - TXIB is a Eastman trademark and is a common plasticizer found in PVC (polyvinyl chloride)

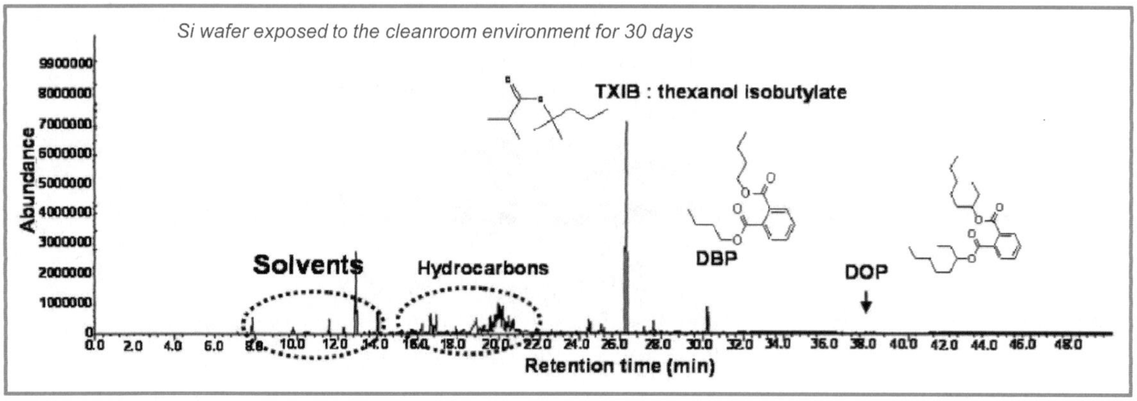

Photo-Ionization Systems

- Portable Photo-ionization Detectors (PID) systems contain built-in correction factors for 100's of compounds.
- This tool is not specific and does not identify the individual volatile material.
- The ppbRAE 3000 measures the presence of VOC's and inorganic gases from 1 ppb-10,000 ppm with widely varying sensitivities.
- System calibrated using standard gas mixture or a specialty custom gas mixture
  - Specifications (Mixture):
    - 10 ppm Isobutylene, Balance Air

ppbRAE 3000

SEMATECH Surface Preparation and Cleaning Conference

Comparison of HF and HCl cross-contamination between different Entegris FOUP platforms and Cu-coated wafers

April 19th, 2016 in Santa Clara (CA), USA

*Fernando Herrán, CEA-Leti, presenter, *fernando.herran@cea.fr*
Hervé Fontaine, CEA-Leti, *herve.fontaine@cea.fr*
Paola González-Aguirre, Entegris, *paola.gonzalez@entegris.com*
Carlos Beitia, CEA-Leti, *carlos.beitia@cea.fr*
Jim Ohlsen, Entegris, *jim.ohlsen@entegris.com*
Jorgen Lundgren, Entegris, *jorgen.lundgren@entegris.com*

CEA–Leti, MINATEC Campus, 38054 Grenoble cedex 9, France
Entegris, SAS, Parc Centr'Alp Ouest, 196 rue du Rocher de Lorzier, 38430 Moirans, France

OUTLINE

1. **Introduction**
2. **Issues/objectives**
3. **Experimental protocol**
4. **Results**
   - HX adsorption by the FOUPs
   - HX outgassing from the FOUPs
   - HX transfer to Cu-coated wafers
5. **Conclusions**

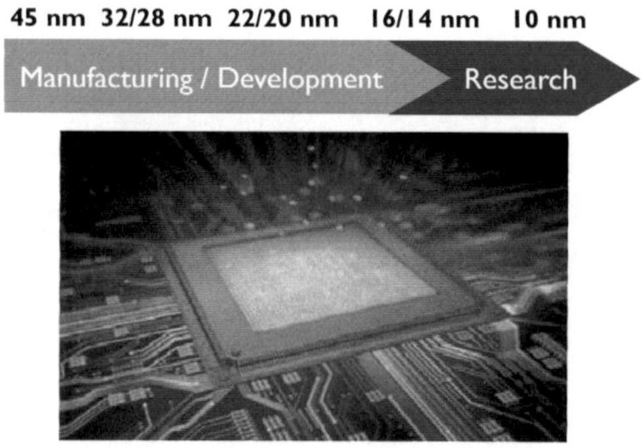

INTRODUCTION

AMC sorption from CR air, released from stored wafers, equipment connection…

Subsequent outgassing, transfer to wafers & potential defectiveness

*Air → FOUP → Wafers

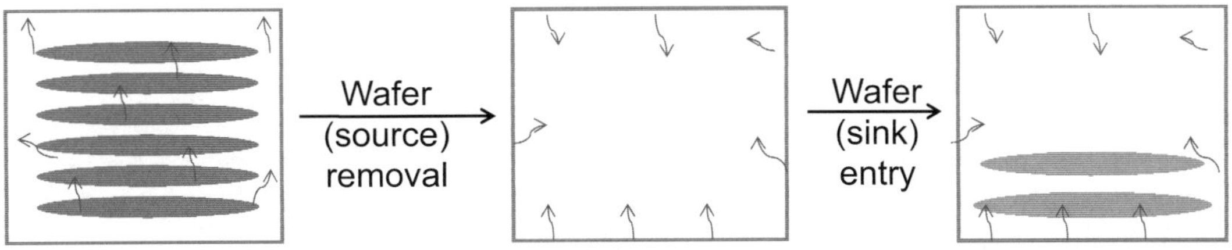

FOUP contamination by wafer outgassing

Contaminant outgassing FOUP → air up to equilibrium

Contamination transfer FOUP → Air → Wafer

- Solution-diffusion model (polymer membranes): molecular transfer governed by gas **solubility** and **diffusion** in polymers
- New generation FOUPs must minimize the impact of the AMCs (not only particles) onto the wafers

*H. Fontaine et al., Solid State Phenomena, 2008

ISSUES/OBJECTIVES

- Among AMCs, **HX** are identified as **root cause of defectiveness** where **moisture** and **time** play a critical role as well:

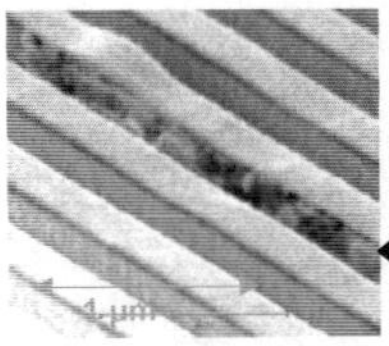

Corrosion of a Cu via

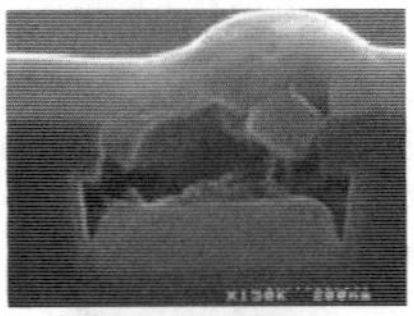

Al lines corrosion

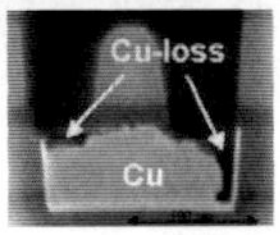

TiF_x crystal growth on TiN

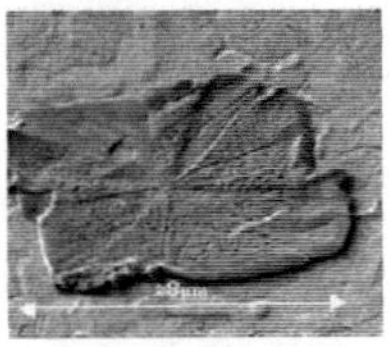

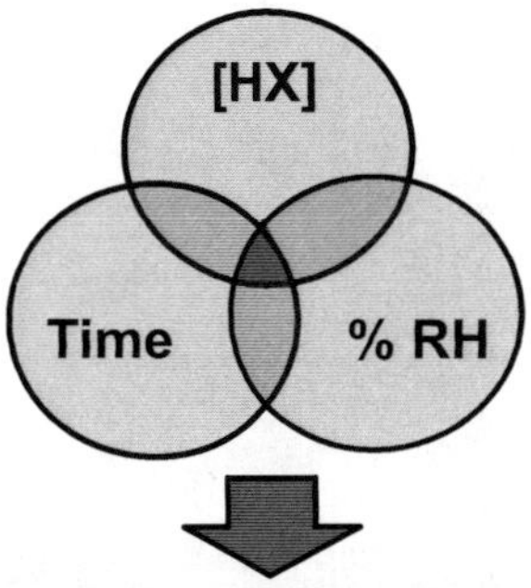

Wafer DEFECTIVINESS

<u>Objectives</u>: to compare two critical molecules (**HF*/HCl**) and **three Entegris FOUP models** in terms of:

- **HX contamination, sorption & subsequent release**
- **HX cross-contamination on stored wafers**

*P. Gonzalez-Aguirre et al., Microelectronic Engineering, 2013

cea tech **EXPERIMENTAL PROTOCOL** Entegris leti

☑ SPECTRA PC | ☑ SPECTRA PC/CP | ☑A300 EBM/CNT

Polycarbonate (ref.)

Polycarbonate/C-powder

EBM/C-nanotubes

FOUP INTENTIONAL CONTAMINATION

Contamination phase; t = 24h

10 µL droplet, evaporation within FOUP's volume:

- **HF 2% → 9.8 ppmv**
- **HCl 3,7% → 10.3 ppmv**

SPCC 2016: CEA/LETI-ENTEGRIS | 5

cea tech **EXPERIMENTAL PROTOCOL** Entegris leti

0) FOUP CONDITIONING
Clean room stabilization
(21°C, 40% RH) 7 days

1) INTENTIONAL GASEOUS HX CONTAMINATION
10µL- droplet, 24h

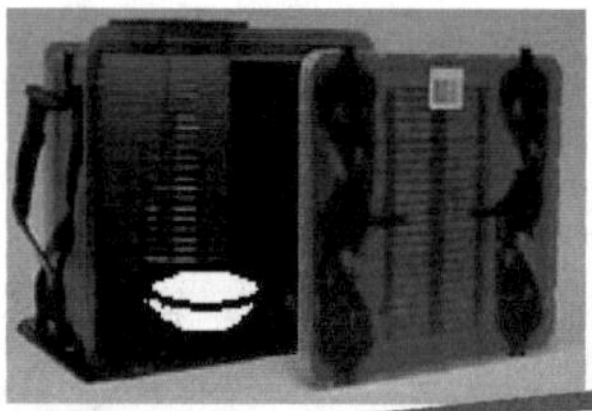

2) PURGE
(N_2 gun) 5 min

3) OUTGASSING MONITORING, IMPINGER-IC
(once per day) 7 days

3) STORAGE of Cu-WAFERS, LPE-IC
HF: t < 12 days
HCl: t < 2 days

IC = Ion Chromatography, LPE = Liquid Phase Extraction

SPCC 2016: CEA/LETI-ENTEGRIS | 6

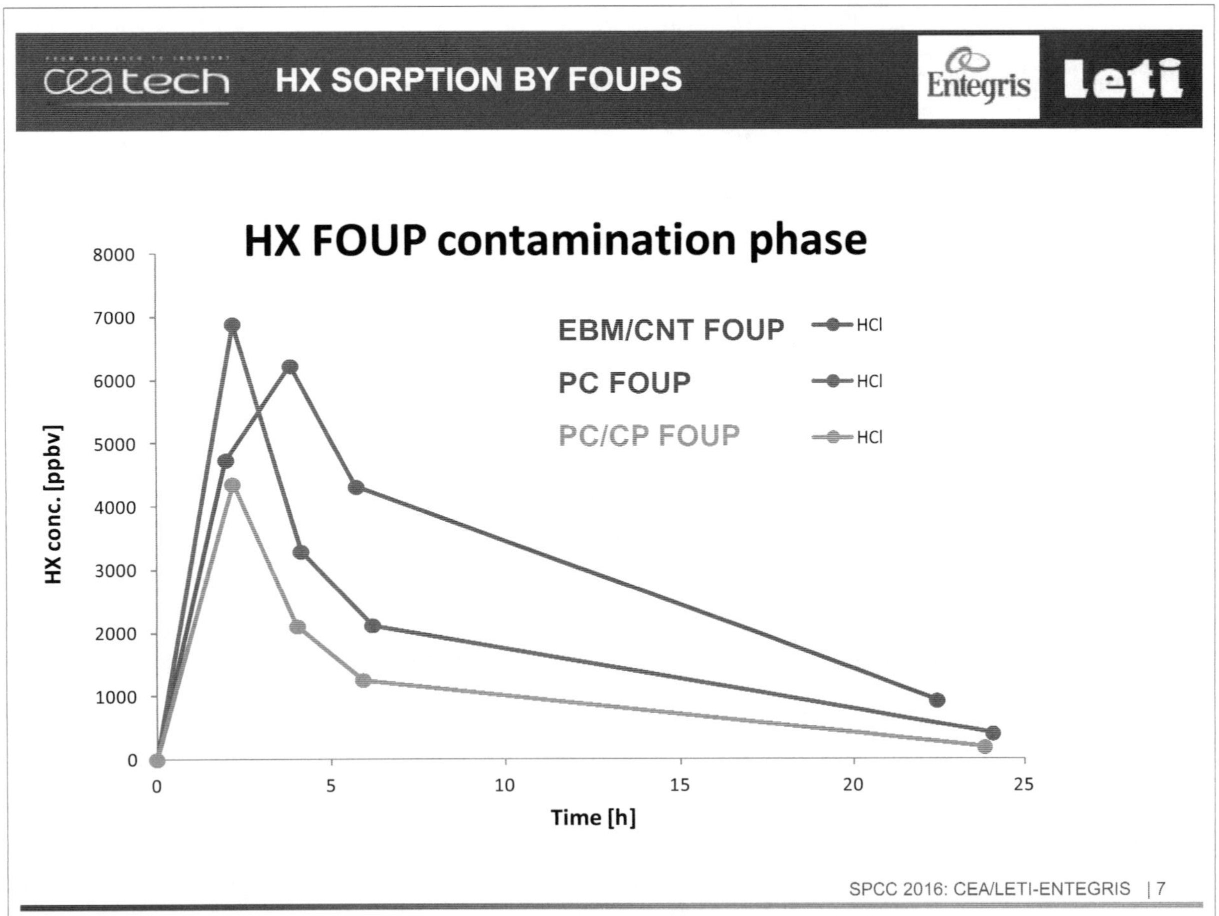
cea tech
HX SORPTION BY FOUPS
Entegris
leti
HX FOUP contamination phase
EBM/CNT FOUP HCl
PC FOUP HCl
PC/CP FOUP HCl
HX conc. [ppbv]
Time [h]
SPCC 2016: CEA/LETI-ENTEGRIS | 7

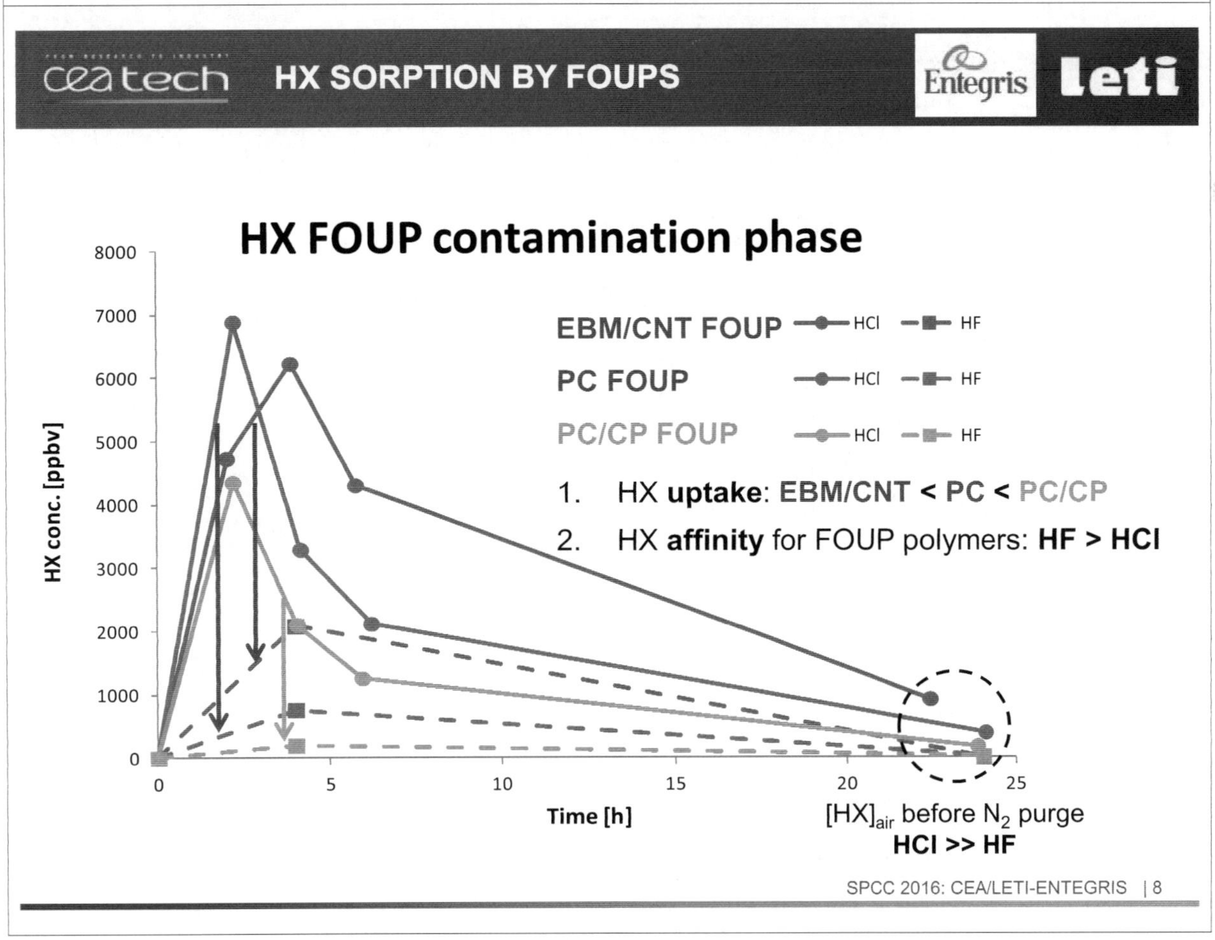
cea tech
HX SORPTION BY FOUPS
Entegris
leti
HX FOUP contamination phase
EBM/CNT FOUP HCl HF
PC FOUP HCl HF
PC/CP FOUP HCl HF
1. HX uptake: EBM/CNT < PC < PC/CP
2. HX affinity for FOUP polymers: HF > HCl
HX conc. [ppbv]
Time [h]
[HX]air before N2 purge
HCl >> HF
SPCC 2016: CEA/LETI-ENTEGRIS | 8

HX SORPTION BY FOUPS

FOUP contamination: wafer outgassing $[HX]_{air}$

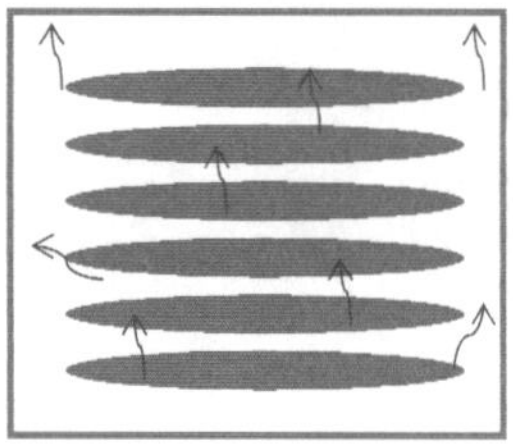

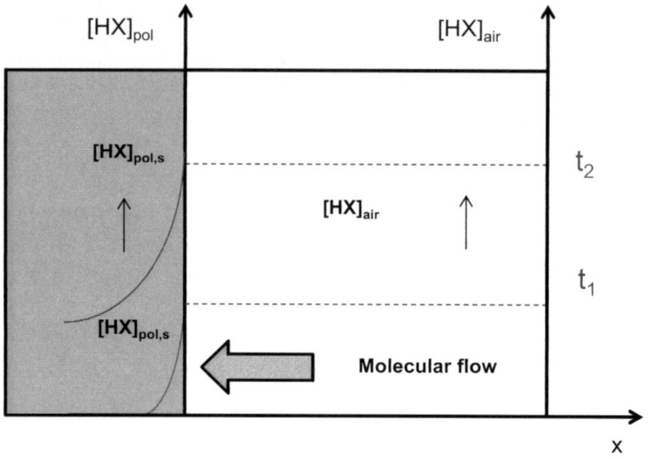

1. $[HX]_{air}$ **increases and penetrates the polymer**
2. **FOUP's contamination: f (D, S)**

$$\begin{cases} [HX]_{pol,s} = \textbf{Solub.}\ [HX]_{air} \\ J_{HX} = -\textbf{Diff.}\ \nabla [HX]_{polymer} \end{cases}$$

HX OUTGASSING BY FOUPS

Before purge [ppbv]:

$[HCl]_{EBM/CNT} \approx 1000$

$[HCl]_{PC} \approx 400$

$[HCl]_{PC/CP} \approx 200$

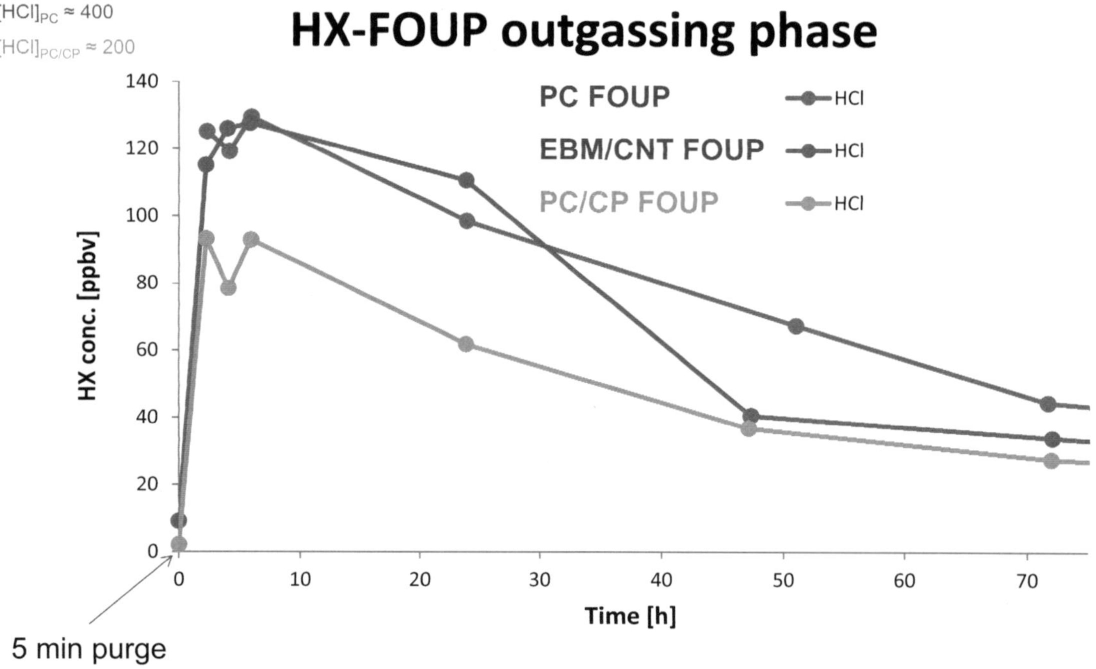

HX OUTGASSING BY FOUPS

Before purge [ppbv]:
$[HCl]_{EBM/CNT} \approx 1000$
$[HCl]_{PC} \approx 400$
$[HCl]_{PC/CP} \approx 200$

HX-FOUP outgassing phase

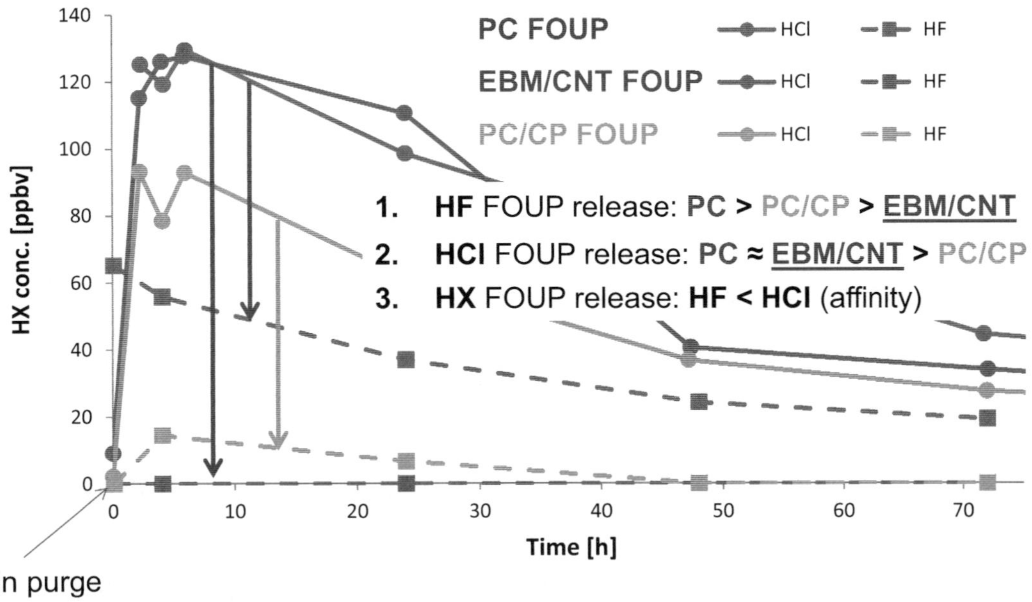

HX OUTGASSING BY FOUPS

✓ HX transfer to Cu-coated wafers??

Empty FOUP: outgassing [HX]air

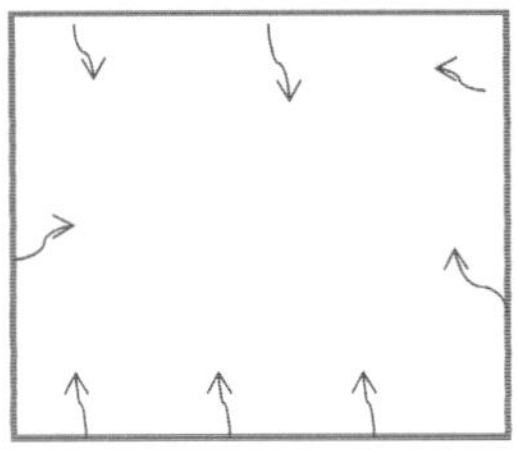

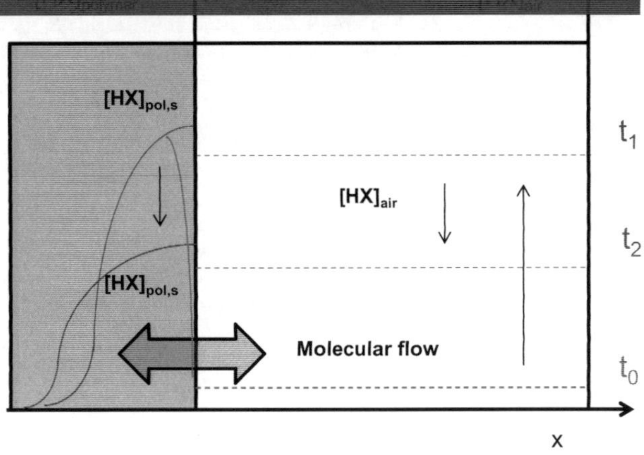

1. $[HX]_{air}$ sharp increase until $[HX]_{pol,s}$ equilibrium → Solub.
2. Outgoing diffusion: $[HX]_{pol,s}$ is reduced → $[HX]_{air}$ pulled down
3. $[HX]_{air}$ is mainly <u>solubility-dependent</u>

HX TRANSFER TO CU-WAFERS

Cu-wafer exposure into HX-contaminated FOUPs

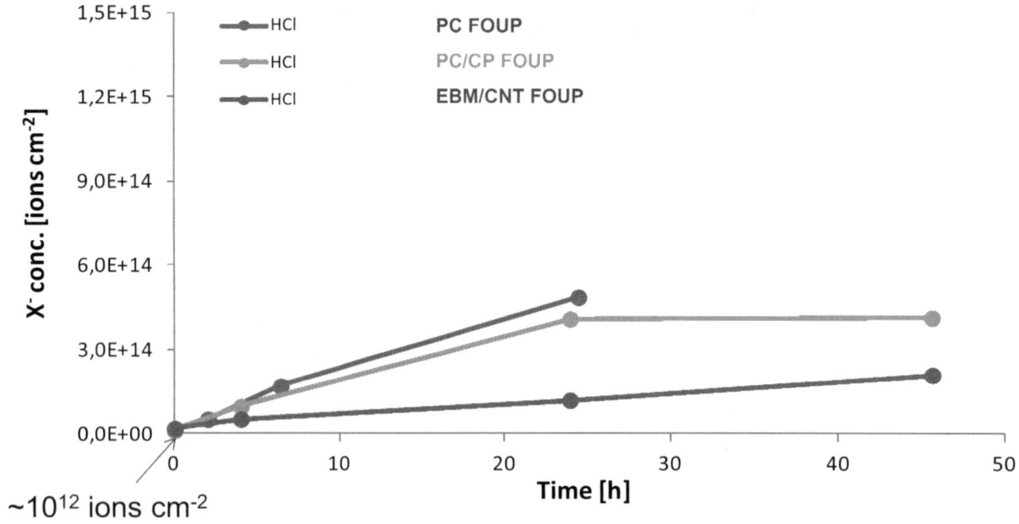

1. **HCl** outgassing rate: **PC ≈ EBM/CNT → though the lowest transfer**

HX TRANSFER TO CU-WAFERS

Cu-wafer exposure into HX-contaminated FOUPs

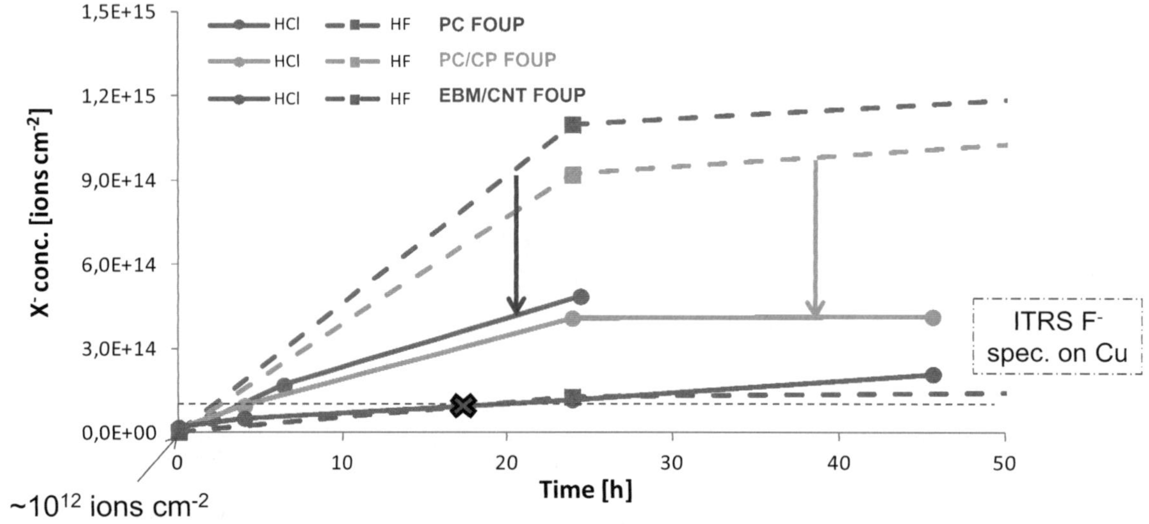

1. **HCl** outgassing rate: **PC ≈ EBM/CNT → though the lowest transfer**
2. **HX** transfer: **PC > PC/CP > EBM/CNT**
3. **HF** transfer ≈ **HCl** transfer for **EBM/CNT** only

HX TRANSFER TO CU-WAFERS

Wafers within FOUP:
$[HX]_{air} \rightarrow 0$

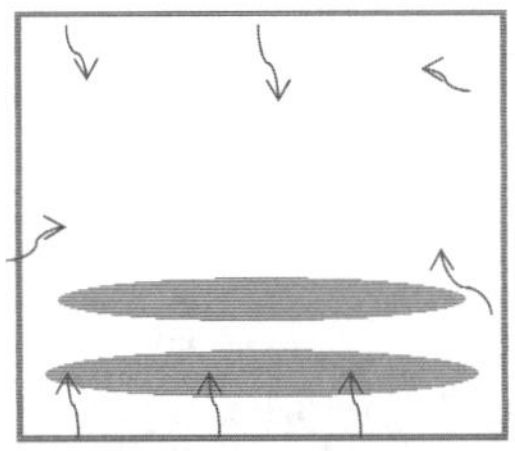

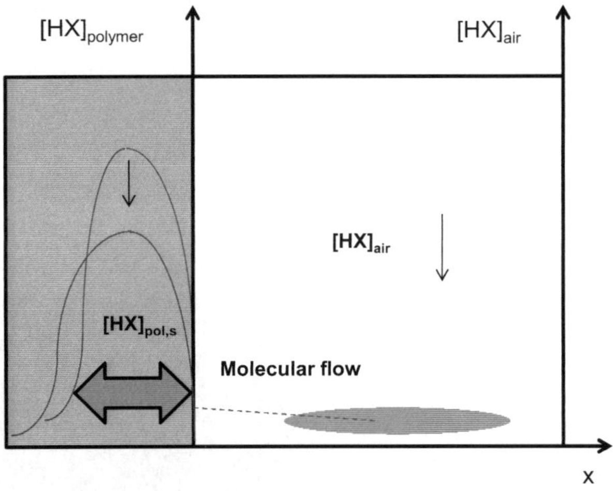

1. **$[HX]_{air} \rightarrow 0$ (HX affinity Cu) then $[HX]_{pol,s} \rightarrow 0$**
2. **HX molecular flow: FOUP's inner surface → copper wafer**
3. **$D_{air} >> D_{polymer}$ → HX transfer to Cu mainly governed by $D_{polymer}$**

CONCLUSIONS

To summarize:

→ **HX affinity** by FOUP's polymers: **HF > HCl**

→ **HX outgassing** after int. contamination: HF < HCl, **f (solubility)**

→ **HX transfer** to wafers: HF > HCl, **f (diffusivity)**

→ **EBM/CNT** FOUP: exhibits a strong HCl outgassing rate but low transfer to copper wafers (**low S and D**)

→ **FOUP's performance** in terms of AMC cross-contamination must be assessed by **wafer exposure**

→ Among the tested FOUPs, **EBM/CNT** is the most efficient to limit **HF and HCl** contamination **transfer** to wafers → **reduced wafer defectiveness is expected**

Thanks for your attention

Leti, technology research institute
Commissariat à l'énergie atomique et aux énergies alternatives
Minatec Campus | 17 rue des Martyrs | 38054 Grenoble Cedex | France
www.leti.fr

INSTITUT CARNOT CEA LETI

Surface Preparation and Cleaning Conference
April 19-20, 2016, Santa Clara, CA, USA

Effect of TMAH cleaning solution for removal of contaminants from EUV mask surface

Min-Su Kim[a], In-Chan Choi[a], Hyun-Tae Kim[a], Sung-Hae Jang[a] and Jin-Goo Park[a, b,†,]

[a] Department of Bio-Nano Technology and
[b] Materials Engineering, Hanyang University, KOREA

†+82(0)31 400 5226
jgpark@hanyang.ac.kr
Hanyang Univ., Ansan, 426-791, KOREA

NEMPL Hanyang University

Nano-Bio Electronic Materials and Processing Lab.

Outline

01 Introduction
- Issues on contaminants on EUV mask

02 Research Objective

03 Experimental Procedure

04 Results and Discussion
- Particle removal on EUV mask surface
- Carbon contamination removal on EUV mask surface

05 Summary

NEMPL Hanyang University

Nano-Bio Electronic Materials and Processing Lab.

Scaling Scenario : Lithography

	2010	2011	2012	2013	2014	2015	2016	2017	2018	2019	2020
MPU	32nm	22nm					16nm			11nm	
Logic	32nm			22nm			16nm			11nm	
DRAM		2Xnm		2Ynm / 1Xnm				1Ynm or Emerging Memory			
NAND	2Xnm		1Xnm	1Ynm or 3D		3D			3D or New		

Technology

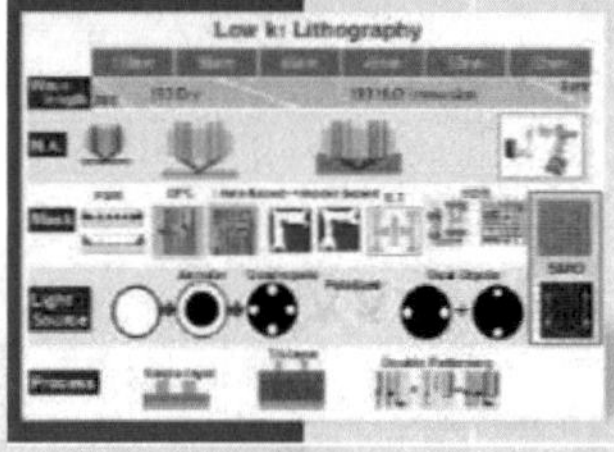

- ArF Immersion
- ArF Immersion DP(SADP, LELE, LLE), NTD, SMO
- ArF Immersion Low-k1 Lithography DP Multiple SADP(1/4) , Spacer & Cut
- EUV Lithography → EUV Extended (DP...)
- Direct Writing / EB, Nano Imprint & Directed Self Assembly

*DP Double Patterning
*SADP Self-Aligned Double Patterning
*LELE Litho Etch Litho Etch
*LLE Litho Litho Etch
*NTD Negative Tone Development
*SMO Source and Mask Optimization
*MP Multiple Patterning

G. Chung, "Partnership to build a better future, and leading edge collaboration", SPCC (2012)

Issues of EUV mask contamination

❖ EUV mask contamination

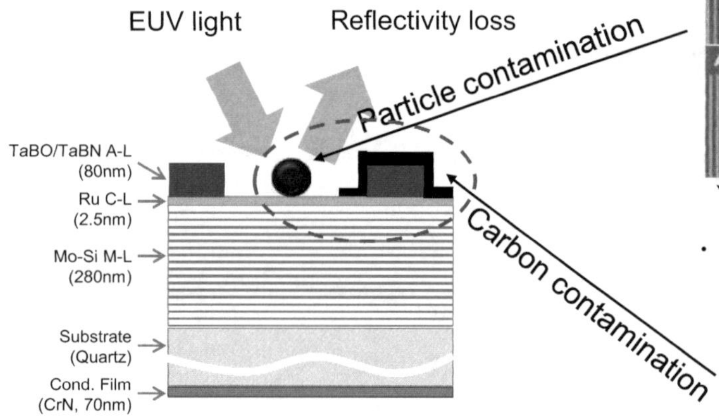

- **Particles: pattern defects on Si wafer**
- **Carbon contaminant: pattern CD change**

- Pattern defect on wafer by particle on EUV mask

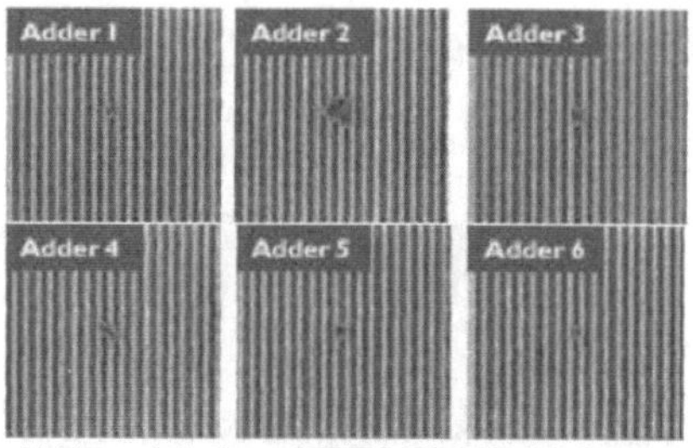

Y. Hyun, et al., Proc of SPIE 9422, 94221U1 (2015)

- CD change by carbon contamination on EUV mask

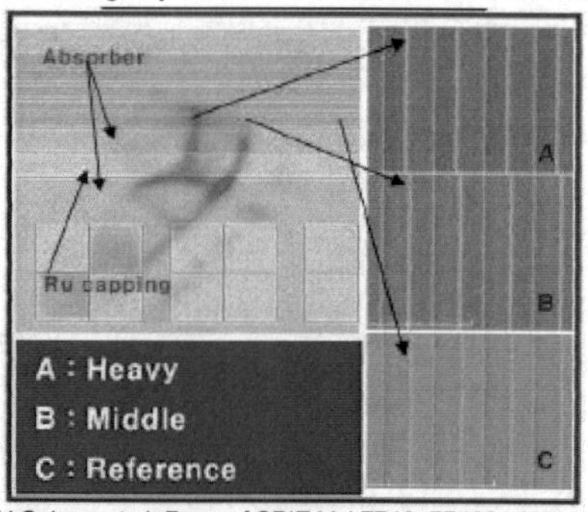

H.S. Lee, et.al, Proc. of SPIE Vol.7748, 774804 (2010)

Particle contamination on EUV mask

❖ Schematic of particle contamination on EUV mask

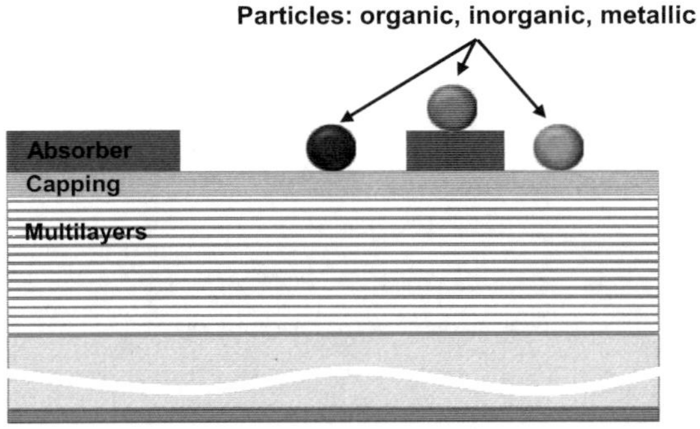

❖ Critical particle size w/ various materials on 16 nm HP node

Simulation condition
- Fast Litho tool
- Pseudo-Spectral Time Domain (PSTD) method
- Parameters: n, k, thickness

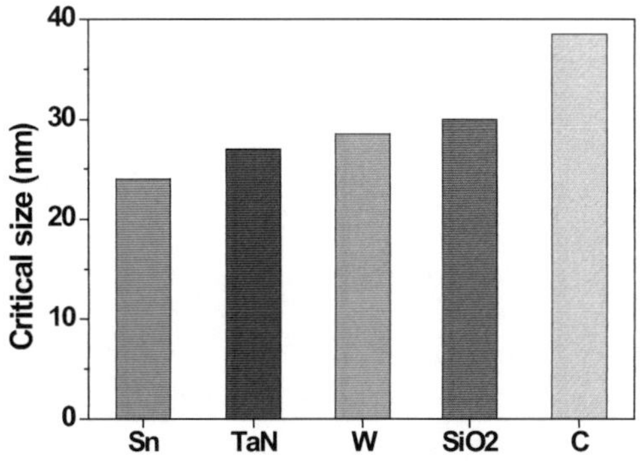

- Various type of particle can be deposited on the capping or absorber layer
- All types of particles should be removed from EUV mask surface (Capping, Absorber)

Issue of carbon contaminant removal

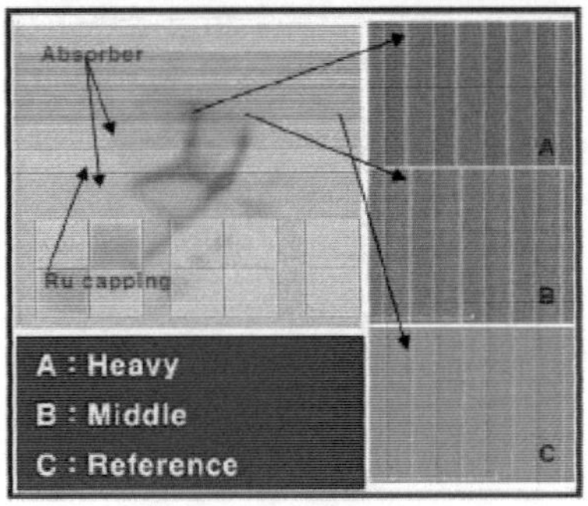

Contamination

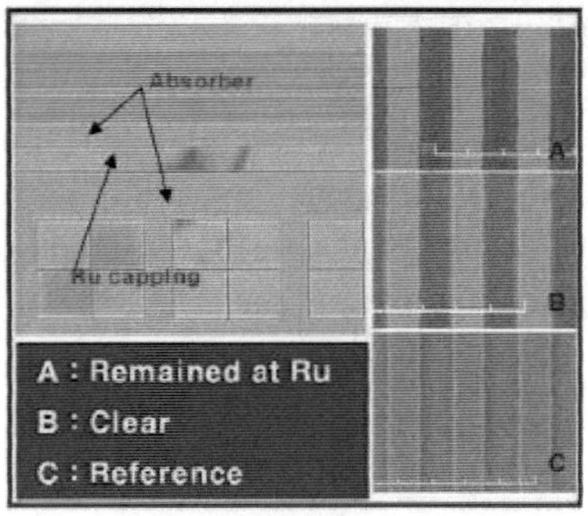

Cleaning

H.S. Lee, et.al, Proc. of SPIE Vol.7748, 774804 (2010)

Carbon contaminant was removed from ARC surface but not from Ru surface

Why carbon contaminant not removed from Ru surface ?

- Carbon contaminants on Ru surface have higher density and thickness than on ARC surface ?
- Carbon contaminants have higher adhesion with Ru than ARC surface ?

H. Lee et al., ECS Transactions, **58** (6), 93 (2013)

Research objective

Surface interactions between contaminants and EUV mask surfaces

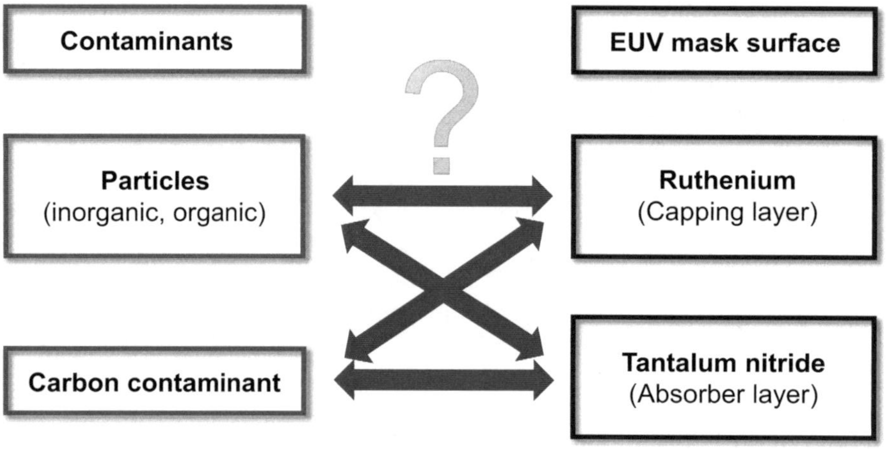

Experiments

❖ **Particle removal test**

- 2 kinds of particles
  - Silica, PSL standard particles (100 nm, Corpuscular, USA)
- 3 kinds of surfaces
  - Si wafer, TaN and Ru coated wafer (2 cm x 2 cm)
- Particle deposition (spin method, particles in DIW)
- Akrion 0.84 MHz megasonic cleaning with DIW, dNH_4OH and dTMAH

❖ **Carbon contaminant removal test**

- Hydrocarbon film deposition on TaN and Ru surface
  - Using PECVD, 40 nm thickness
- Hydrocarbon removal using solvent based cleaning solution
  - Dimethyl sulfoxide (DMSO) 80% with NH_4OH or TMAH (0.1M), 65 ℃, 20min (dipping)

❖ **Analysis**

- Particle removal efficiency measurement
  - Optical microscope (dark field mode, LV-100D, Nikon, Japan)
- Chemical bonding analysis using FTIR (Nicolet iS50, Thermo Scientific, USA)
  - Multiple internal reflection (MIR) and attenuated total reflection (ATR) method
- Contact angle analyzer (Phoenix, SEO, KOREA)

Silica particle removal with various surfaces

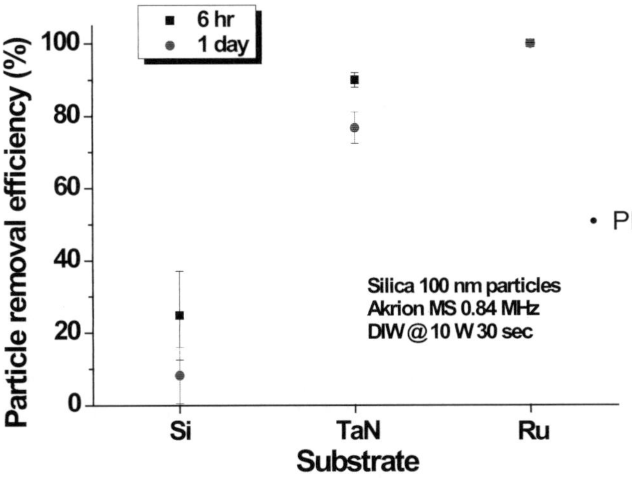

- PRE of silica particles @ various surfaces

  Si ≪ TaN < Ru

- **Silica particle removal efficiency was very lower @ Si surface even 6hr aging time**
- **Silica particle was easily removed from Ru surface**

  → Why PRE is different with surface materials?

Surface interaction (hydrogen bonding)

❖ **Hydrogen bonding between Si surface and oxide particle**

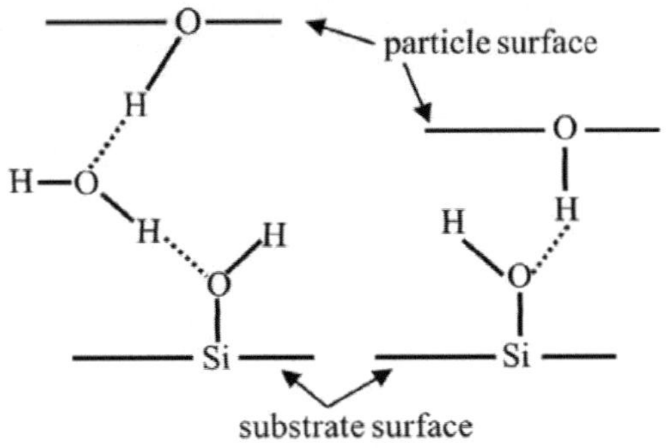

Hydrogen bonding with silica ($F_{H\text{-}bond}$)	**600 nN**
Van der Waals force with silica (F_{vdW})	**30 nN**

Hydrogen bonding ≫ van der Waals force

X. Wu, et al., *J. of Appl. Phys.* **86** (3) (1999)

- **Si surface and inorganic oxide particles can interact to form hydrogen bonds**
- **Adhesion force of hydrogen bonding is much larger than van der Waals force**

Hydrogen bonding analysis (MIR-FTIR)

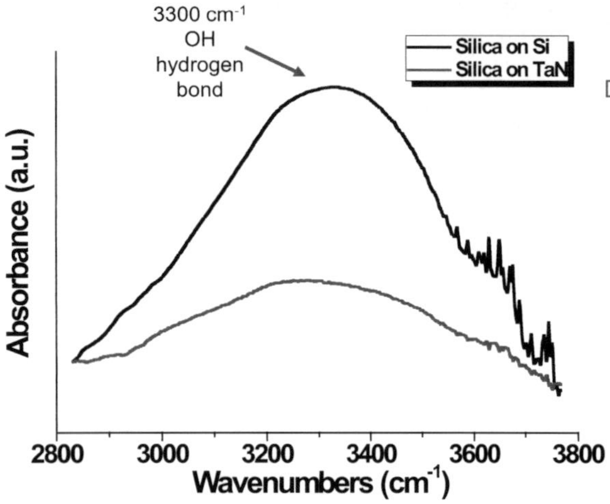

MIR-FTIR method

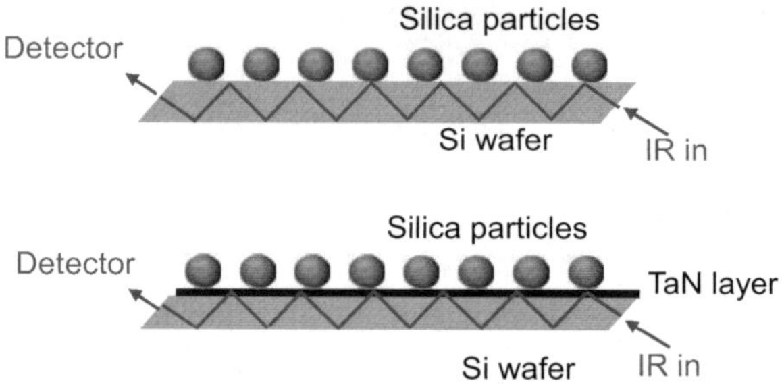

Surface	Hydrogen bonding with silica
Si	Strong
TaN	Weak
Ru	None (only F_{vdW})

- **Hydrogen bond peak (3300 cm^{-1}) was measured using MIR-FTIR method**
- **OH peak is much higher @ Si than TaN surface**
  → PRE was much lower @ Si surface

PSL particle removal with various surfaces

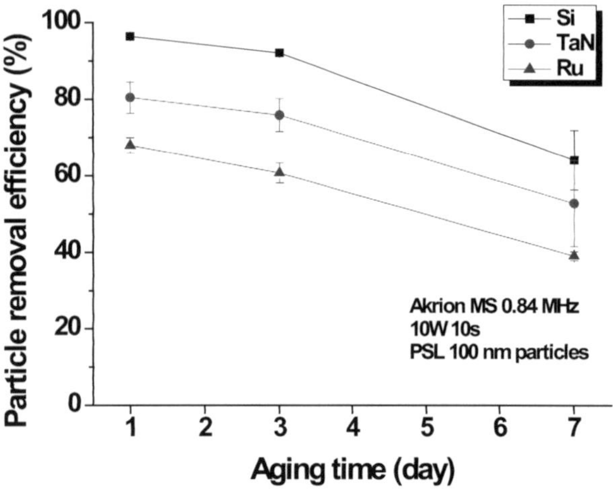

Particle	Bonding with Ru surface
Silica	F_{vdW}
PSL	Metal-carbon bond

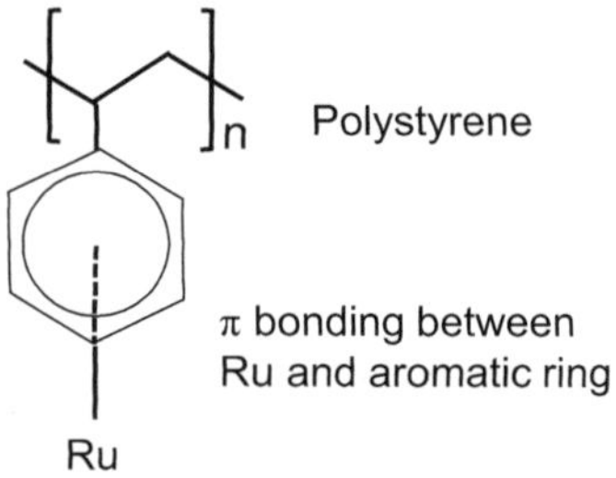

R. A. Zelonka, et al., Can. *J. Chem.* **50** (1972)

- **In case of PSL particle, PRE was lower @ Ru surface than Si and TaN surface**
- **Ru (transition metal) can form chemical bonding with aromatic ring of PSL particle**
  → Increase adhesion force

Effect of TMAH on PSL particle removal

❖ PRE of silica on Ru surface

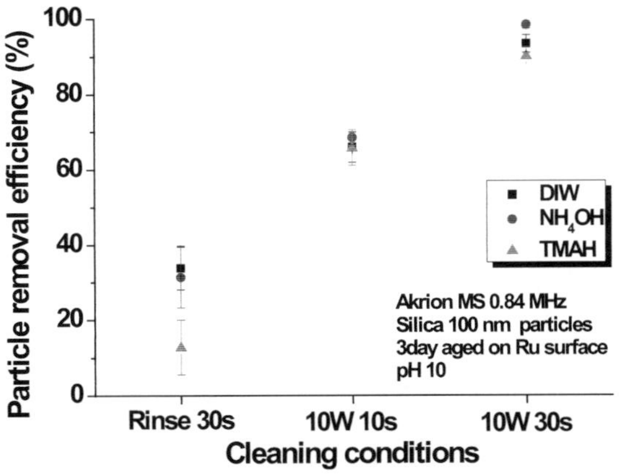

❖ PRE of PSL on Ru surface

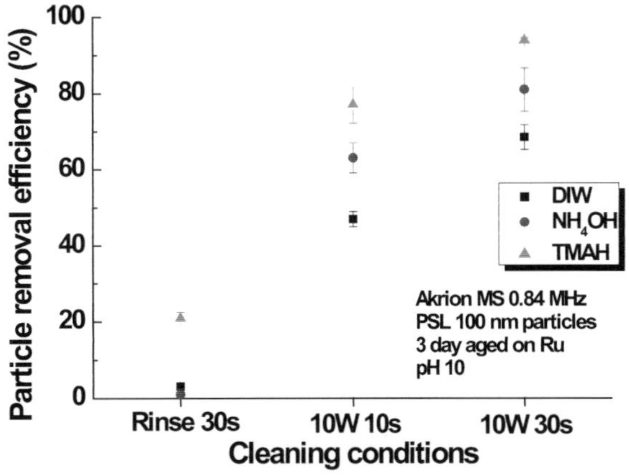

- No effect of cleaning solution on silica particle removal @ Ru surface
- TMAH cleaning with megasonic showed higher PRE on PSL particle from Ru surface
  - → TMAH can break the metal-carbon bonding as solvent

TMAH is very effective to remove organic particle from Ru surface

Hydrocarbon film deposition

❖ RF-PECVD (SRN-501, Sorona, Korea)

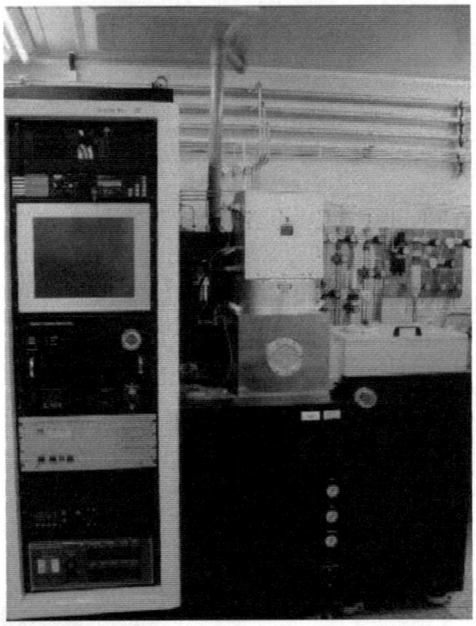

Process condition	
Gas ratio (CH_4 : Ar)	1 : 1
Plasma power	240 W
Temperature	30 ℃
Chamber pressure	0.6 torr

- Density of hydrocarbon film (1.3 g/cm^3)

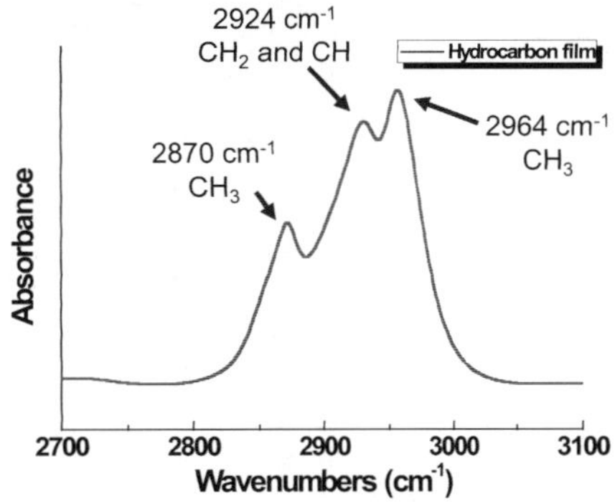

Hydrocarbon removal from TaN and Ru surface

❖ **ATR-FTIR analysis**

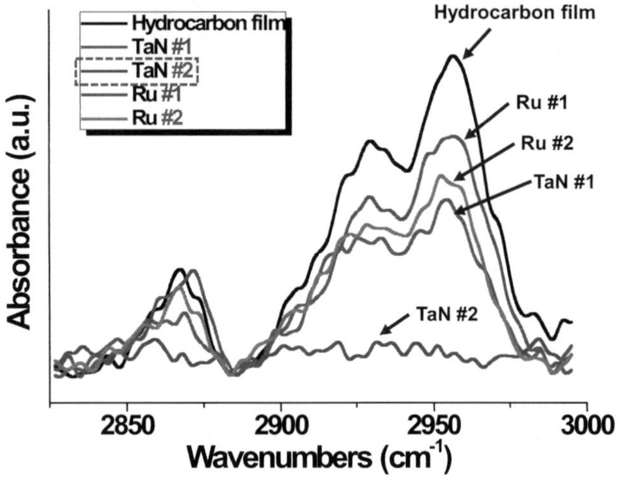

	Cleaning solution
#1	DMSO (80%) + NH_4OH (0.1 M)
#2	DMSO (80%) + TMAH (0.1 M)

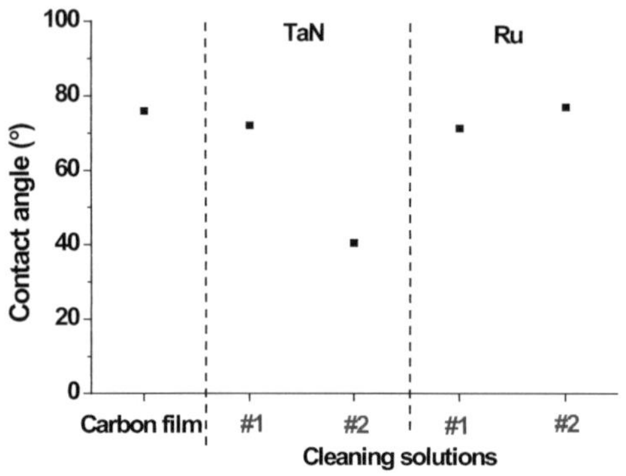

- TMAH can help remove hydrocarbon with DMSO
- Carbon peaks of hydrocarbon film was removed at TMAH added DMSO cleaning solution from TaN surface
- Hydrocarbon film was not removed from Ru surface @ TMAH with DMSO cleaning solution

Hydrocarbon removal from TaN and Ru surface

❖ **Surface images after cleaning process (OM)**

	#1 (DMSO + NH_4OH)	#2 (DMSO + TMAH)
TaN surface	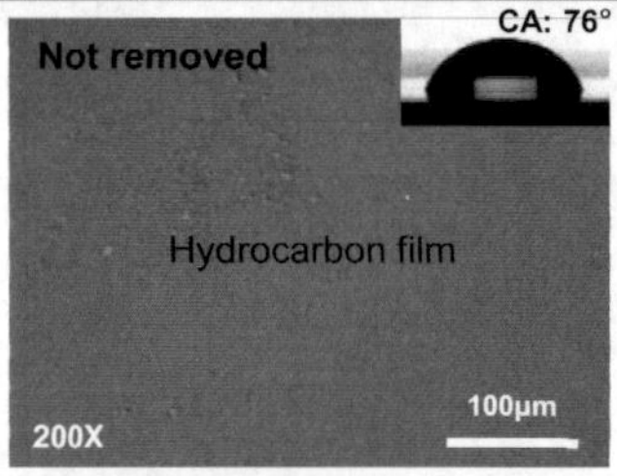	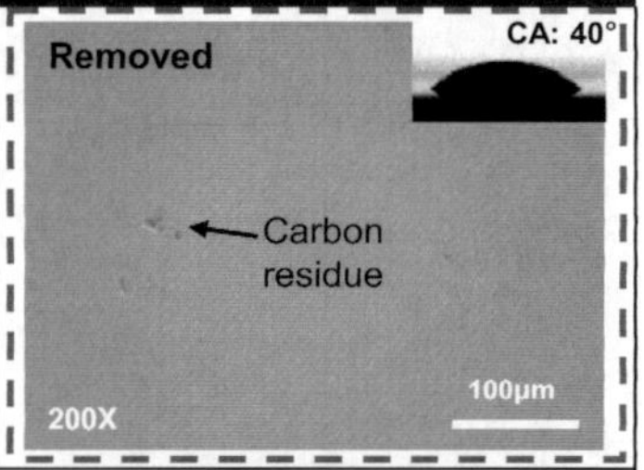
Ru surface	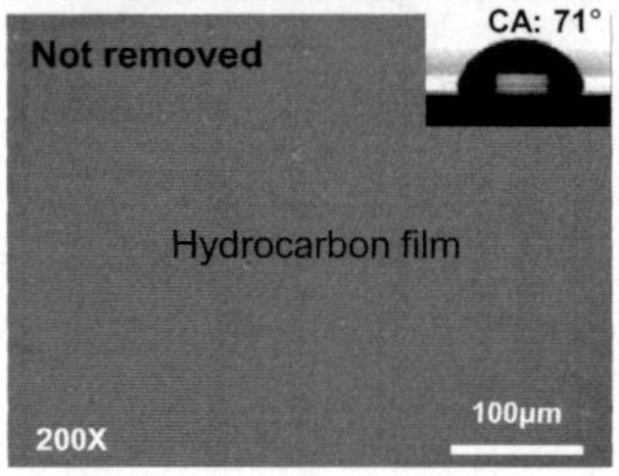	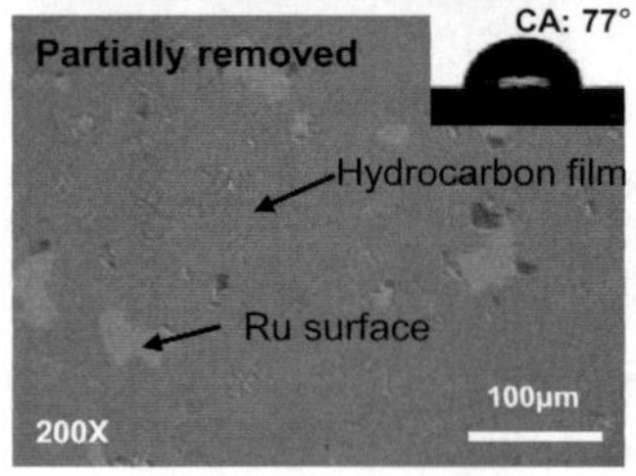

Hydrocarbon film is still remained on Ru surface (partially removed)

Mechanism of hydrocarbon removal

❖ **Hydrocarbon removal in DMSO with TMAH cleaning solution**

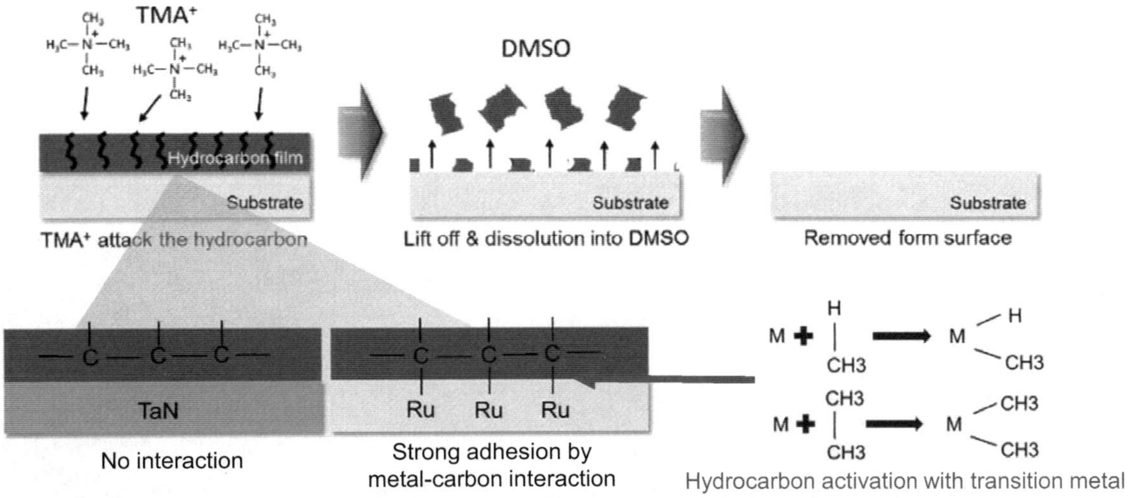

M. R. A. Blomberg, et al., *J. Am. Chem. Soc.*, **113** (2), 424 (1991)

More difficult to remove hydrocarbon film from Ru surface by transition metal – carbon interaction than those from TaN surface

Summary

- Need to understand the surface interaction between contaminants and substrates in EUV masks

- Silica & PSL particle removal
  - Hydrogen bonding is dominant on oxide particle removal from surface
    → PRE of silica : Si << TaN < Ru
  - PRE of PSL is lower @ Ru surface due to metal – carbon interaction
    → **TMAH is effective to enhance organic particle removal from Ru surface**

- Hydrocarbon film removal
  - DMSO with TMAH cleaning solution can remove hydrocarbon film from TaN surface
  - More difficult to remove carbon contaminant from Ru surface
    → **Transition metal – hydrocarbon interaction between Ru and hydrocarbon film**

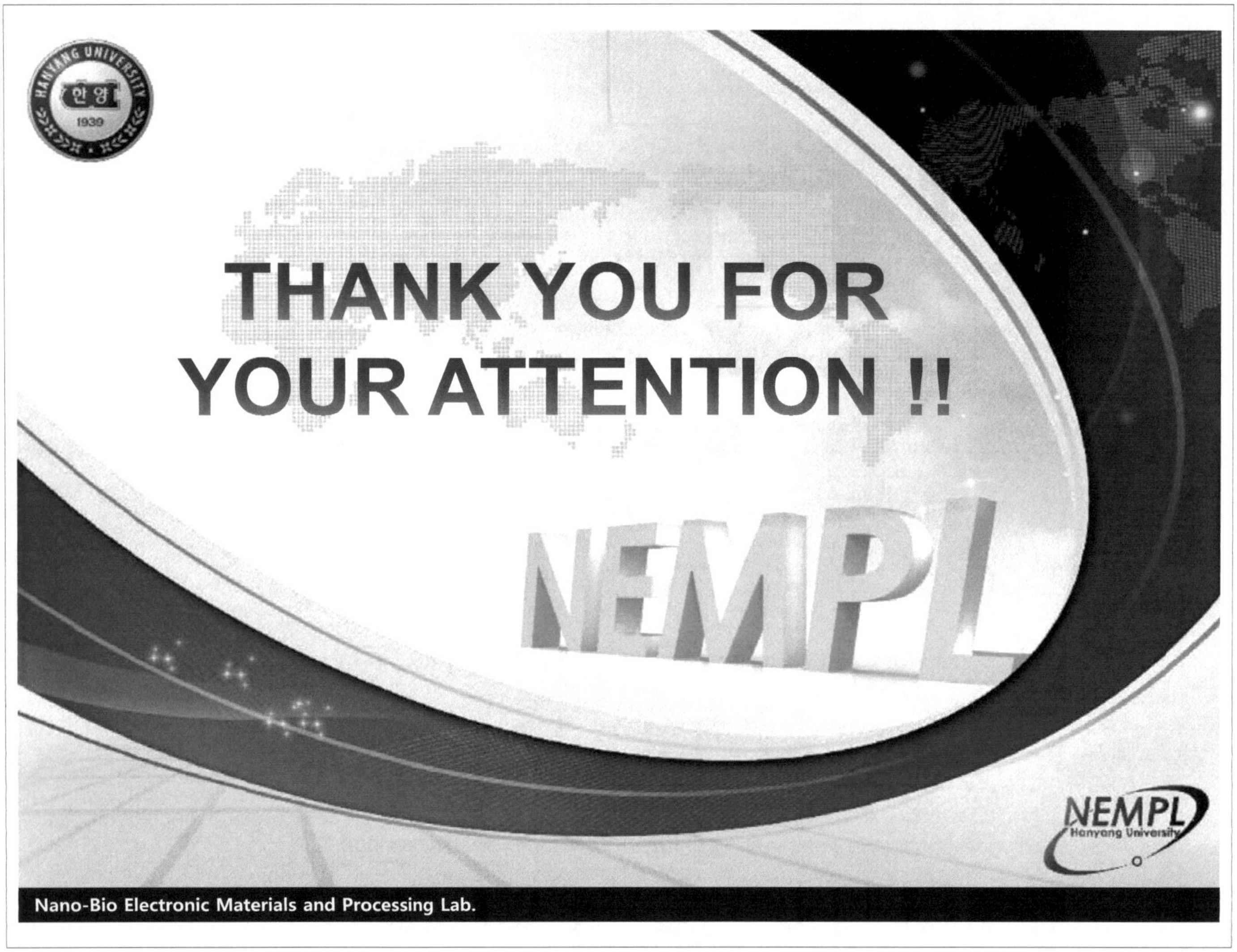
HANYANG UNIVERSITY
한양
1939
THANK YOU FOR
YOUR ATTENTION !!
NEMPL
NEMPL
Hanyang University
Nano-Bio Electronic Materials and Processing Lab.

PALL Pall Corporation

Better Lives. Better Planet. SM

Advanced Contamination Control using Novel Polyarylsulfone Membrane Technology

Patrick Connor (patrick_connor@pall.com, 516-801-9249), Rao Varanasi, Lawrence Johnson, Jian Tan, David Huang, Tony Shucosky, Katsuhiko Tokuno, *Pall Corporation*

Archita Sengupta, *Intel Corporation*

Presented at Surface Preparation and Cleaning Conference 2016

Santa Clara, CA

Outline

- Background
- Development of Evaluation Program
- Test Program and Results
- Supporting Fab Application Results
- Assessment of Results and Conclusions
- Next Steps
- Acknowledgments

Background: Rationale for Study

- ➢Filtration must keep up with "Paradigm Shift" in Defect Tolerance for Next generation Technology Nodes
- ➢Customized Filtration is needed to Aggressive Chemistry/Processes
- ➢ Innovation needed for <=5nm Metals/Particles/Organics/NVR removal for next generation need

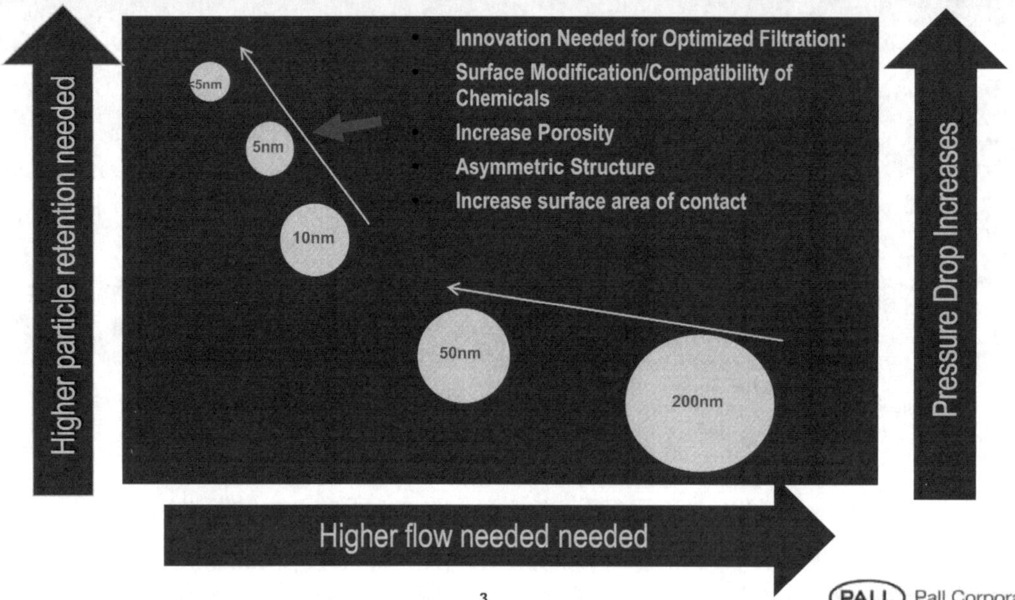

3

PALL Pall Corporation

Core Membrane Properties: PTFE and HAPAS

PTFE vs HAPS comparison	PROS	CONS
$(CF_2 - CF_2)_n$ 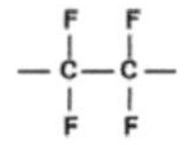 PTFE (PFOA free) 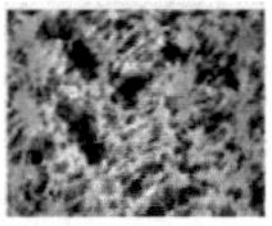	Chemically most inert	High unit cost
	High Temperature Resistance (185°C)	Most expensive surface modification
	High Purity	High temperature aggressive service requires special package design
	Primary retention by surface size exclusion	Minimal depth capture, surface loading only
	Long term stability	Typically low flux rates
	Available semi-phobic	Completely to semi-phobic
	Aggressive chemistry application	Disposability an isue in some countries
$[O-Ø-C(CH_3)_2-Ø-O-Ø-SO_2-Ø]_n$ 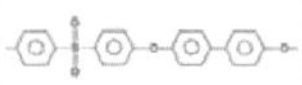 HAPAS Highly Asymmetric Polyarylsulfone 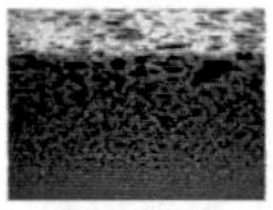	Good chemical resistance in aqueous chemistries pH 1 -14	Limited solvent and conc acid, oxidizer resistance
	Dual retention mechanisms: mechanical sieve final filter with integral tortuous path prefilter	Packaging material may limit thermal or long term chemical resistance
	Highest flux rate	
	Better Cost of Ownership as alternative for non-aggressive chemistry	
	High surface energy w/o surface modification – helps wetting of medium	
	Good temperature resistance (90°C)	
	High dirt holding capacity	

- The present study was undertaken to explore the suitability of a filter employing **highly asymmetric polyarylsulfone** ("HAPAS") medium with other components of high density polyethylene (HDPE) in a variety of aggressive acids, bases, and blends
  - The 5 nm rated Pall Ultipleat® SP DR product was the model chosen for evaluation

4

PALL Pall Corporation

Considering HAPAS vs. PTFE for Critical Filtration

- The results show that a next-generation filter containing HAPAS medium has applicability in a wide variety of dilute and aggressive chemicals
  - But how does it compare to an all-perfluoropolymeric filter? The table below considers various characteristics:

Property	HAPAS	PTFE
Finest available rating	2 nm	5 nm
Flow, psid/GPM-10" equiv.	0.7	0.75 - 1.0
Unit Cost	X	3X - 7X
Compatibility	Acceptable for a wide range of chemicals, up to 70°C	Acceptable across most chemicals, up to 185°C

- **The Ultipleat SP DR product employing HAPAS medium provides a suitable alternative on the basis of several characteristics**

5

PALL Pall Corporation

Advantages of Highly Asymmetric Polyarylsulfone ("HAPAS") Filtration Medium

Asymmetric   Isotropic (for comparison)

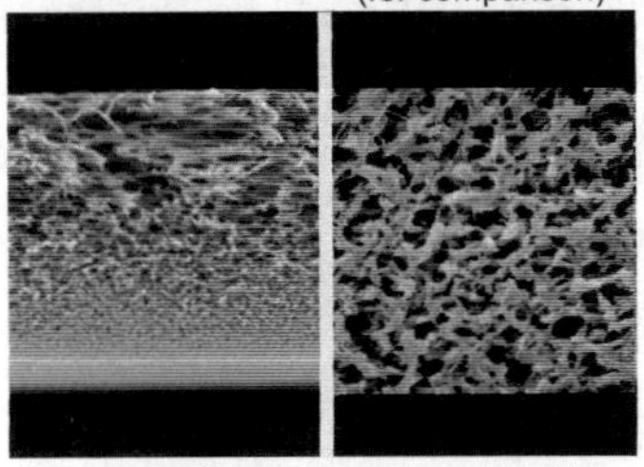

Cross-section – upstream at top

Medium available as fine as 2 nm (5 nm shown)

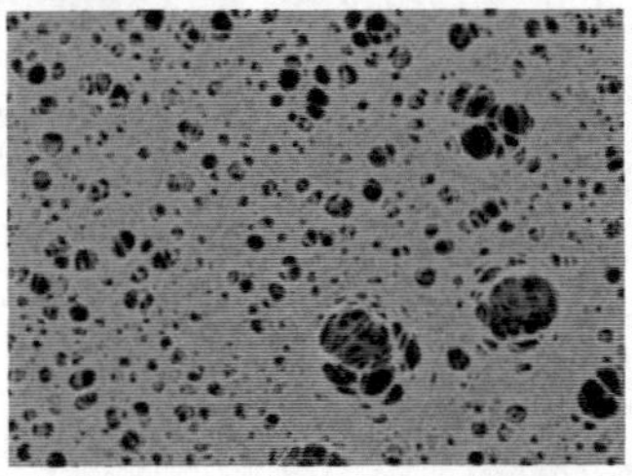

Large pore face – upstream

Fine pore face – downstream

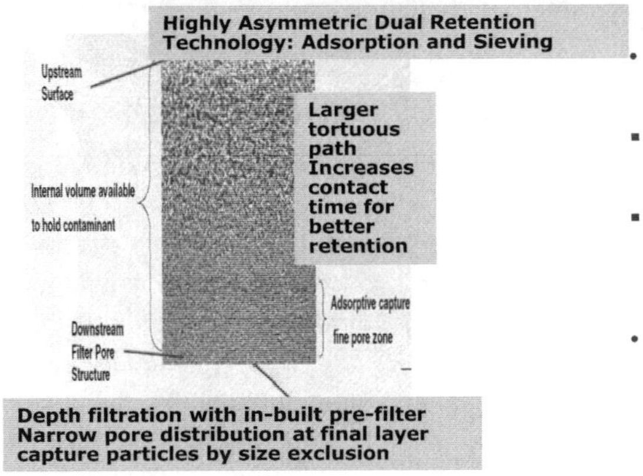

- Longer on-stream life: Higher Voids Volume for greater dirt holding capacity
- Highly asymmetric structure: Improved pressure drop characteristics compared to isotropic membranes
- While provided in prewet fashion: possesses relatively high critical wetting surface tension
  - Diminishes dewetting potential in aqueous application
- Superior Flow Rates w/ lower pressure drops
  - High bath turnover rate for rapid particle removal

6

PALL Pall Corporation

Planned Program of Study

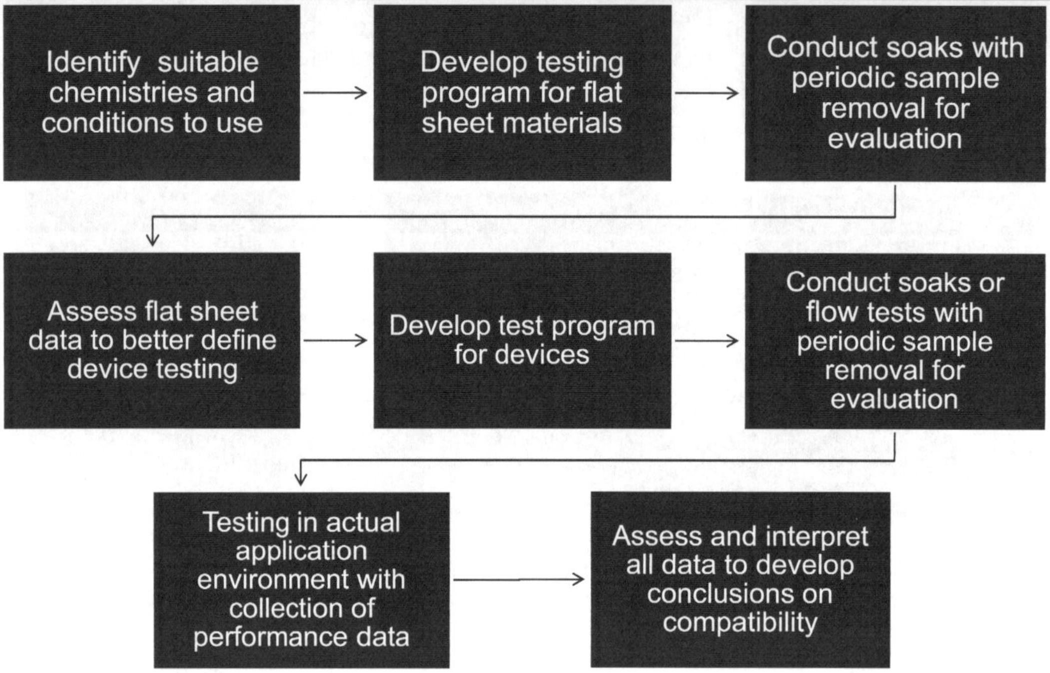

7 PALL Pall Corporation

Targeting Candidate Chemistries and Conditions to Use

Chemical / Chemistry	Reason for Interest	Conditions
Hydrogen Peroxide (H_2O_2)	Occurs widely in cleaning, etching, and stripping chemistries	5% - 30%, room temperature (R.T.)
Tetramethylammonium hydroxide ("TMAH")	Widely used in cleaning, etching, and PR developer chemistries	2% - 25%, R.T.
Hydrofluoric Acid (HF)	Widely used in cleaning and etching, sometimes at high conc.	49%, R.T.
Isopropanol ("IPA")	Frequent use in drying; model of organic content	100%, R.T and 70°C
SC1, SC2 (ammonia or HCl + H_2O_2 + H_2O)	Aggressive blends, often used at elevated temperatures	SC1: 1:1:5; SC2: 1:5:10, R.T. (sheet), 60°C (device)
Hydrochloric Acid (HCl)	Strong acid model; used in blends	10% - 37%, R.T.
DI Water	Cleaning agent	100%, 70°C
Proprietary Blends	Used for BEOL cleaning / stripping; indicator for organics	Specialized

8 PALL Pall Corporation

Suitable Test Methods for Flat Sheet Materials

- Samples of HAPAS medium were produced in coupon and disk forms; samples of HDPE support material were produced as coupons
- Collections of disks and coupons were immersed in various chemicals selected
  - Sealed in vessel and subjected to elevated temperature, where applicable
- Samples removed periodically and subjected to physical and chemical testing
  - Tensile measurements of coupons via Dynamic Mechanical Analyzer ("DMA")
  - Flow rate measurement, using water, at defined ΔP, for disks
  - FTIR spectra of coupons obtained
- Where applicable and suitable, HAPAS medium exhibiting retention of other properties subjected to 5 nm gold nanoparticle ("GNP") retention testing
  - Generally restricted to fresh HAPAS and longer duration soaks

9

PALL Pall Corporation

Summary of Tensile, Flow, and IR Spectroscopic Test Results For Flat Sheet Materials—Three Month Exposure

Media 5 nm HAPAS							S/D layer	
Chemical	Conc	temp	Water flow	Retention (LRV at 5 nm)	Mechanical Property (DMA)	Chemical Composition (FTIR)	Mechanical Property (DMA)	Chemical Composition (FTIR)
HCl	37%	20C	↔	↓	↔	↔	↔	↔
	20%		↔		↔	↔	↔	↔
	10%		↔	↔	↔	↔	↔	↔
H2O2	30%	20C	↑↑	↓↓	↔	↔	↔	↔
	20%		↑↑	↓↓	↔	↔	↔	↔
	10%		↑↑	↓↓	↔	↔	↔	↔
	5%		↑↑	↓↓	↔	↔	↔	↔
SC1	*	20C	↔	↔	↔	↔	↔	↔
*NH_4OH:H_2O_2:H20 – 1:1:5								
SC2	†	20C	↔	↔	↔	↔	↔	↔
†HCl:H_2O_2:H20 – 1:5:10								
IPA	100%	20C	↔	↔	↔	↔	↔	↔
IPA	100%	70C	↑	↓	↓	↔	↓	↔
HF	49%	20C	↔	↓	↔	↔	↔	↔
TMAH	25%	20C	↔	↔	↔	↔	↔	↔
	10%		↔	↔	↓	↔	↔	↔
	2%		↔	↔	↔	↔	↔	↔

↓↓ ↑↑	significant change	↓ ↑	slight change	↔	no significant change

Results suggest compatibility with most chemicals used

10

PALL Pall Corporation

Assessment of Flat Sheet Results and Continuing on to Device Testing

- The decision made to soak actual filters in <u>the most concentrated version of chemical</u>
  - Including 30% H_2O_2, essentially as a negative control
- Expansion of program to test in (1:1:5) SC1 and (1:5:10) SC2 at 60°C with flow
- Evaluation plan:
  - Collect baseline data on filters
    - Non-destructive integrity test
    - ΔP vs. flow rate (using water)
    - Rinse test in water for particles
  - Soak filters for periods up to 3 months with periodic repeat of tests
  - Post-soak check of robustness via pulsation testing in water

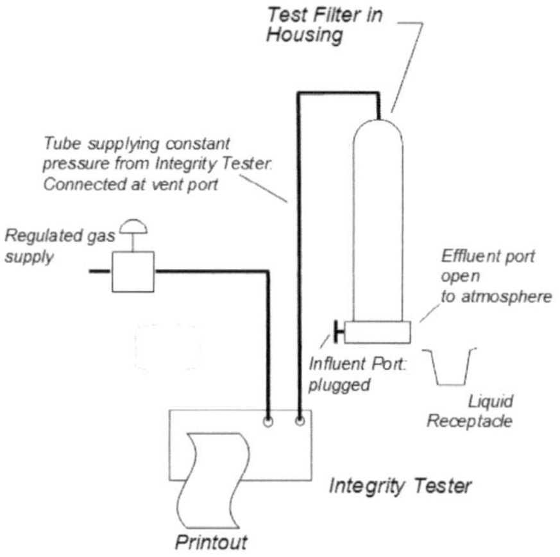

Schematic of Integrity Test Set-up

Pall Corporation

Integrity Test and Pressure Drop *vs.* Flow Rate with Respect to Time for Target Chemicals

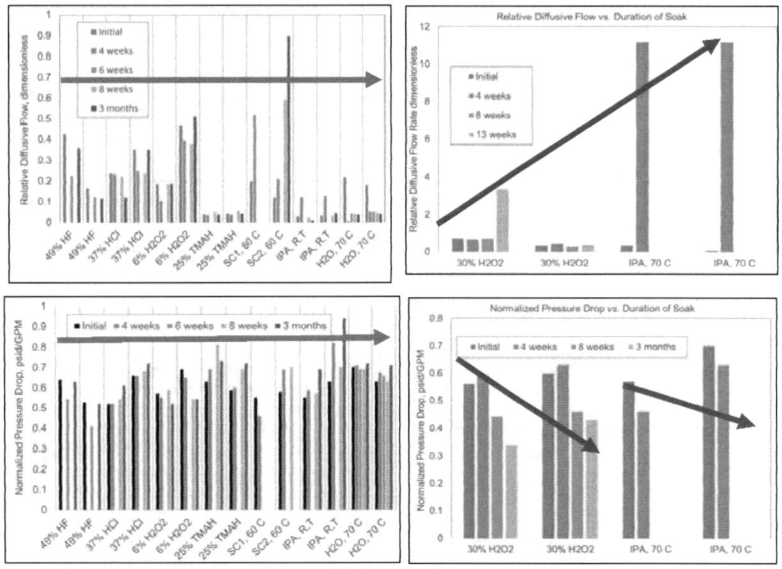

Diffusive flow and pressure drop testing revealed the following:

- For most chemicals, parameter changes within acceptable limits
  - Suggests ***compatibility*** over soak time used
- **30% H_2O_2** sample shows integrity loss at 3 mo, and major drop in ΔP at 8 wk.
  - Some affect—***limited compatibility***
- Loss of integrity and ΔP drop at 4 wk for **70°C IPA**
  - This suggests ***incompatibility***

Pall Corporation

Comparative Initial and Post-Contact "Rinse-in" Results in DIW for Selected Cases

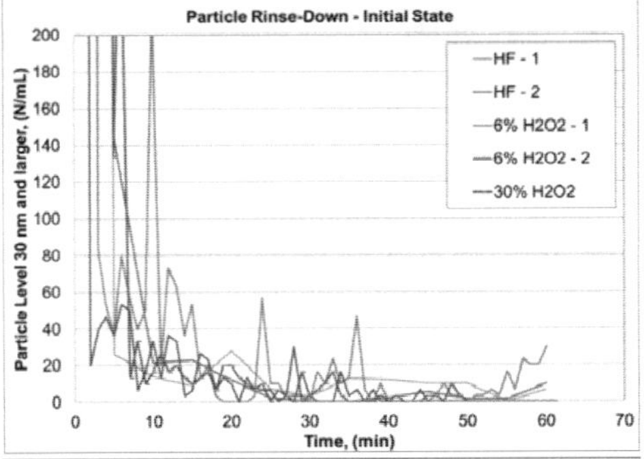

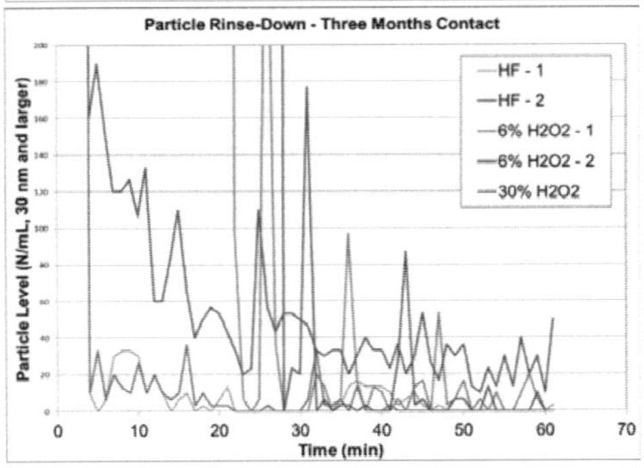

Particle rinse tests in water pre- and post-soak can be used to evaluate potential changes resulting from contact:

- Filters soaked in 49% HF and 6% H_2O_2 recover background levels
  - **Suggests minimal effect from long-term contact, and compatibility with these chemicals**
- By comparison, filter soaked in 30% H_2O_2 exhibits relatively high level after 60 minute rinse
  - **Suggests limited compatibility**

Comparative Initial and Post-Contact "Rinse-in" Results in DIW for Selected Cases, *cont.*

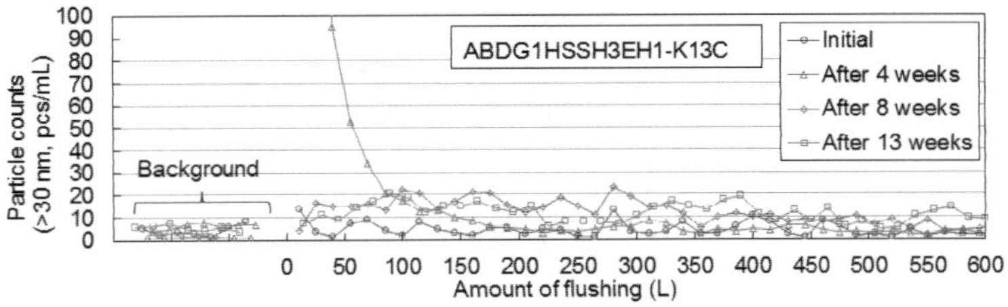

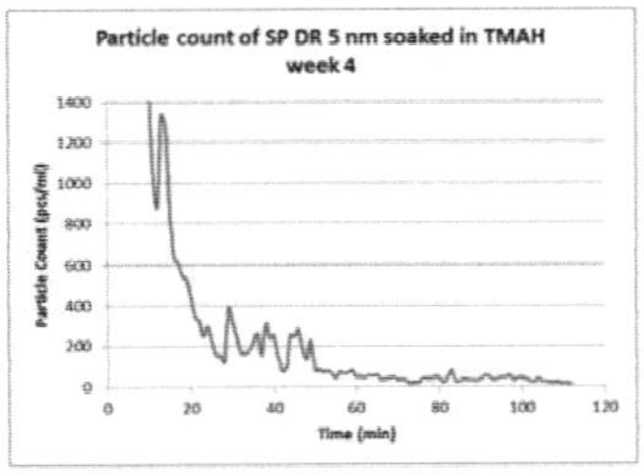

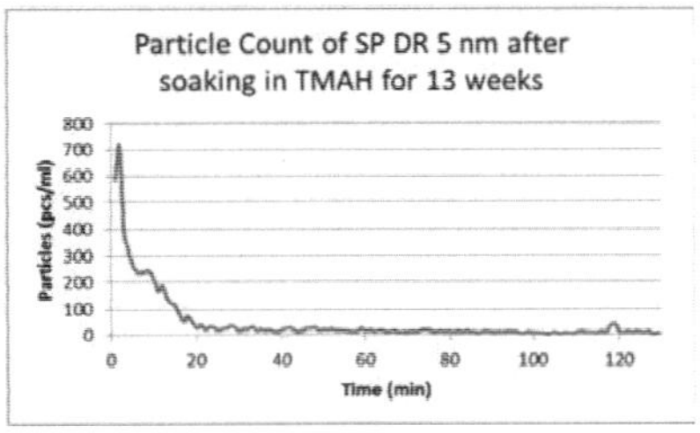

- Filter contacted with flowing **SC2** (1:5:10) at 60°C exhibits **minimal change in transient particles over time**
- Filter soaked in **TMAH** appears to **achieve low particle level *quicker* after longer soak duration**

Results confirm absence of degradation of the filter in these chemistries and confirm compatibility

Durability Tests on IPA (20°C) and TMAH Soaked Filters

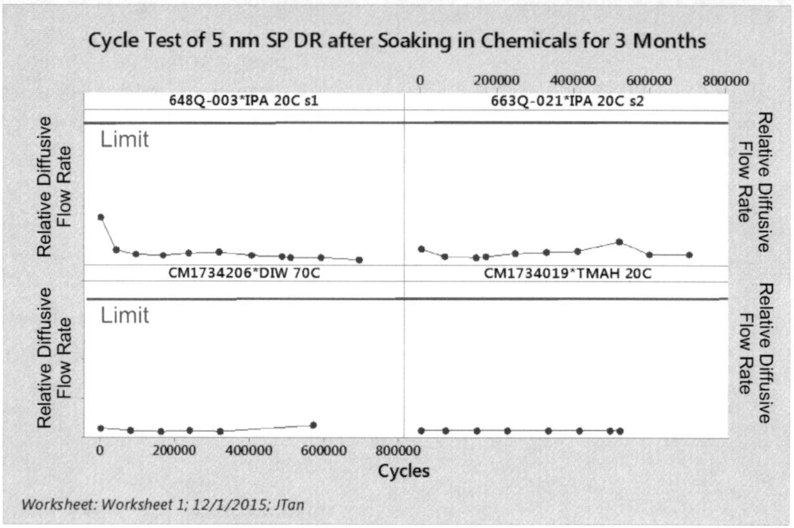

This test provides supplemental information on post-soak filter behavior by subjecting them to additional, cyclic stress

- ***Stressed filters show no loss of integrity even after over 500k cycles, further confirming compatibility in R.T. IPA and 25% TMAH***

15

PALL Pall Corporation

Side by Side Comparison of PTFE vs HAPAS performance

First 30 days Defect monitor data at IDM Advance technology Node WE tools

TOTAL_ADDERS: Chemistry A: PTFE vs HAPAS

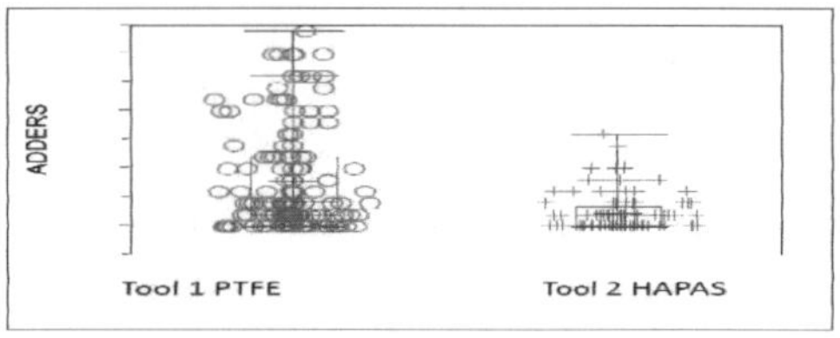

TOTAL_ADDERS: Chemistry B: PTFE vs HAPAS

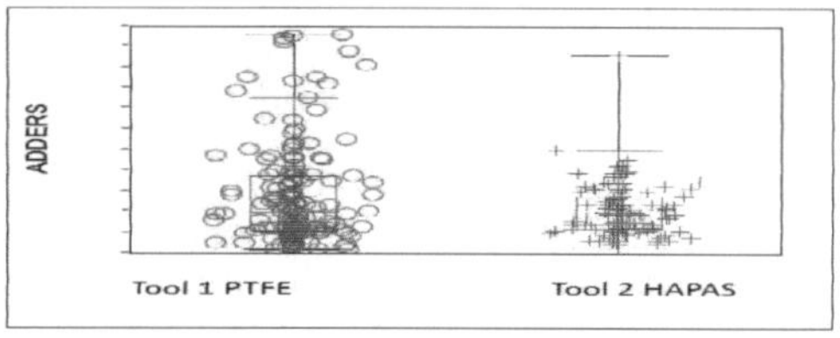

TOTAL_ADDERS: Chemistry C: PTFE vs HAPAS

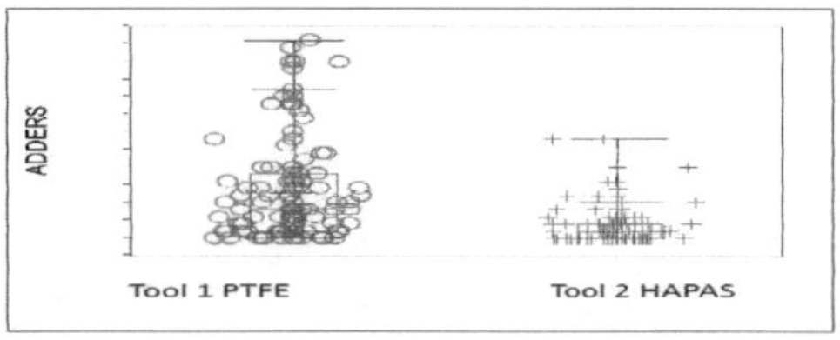

TOTAL_ADDERS: Chemistry D: PTFE vs HAPAS

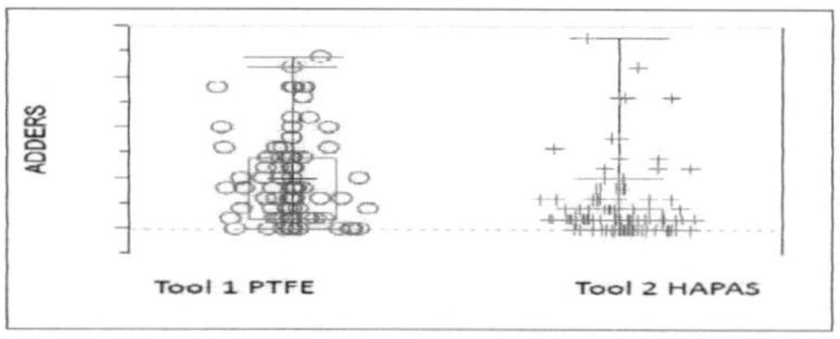

Data Reference: Ryan Mackiewicz, Intel Corporation.

intel

16

PALL Pall Corporation

Assessment of Results and Conclusions

- Test program was effective in showing differences in behavior of both components and complete filters tested in various chemicals
  - Flat sheet tests were useful for guiding device testing
- Test Results show that the Ultipleat SP DR product with HAPAS medium can be applied in a variety of WEC chemicals, including concentrated versions of TMAH, HF, and HCl, dilute H_2O_2, and IPA at room temperature, and SC1 and SC2 at 60°C
  - The product exhibited incompatibility with 30% H_2O_2, and with IPA at 70°C
- Lab test results are supplemented by real time field qualification data on advanced technology node tools showing value of employing the HAPAS product in actual WEC application .

Overall, we find that a filter product containing HAPAS medium indeed provides a very useful technically viable and lower cost-of-ownership alternative to all-perfluoropolymeric filters for a wide range of applications

Next Steps

- Expansion of program to broader array of specialty chemicals and conditions
  - Especially to include more blends, including organics
  - Extension to higher temperatures
- "Fine-tuning" of testing
  - Exploration of additional evaluation methodologies
  - Shorter duration-increased interval testing for certain chemicals
- Incorporation of more flow-based tests

PALL Pall Corporation

Acknowledgements

- PALL: Thank you so much to Ibrahim Mohamed-Ali, Mai Nguyen, Kei Lau, and Pauline Adejo of the Port Washington team, Virgil Briggs, Mike Gofkowski, Shawn Hubbard, Yi Wang, Valerie Pritchard, John Tomancik, and Dick Stoyell of the Cortland team, and Shusaku Daikoku, Mitsutoshi Nakagawa, Tomoyuki Takakura, and Shuichi Tsuzuki of the Tsukuba team for all the extensive lab work and additional assistance provided during the project
- Intel: Thank you so much to Ryan Mackiewicz And Ameya Bapat for providing field data

Thank you for listening! ☺

Electron beam generated plasmas:
Ultra cold sources for low damage, atomic layer processing

Scott G. Walton
Plasma Physics Division, Naval Research Laboratory
Washington, DC 20375

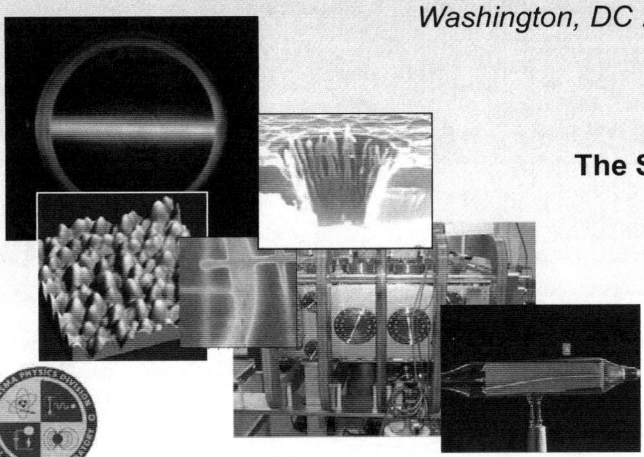

The Surface Preparation and Cleaning Conference (SPCC)
April 19-20, 2016
Santa Clara, CA

Motivation

As material demands evolve toward the single nanometer-scale dimensions, one would like to systematically modify *one and only one monolayer* at a time, without "damaging" other layers

Atomic layer etching/deposition

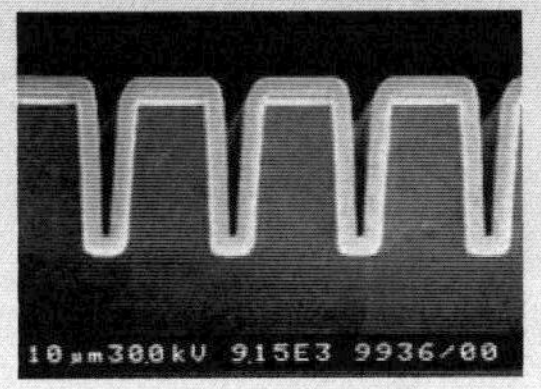

2-D material processing

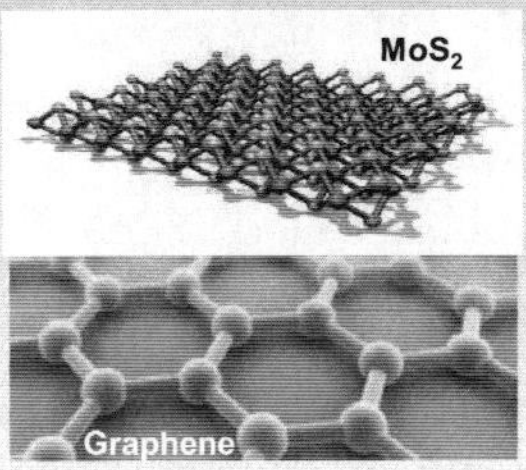

Polymer processing

Plasma Requirements

- **Precise control over the *flux of species* and the *ion energy* at surfaces during processing**
- **For very thin materials (e.g. 2-D materials), the energy of incident ions should be low as possible to minimize damage while processing**

A. Agarwal and M.J. Kushner, J. Vac. Sci. Tech. - A, 27, 37 (2009)
K.J. Kanarik, G. Kamarthy, R.A. Gottscho, Solid State Technology, Volume 55, Issue 3, (2012)
G. S. Oehrlein, D. Metzler, and C. Lia, ECS J. Solid State Sci. Technol. 4(6),) N5041-N5053 (2015)

2

Electron Beam Generated Plasmas

- Unique control over species production
- Inherently low electron temperature and thus, uniquely low ion energies

3

Electron beam generated plasma processing system

Large Area Plasma Processing System (LAPPS)*

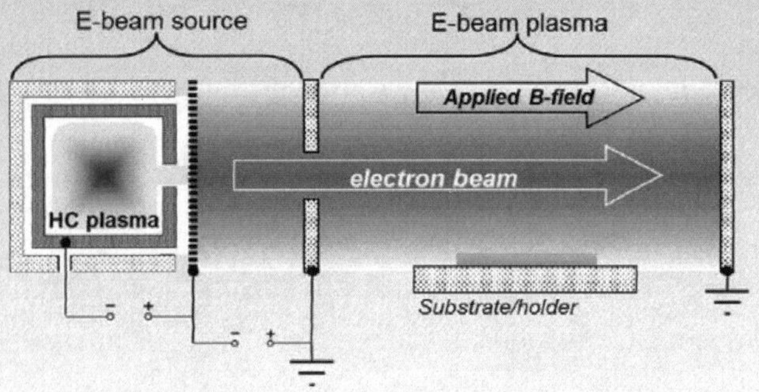

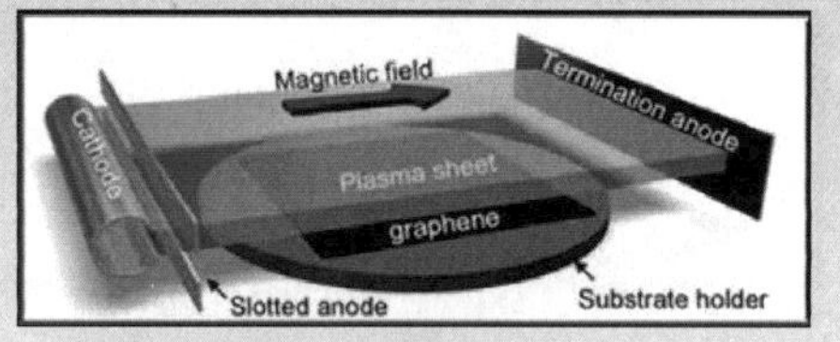

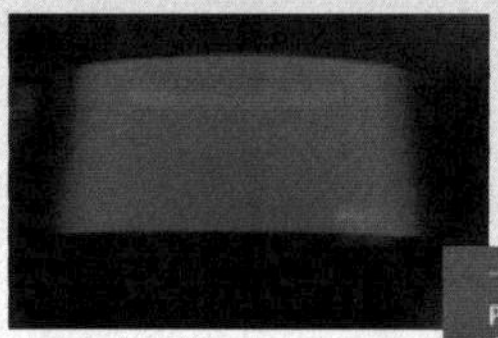

Basic Operation

- High energy beam injected into background
- Creates Plasma
  - Ionizes: Charged Particles (ions and electrons)
  - Excites: Species emit photons
  - Dissociates: Reactive Radicals

*Meger et al., US patent no. 5,874,807 (Feb. 1999)

S. G. Walton, et al., ECS Journal of Solid State Science and Technology, **4(6)** N5033-N5040 (2015)

4

Platforms for processing

Source is decoupled from reactor geometry

- Flexible design
- Unique geometries

Magnetized or not

PEPVD*

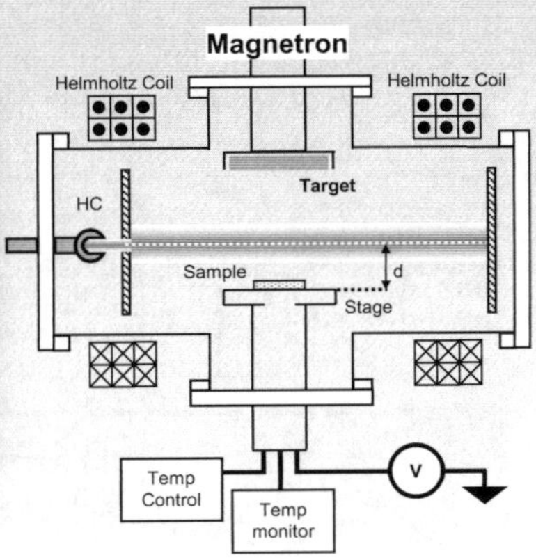

Roll to Roll**

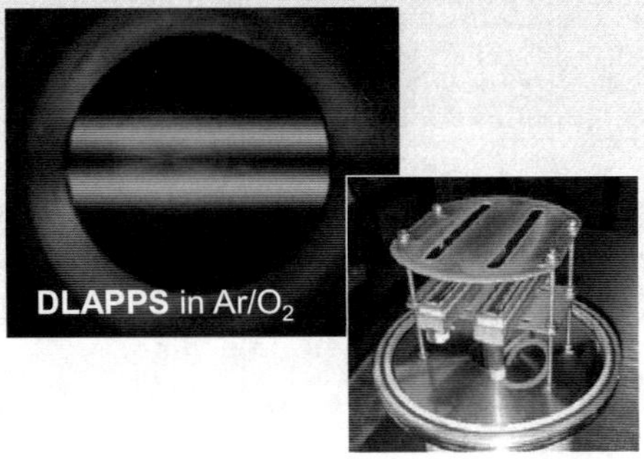

* C. Muratore, et al., J. Vac. Sci. Technol. A, 24(1), 25 (2006)

** S.G Walton, et al., Plasma Proc. and Poly. 5, 453 (2008)

5

Platforms and processing

- Scaling is straightforward to accommodate large area processing
  - **width** scales with e-beam source width
  - **length** scales with beam energy and pressure

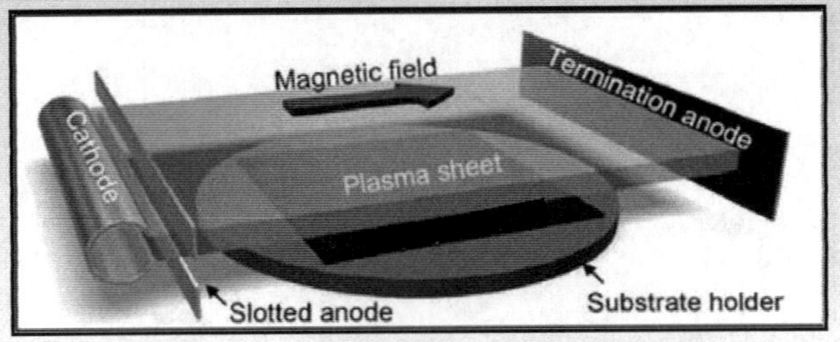

Large area rectangular chamber (1 m^2 capability)

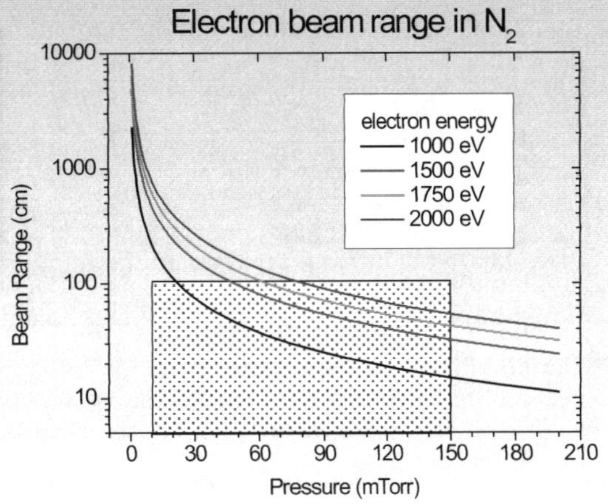

6

Platforms and processing

- Source is scalable to large areas without losses in uniformity

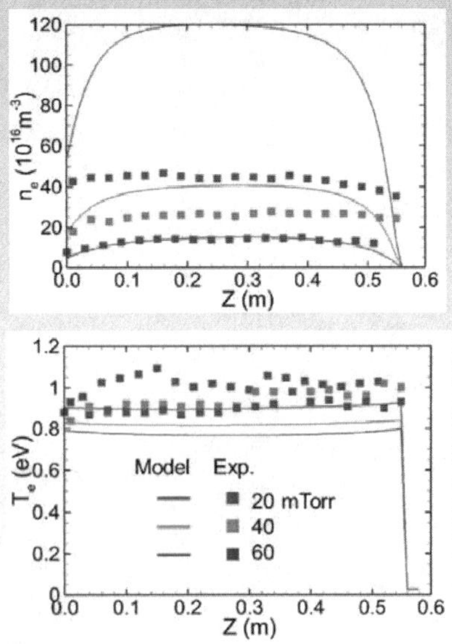

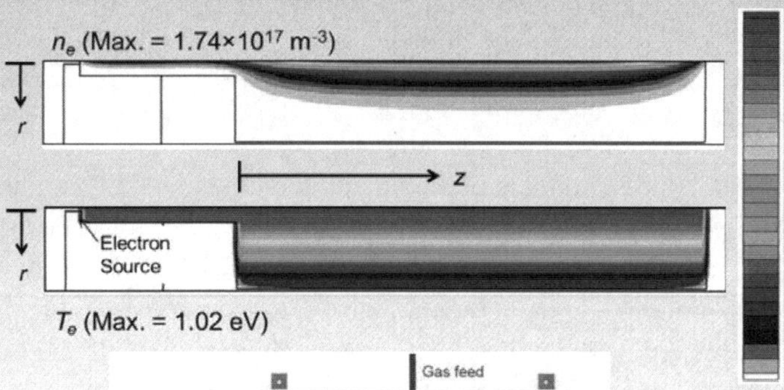

S. Rauf et al., AVS International Symposium, San Jose, CA (October 18-23, 2015)

7

Plasma generation with high-energy electron beams

- In discharges,
  - electric fields to energize plasma electrons
  - small fraction of electrons ionize gas
  - most energy is used to excite the gas
- The injection of a 2 keV beam into the background gas will directly ionize and dissociate the gas.
  - more *efficiently* ionize
  - no threshold determination
  - ionize all species equally
- The plasma (or secondary) electrons have much a lower energy
  - are not required to sustain the plasma
  - cooled electrostatically and through inelastic collisions

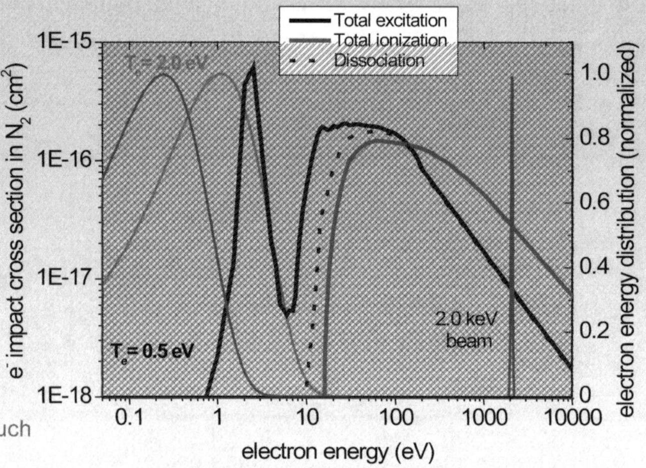

... the resulting plasmas have unique features

8

Unique Features of Electron Beam-Generated Plasmas

Simple Control of plasma density and species production

$$\frac{dn_i}{dt} = S_i - L_i = 0 \qquad S_i = k\, I_{beam}\, P_i\, \sigma_i$$

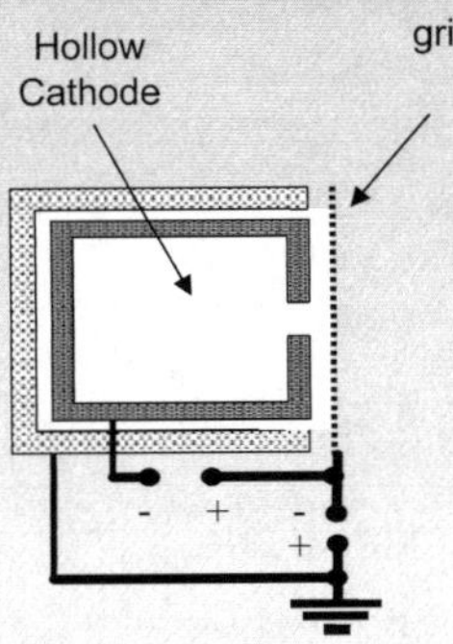

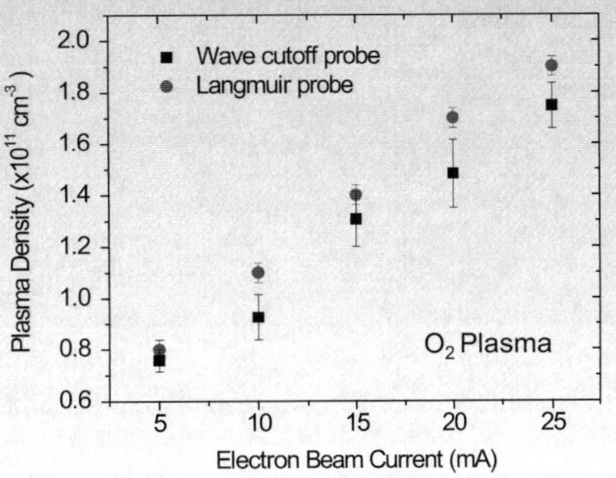

Plasma density scales with beam current

E.H. Lock, et al., Plasma Sources Sci. Technol. 17, 025009 (2008)

S.G. Walton, et al. J. Appl. Polym. Sci. 117, 3515 (2010)

D.R. Boris, et al., Plasma Sources Sci. Technol. 20, 025003-7 (2011)

9

Unique Features of Electron Beam-Generated Plasmas

Species generation is proportional to the relative concentrations of the working gases

$$S_i = k\, I_{beam}\, P_i\, \sigma_i(E)$$

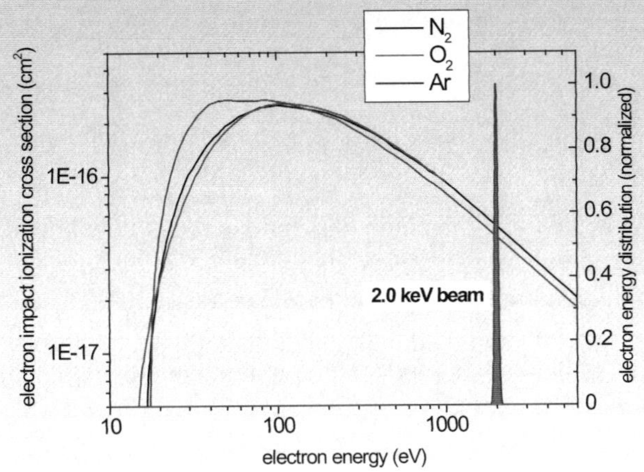

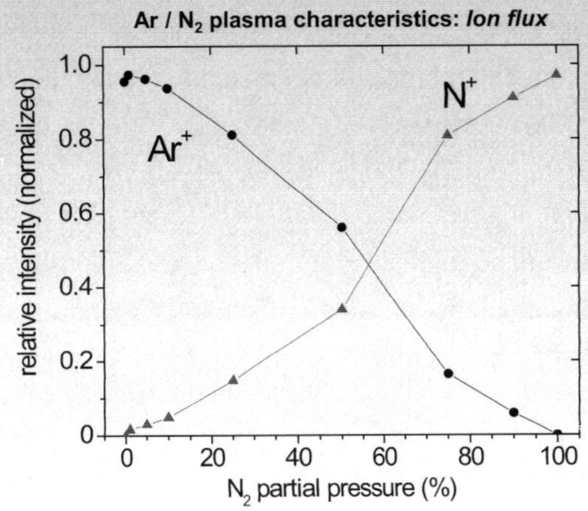

S.G. Walton, et al., Surf. Coat. Technol., 186 (1-2), 40 (2004)

10

Unique Features of electron beam generated plasmas

T_e is low and dependent on background gas

EEDF in beam channel

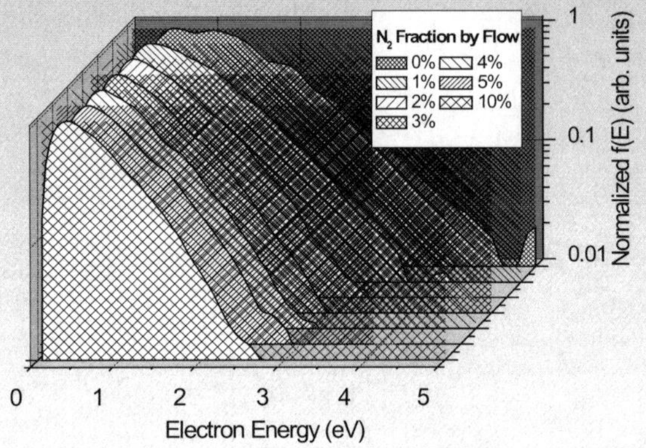

T_e in beam channel

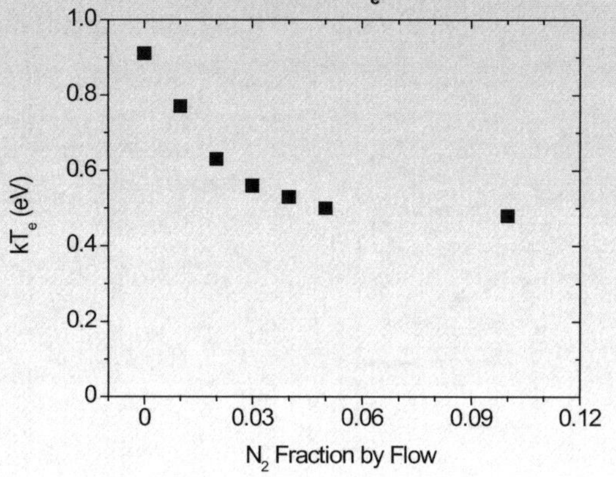

EEDF cools with increasing nitrogen concentration

Leading to a corresponding drop in T_e

D. R. Boris, et al., Plasma Sources Sci. Tech. 22, 065004-6 (2013)
G. M. Petrov, et al., Plasma Sources Sci. Tech. 22, 065005 -8 (2013)

11

Unique Features of electron beam generated plasmas

Very low T_e provides very low ion energy

IED at Electrode surface

Plasma Potential:

$$V_p = T_e \ln(M_i/2\pi m_e)^{1/2}$$

Ion Energy:

$$E_{ion} = V_p - V_{sb}$$

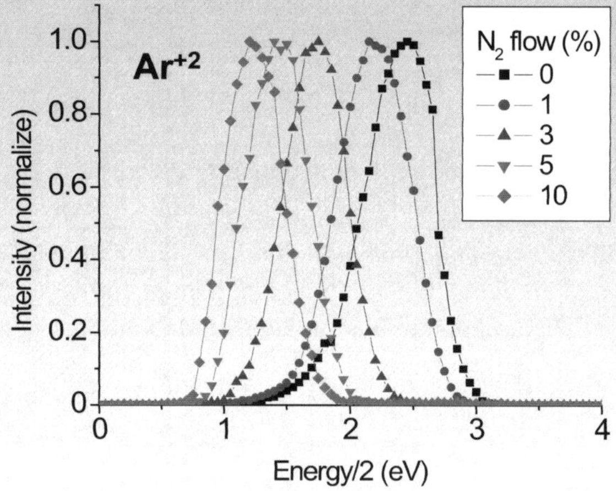

D. R. Boris, et al., Plasma Sources Sci. Tech. 22, 065004-6 (2013)
G. M. Petrov, et al., Plasma Sources Sci. Tech. 22, 065005-8 (2013)

12

Unique Features of electron beam generated plasmas

Electron Beam Generated Plasmas have a fundamentally low T_e (even at high plasma densities) and thus provide a large flux of low energy ions

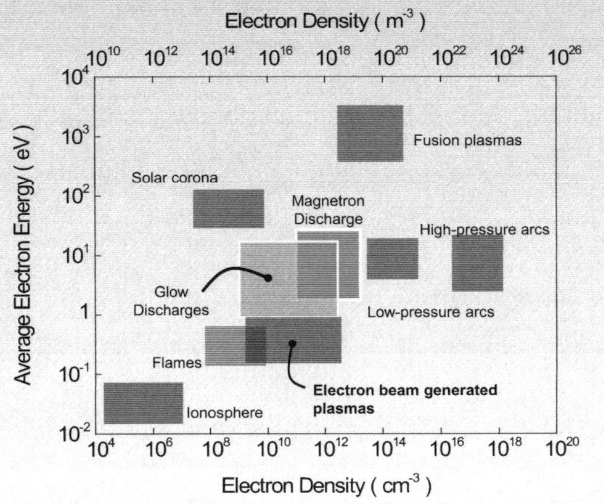

Plasma sources (discharges) used in materials processing

Type	T_e (eV)	n_e (cm^{-3})	KE_{ion} (eV)
ICP	4	10^{11}	20
ECR	4	10^{11}	20
DBD	2 - 10	10^{13}	10 - 100
RIE	8	10^{10}	40
DC Diode	2 - 10	10^{10}	10 - 100
CCP	1 - 5	10^9 - 10^{10}	5 - 25
Electron Beam	**0.3 - 1**	**10^9 - 10^{12}**	**1.5 - 5**

From: S.G. Walton and J.E. Green, "Plasmas in Deposition Processes," In *Handbook of Thin Film Deposition Technology: 3rd Edition,* ed. Peter Martin, Holland: Elsevier (2009).

13

Processing implications

Electron beam generated plasmas provide:

- Unique control over the production of species and their transport to surface
- n_e is high (10^{10} - 10^{11} cm^{-3}); T_e is very low ($\approx$ 0.5 eV)
- In the absence of any biasing, a mix of reactive species in concert with a large flux of very low energy (0 - 5 eV) ions

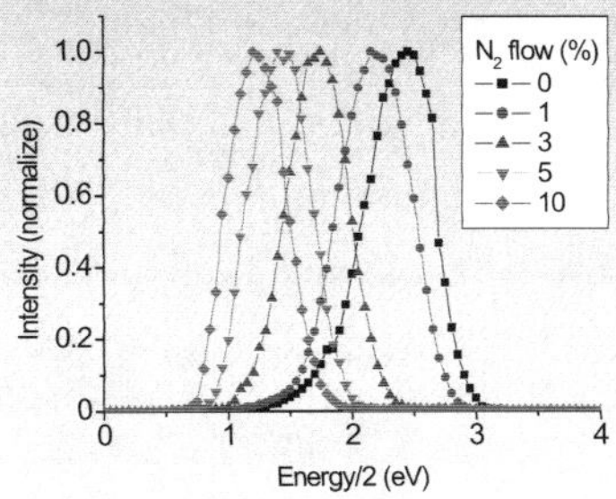

Ion-induced surface processes*

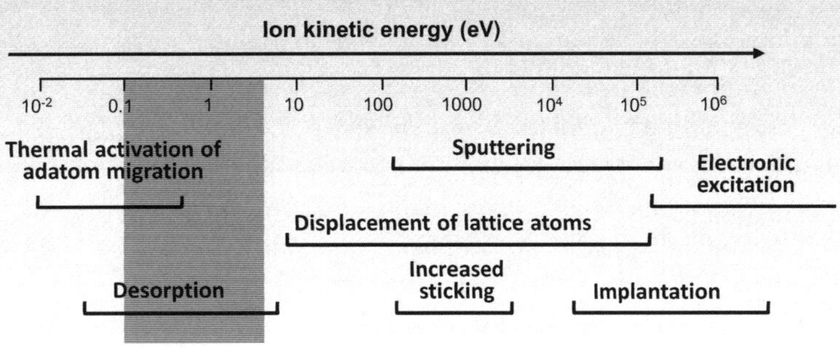

* Adapted From: Grill, A., *Cold plasma in materials fabrication: From fundamentals to applications*, Piscataway: IEEE Press, 1994.

14

Processing implications: Low damage

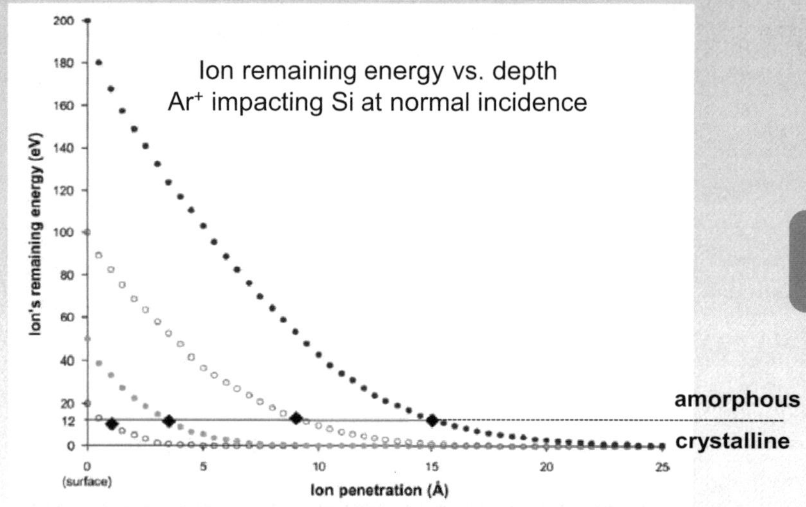

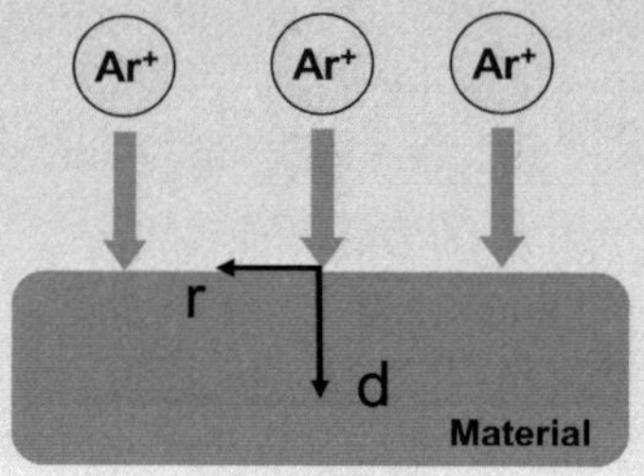

Low ion energy is valuable in maintaining crystallinity

D.B. Graves and D. Humbird, Applied Surface Science 192, 72–87 (2002).

15

Target Processing Applications

Chemical Modification of Ion Energy Sensitive Materials

- Chemical modification without physical changes (e.g. surface roughening)
- Polymers or other "soft" materials
- *Sensors and biomaterials applications*

Atomic Layer Modification

- Graphene Oxide Reduction – *reduction to graphene without damage*
- Graphene (or other 2-D materials) functionalization - *etching is not an option*
- *Sensors and electronic applications*

Atomic Layer Etching

- Selective removal of only one monolayer at a time
- Polymers (e.g. photoresist)
- Silicon
- Metal and/or silicon oxide/nitrides
- *Microelectronics applications*

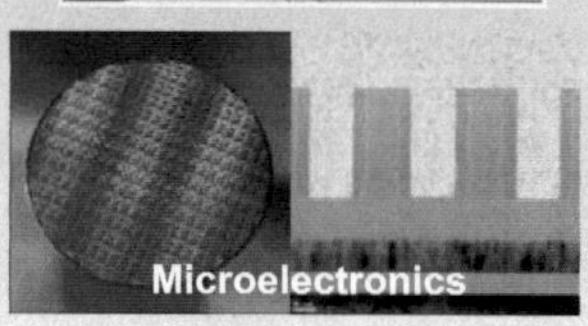

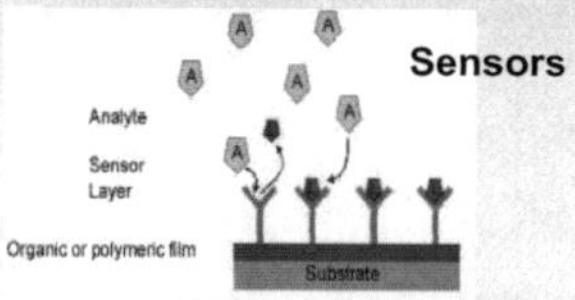

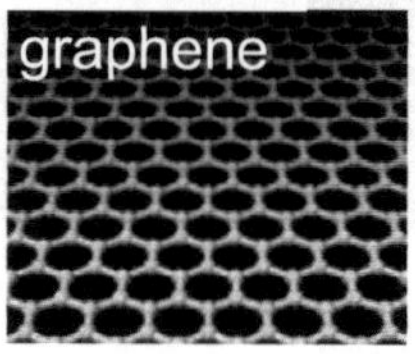

16

Graphene Processing

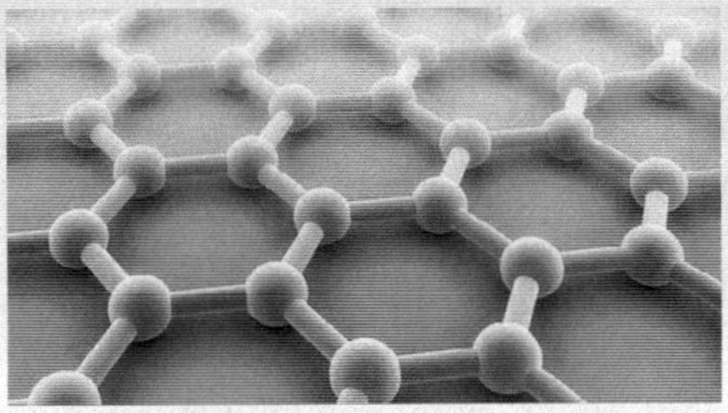

17

Functionalized Graphene: Chemical Modification

Graphene on Si/SiO$_2$ *

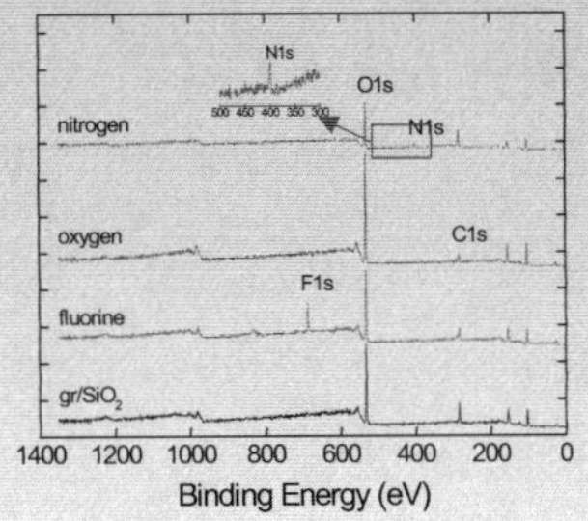

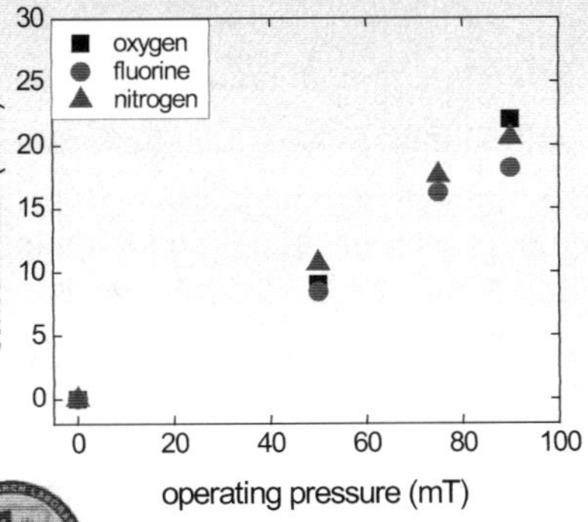

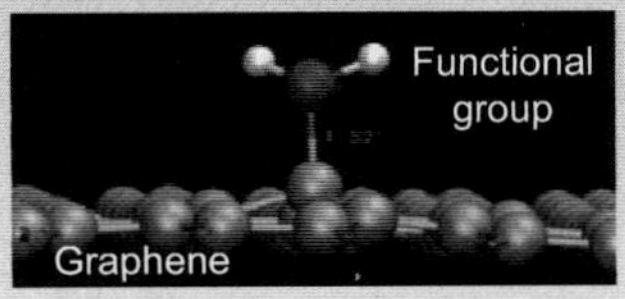

Controllably introduce functional groups

Epitaxial Graphene on SiC **

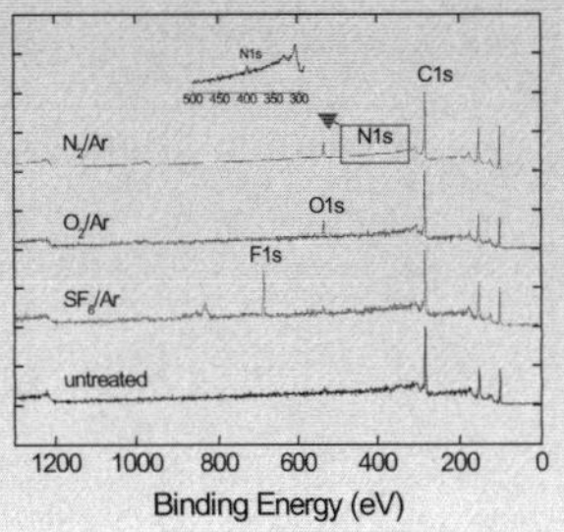

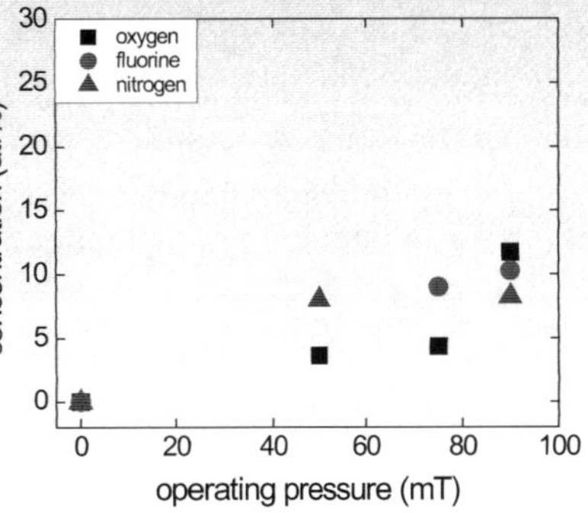

* M. Baraket, et al., Appl. Phys. Lett. 96 231501 (2010).

** S.G. Walton, et al., Materials Science Forum (2012); 16. S.C. Hernández et al, Surf. Coat. Tech. 241, 8 (2014)

18

Functionalized Graphene; Functional Surfaces

- Exposure to ammonia-containing plasmas leads to the incorporation of both primary amines (NH_2) and N
- Surface reactivity increases and sheet resistance remains reasonably low

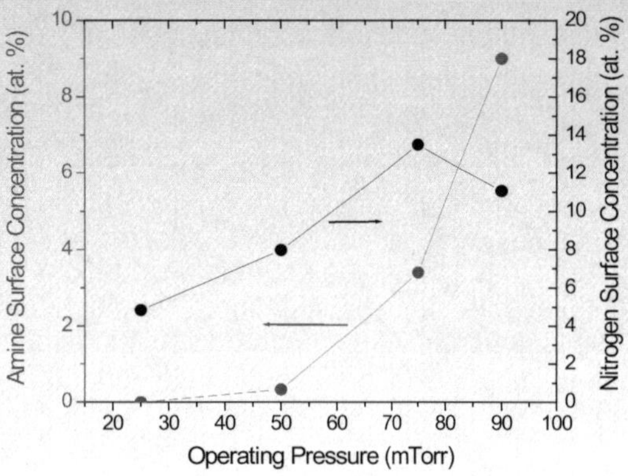

Primary amine coverage

Sheet resistance and reactivity

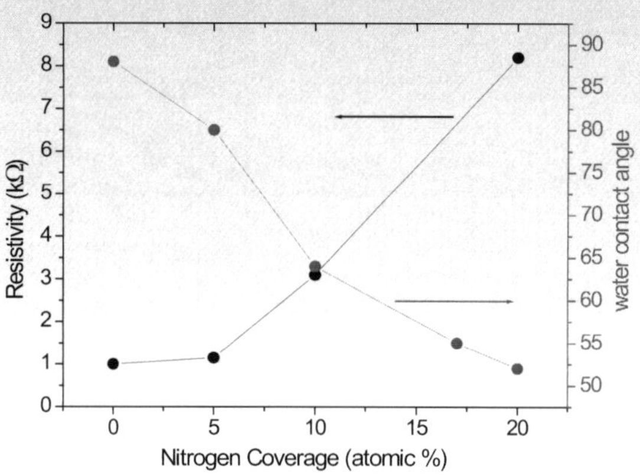

Functionalized Graphene: Enhanced atomic layer deposition (ALD)

ALD of Al_2O_3 on fluorinated epitaxial graphene

F-functionalized un-functionalized

- 20nm of Al_2O_3
- Conformal oxide layer only on the functionalized regions

ALD of HfO_2 on fluorinated CVD graphene*

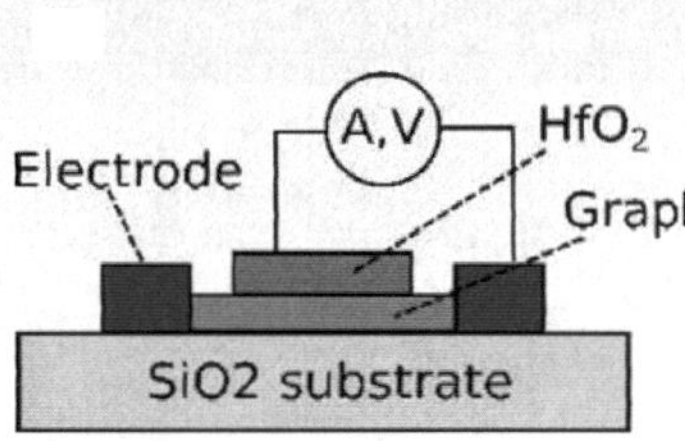

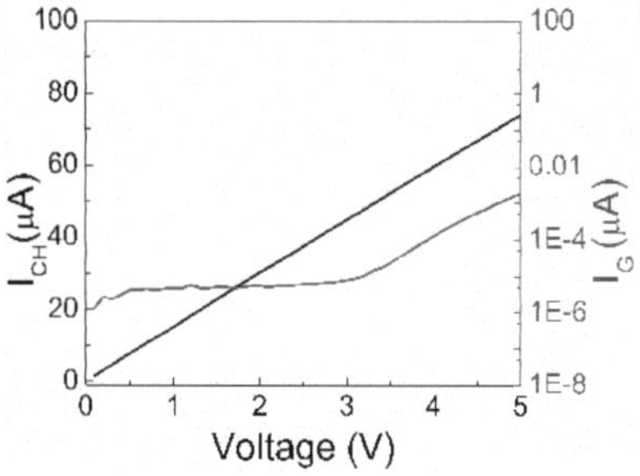

Large difference between source-drain current (I_{CH}) and source-gate current (I_G) indicate successful isolation of graphene

* In collaboration with IBM: A. V. Jagtiani, et al,J. Vac. Sci. Technol. A 34, 01B103 (2016).

Functionalized Graphene: Controlling the properties at contacts

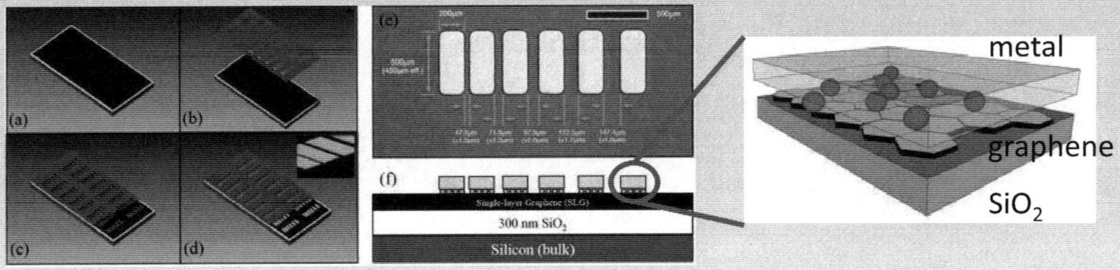

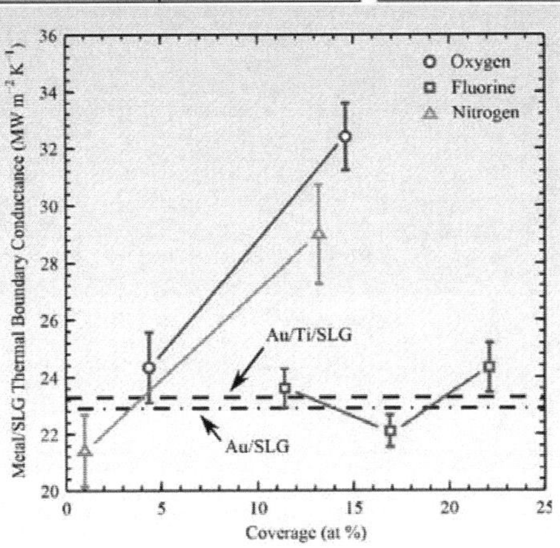

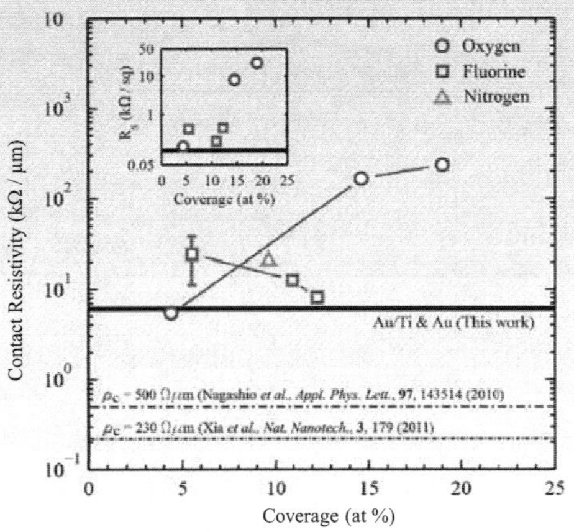

University of Virginia Engineering
Developing Leaders of Innovation

In collaboration with Prof. Patrick Hopkins (UVA)

P.E. Hopkins et al., Nano Lett. 12, 590 (2012) ;B. M. Foley et al., Nano Lett. 15, 4876 (2015)

21

Chemical Patterning of Graphene

Direct chemical patterning of graphene for greater control over material properties and reactivity

Physical masking technique + plasma functionalization

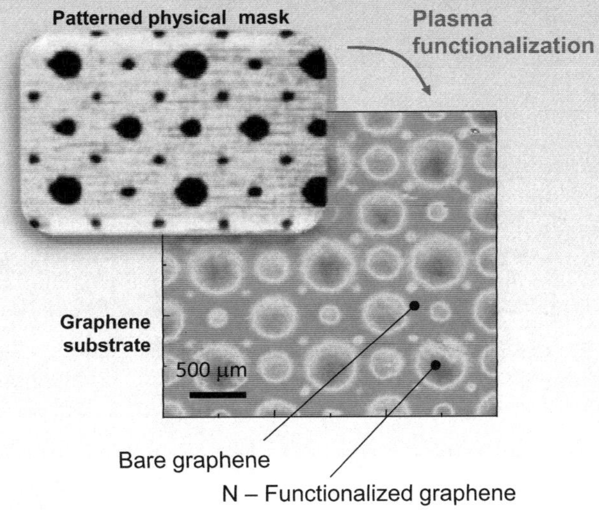

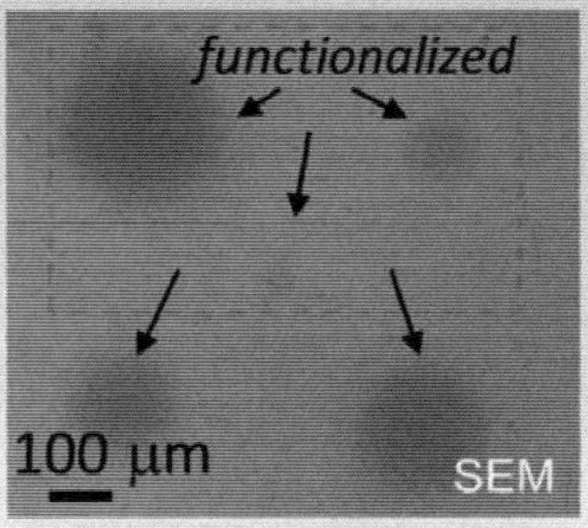

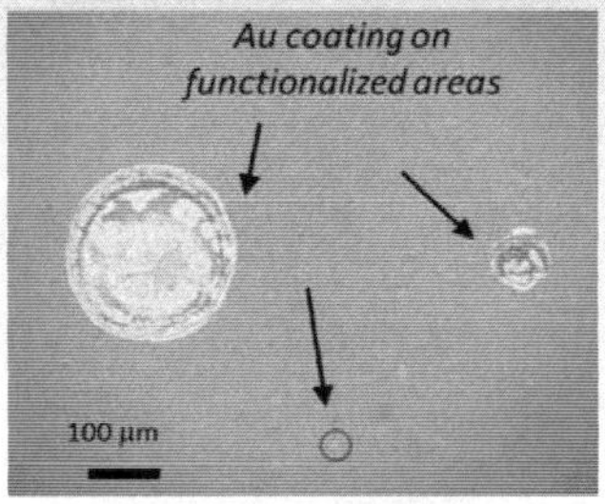

electrodeposition

S. C. Hernández et al., Carbon 60, 84 (2013)

22

Chemical Patterning of Graphene: Gradients

Producing chemical gradients on graphene

- Canopy mask + plasma functionalization to produce chemical gradients
  - Continuous in gradient direction (x) and uniform laterally (y)
  - Flexibility in chemistry to grade the surface reactivity

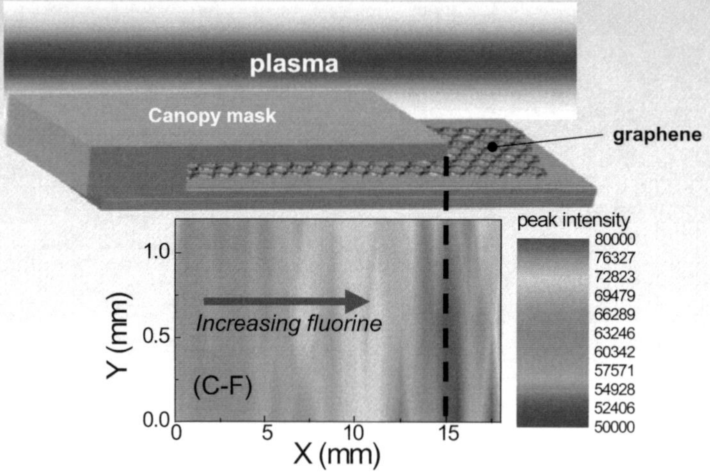

S. C. Hernández et al, ACS Nano 7, 4746 (2013)

23

Chemical Patterning of Graphene: Gradients

Oxygen gradient *for "pulling"* H_2O

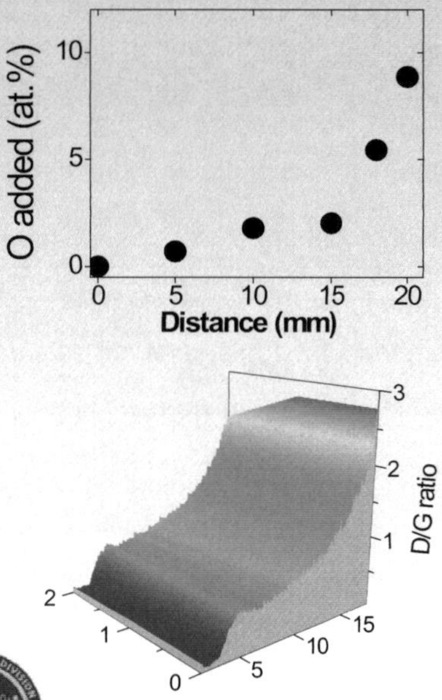

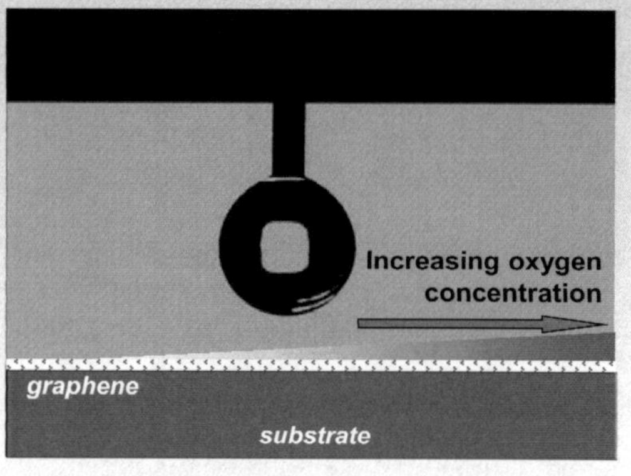

S. C. Hernández et al, ACS Nano 7, 4746 (2013)

24

Atomic Layer Etch

25

CNT Etching: A comparative study

Compare the response of semiconducting CNT's to CH_4 plasmas

Before exposure | After exposure

e-beam generated plasma + 10 V surface bias

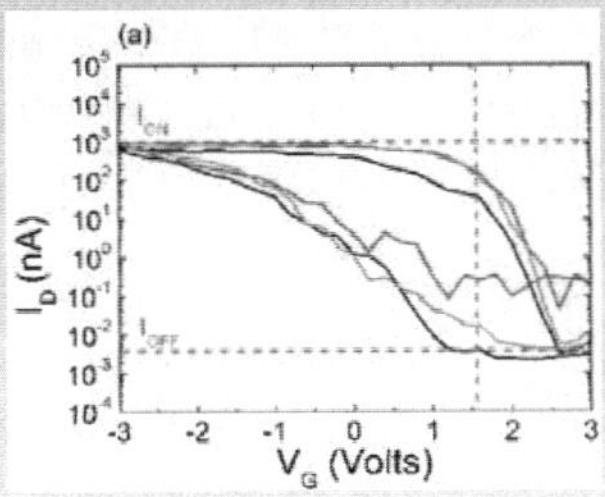

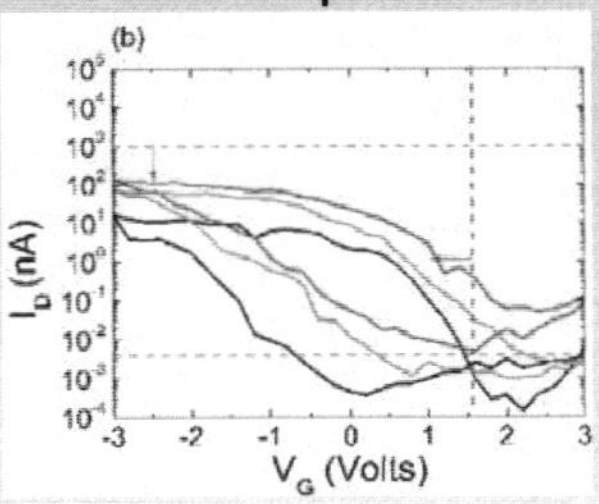

- I_{on} decreases
- I_{off} remains similar
- Semiconducting properties generally retained

Low power ICP etch tool + 25 W surface bias

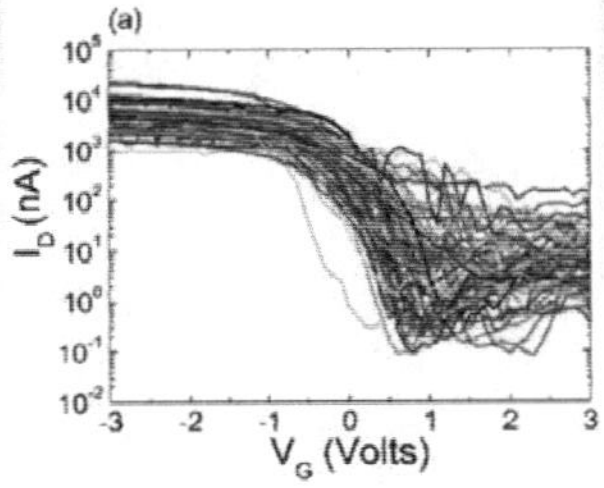

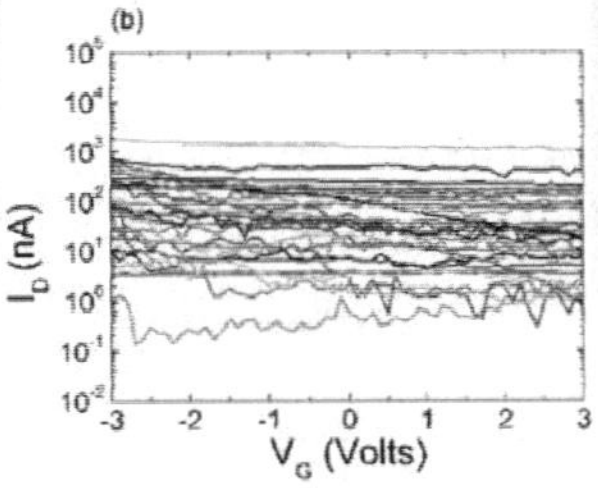

- I_{on} decreases
- I_{off} increases
- Semiconducting properties *not* generally retained

IBM

In collaboration with IBM: A. V. Jagtiani, et al,J. Vac. Sci. Technol. A 34, 01B103 (2016).

26

Etch Stop on Graphene: A comparative study

Compare the blanket etching of SiN (on graphene) with Ar/SF_6 plasmas

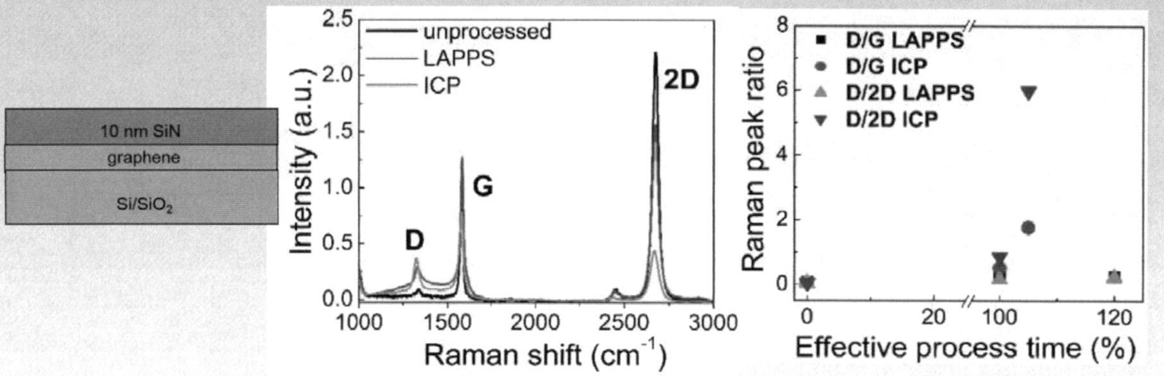

Results show less damage (D/G ratio) to the graphene after e-beam plasma etch compared to ICP etching

IBM

In collaboration with IBM: A. V. Jagtiani, et al, J. Vac. Sci. Technol. A 34, 01B103 (2016).

27

Summary

Electron Beam Generated Plasmas

- Unique control over the production of species and their transport to surface
- n_e is high (10^{10}- 10^{11} cm^{-3}); T_e is very low ($\approx$ 0.5 eV)
- Large flux of very low energy (0 - 5 eV) ions
- Well-suited to advance unique processing applications
  - Ion energy-sensitive materials
  - Atomic Layer Etch (or Deposition)
  - Processing of very thin (~ nm) or atomically thin (2-D) materials

28

Thank you

scott.walton@nrl.navy.mil

This work is supported by the Naval Research Laboratory Base Program

Powered by Naval Research

TM/SM

29

Select Publication List

- M. S. Osofsky, S. C. Hernández, A. Nath, V. D. Wheeler, S.G. Walton, C. M. Krowne, and D. K. Gaskill, "Functionalized graphene as a model system for the two-dimensional metal-insulator transition," Sci. Rep. 6, 19939; doi: 10.1038/srep19939 (2016).
- G.M. Petrov, D.R. Boris, Tz. B. Petrova, and S.G. Walton, "One-dimensional Ar-SF6 hydromodel at low-pressure in e-beam generated plasmas," J. Vac. Sci. Technol. A 34, 021302-12 (2016).
- A. V. Jagtiani, H. Miyazoe, J. Chang, D. B. Farmer, M. Engel, D. Neumayer, S.-J. Han, S. U. Engelmann, D.R. Boris, S.C. Hernández, E.H. Lock, S.G. Walton, E.A. Joseph, "Evaluation of plasma damage to atomic layer carbon materials," J. Vac. Sci. Technol. A 34, 01B103 (2016).
- Dominik Metzler, Florian Weilnboeck, Sandra C. Hernández, Scott G. Walton, Robert L. Bruce, Sebastian Engelmann Lourdes Salamanca-Riba, and Gottlieb S. Oehrlein, "Formation of nanometer-thick delaminated amorphous carbon layer by two-step plasma processing of methacrylate-based polymer," J. Vac. Sci. Technol. B 33, 051601 (2015).
- B.M. Foley, S.C. Hernández, J.C. Duda, J.T. Robinson, S.G. Walton, and P.E. Hopkins, "Modifying Surface Energy of Graphene via Plasma-Based Chemical Functionalization to Tune Thermal and Electrical Transport at Metal Interfaces," Nano Lett. 15, 4876 (2015).
- G. M. Petrov, D. R. Boris, E. H. Lock, Tz. B. Petrova, R. F. Fernsler and S. G. Walton "The influence of magnetic field on electron beam generated plasmas," J. Phys. D: Appl. Phys. 48, 275202-8 (2015).
- C.D. Cothran, D.R. Boris, C.S. Compton, E.M. Tejero, R.F. Fernsler, W.E. Amatucci, and S.G.Walton, "Continuous and pulsed electron beam production from an uninterrupted plasma cathode," Surf. Coat. Technol. 267, 111–116 (2015).
- D. R. Boris, R. F. Fernsler, and S. G. Walton, "Measuring the Density, Electron Temperature, and Electronegativity in Electron Beam Generated Plasmas Produced in Argon /SF_6 Mixtures," Plasma Sources Sci. Tech. 24, 025032 (2015).
- S. G. Walton, D. R. Boris, S. C. Hernández, E. H. Lock, Tz. B. Petrova, G. M. Petrov, and R. F. Fernsler, "Electron Beam Generated Plasmas for Ultra Low T_e Processing," ECS J. Solid State Sci. Technol. 4(6): N5033-N5040 (2015).
- S. U. Engelmann, R. L. Bruce, M. Nakamura, D. Metzler, S.G. Walton, and E. A. Joseph, "Challenges of Tailoring Surface Chemistry and Plasma/Surface Interactions to Advance Atomic Layer Etching," ECS J. Solid State Sci. Technol. 4(6), N5054-N5060 (2015).
- Jonathan R. Felts, Andrew J. Oyer, Sandra C. Hernández, Keith E. Whitener Jr, Jeremy T. Robinson, Scott G. Walton and Paul E. Sheehan, "Direct mechanochemical cleavage of functional groups from graphene," Nature Commun. 6:6467 DOI: 10.1038ncomms7467 (2015).
- E.H. Lock, D.M. Delongchamp, S.W. Schmucker, B. Simpkins, M. Laskoski, S.P. Mulvaney, D.R. Hines, M. Baraket, S.C. Hernandez, J.T. Robinson, P.E. Sheehan, C. Jaye, D.A. Fisher, S.G. Walton, "Dry graphene transfer to polystyrene and ultra-high molecular weight polyethylene – Detailed chemical, structural, morphological, and electrical characterization," Carbon 86, 288 (2015).
- E.H. Lock, R.F. Fernsler, S. Slinker, I. Singer, S.G. Walton, "Global model for plasmas generated by electron beams in low pressure nitrogen," J. Phys. D: Appl. Phys. 47, 425206 (2014).
- S.C. Hernández, V.D. Wheeler, M.S. Osofsky, V. K. Nagareddy, E.H. Lock, L.O. Nyakiti, R.L. Myers-Ward, A.B. Horsfall, C.R. Eddy Jr., D. K. Gaskill, S.G. Walton, "Plasma-Based Chemical Modification of Epitaxial Graphene with Oxygen Functionalities," Surf. Coat. Tech. 241, 8 (2014).

30

Select Publication List

- D. R. Boris, R.F. Fernsler, and S. G. Walton "The spatial profile of density in electron beam generated plasmas," Surf. Coat. Tech. 241, 13 (2014).
- D. R. Boris, G. M. Petrov, E. H. Lock, Tz. B. Petrova, R.F. Fernsler, and S. G. Walton "Controlling the electron energy distribution function of electron beam generated plasmas with molecular gas concentration: part I. experimental results," Plasma Sources Sci. Tech. 22, 065004-6 (2013)
- G. M. Petrov, D. R. Boris, E. H. Lock, Tz. B. Petrova, R.F. Fernsler, and S. G. Walton "Controlling the electron energy distribution function of electron beam generated plasmas with molecular gas concentration: part II. Numerical modeling," Plasma Sources Sci. Tech. 22, 065005 -8 (2013).
- S. C. Hernández, C. J. C. Bennett, C. E. Junkermeier, S. D. Tsoi, F. J. Bezares, R. Stine, J. T. Robinson, E. H. Lock, D. R. Boris, B. D. Pate, J. D. Caldwell, T. L. Reinecke, P. E. Sheehan and S. G. Walton, "Chemical Gradients on Graphene to Drive Droplet Motion," ACS Nano, 7, 6, 4746-4755 (2013)
- S. C. Hernández, F. J. Bezares, J. T. Robinson, J. D. Caldwell, S. G. Walton, "Controlling the local chemical reactivity of graphene through spatial functionalization," Carbon 60, 84-93 (2013).
- V. K. Nagareddy, H. K. Chan, Sandra C. Hernández, Virginia D. Wheeler, Luke O. Nyakiti, Rachel L. Myers-Ward, Charles R. Eddy, Jr, Jonathan P. Goss, Nicholas G. Wright, Scott G. Walton, D. Kurt Gaskill, and Alton B. Horsfall "Improved Chemical Detection and Ultra-Fast Recovery Using Oxygen Functionalized Epitaxial Graphene Sensors," IEEE Sensors Journal 13(8), 2810-2817 (2013).
- Mira Baraket, Rory Stine, Woo K. Lee, Jeremy T. Robinson, Cy R. Tamanaha, Paul E. Sheehan and Scott G. Walton, "Aminated graphene for DNA attachment produced via plasma functionalization," App. Phys. Lett. 100, 233123 (2012).
- Patrick E. Hopkins, Mira Baraket, Edward V. Barnat, Thomas E. Beechem, Sean P. Kearney, John C. Duda, Jeremy T. Robinson, and Scott G. Walton, "Manipulating Thermal Conductance at Metal–Graphene Contacts via Chemical Functionalization," Nano Lett. 12, 590 (2012).
- S. G. Walton, C. D. Cothran, and W. E. Amatucci, "Electron Beam Propagation in Magnetic Fields," IEEE Trans. of Plasma Sci 39(11), 2574 (2011).
- D.R. Boris, S.G. Walton, and R.F. Fernsler, "The LC Resonance Probe for Determining Local Plasma Density," Plasma Sources Sci. Technol. 20, 025003-7 (2011).
- S.H. North, E.H. Lock, C.J. Cooper, J.B. Franek, C.R. Taitt, and S.G. Walton, "Plasma-based surface modification of polystyrene microtitre plates for covalent immobilization of biomolecules," ACS Applied Materials & Interfaces, 2(10), 2884 (2010).
- M. Baraket, S.G. Walton, E.H. Lock, J.T. Robinson, F.K. Perkins, "Electron beam generated plasmas for the functionalization of graphene," Appl. Phys. Lett. 96, 231501 (2010).
- E.H. Lock, D.Y. Petrovykh, P. Mack, T. Carney, R.G. White, S.G. Walton and R. F. Fernsler, "Surface Composition, Chemistry, and Structure of Polystyrene Modified by Electron Beam Generated Plasma," Langmuir 26(11) 8857 (2010).

31

Selectivity in Atomic Layer Etching Using Sequential, Self-Limiting Thermal Reactions

Younghee Lee[1], Craig Huffman[2] and Steven George[1,3]
[1]Depts. of Chemistry & [3]Mechanical Engineering, University of Colorado, Boulder, Colorado 80309.
[2]SUNY Poly SEMATECH, Albany, New York 12203.

Outline

1. Al_2O_3 ALE using HF and $Sn(acac)_2$ or $Al(CH_3)_3$ as metal precursors.
2. Selectivity using $Sn(acac)_2$, $Al(CH_3)_3$, $AlCl(CH_3)_2$ and $SiCl_4$ as metal precursors.
3. Selectivity in ALE based on temperature.

Requirements for Sequential, Self-Limiting Thermal Reactions for ALE

Need **spontaneous**, sequential, **self-limiting** thermal reactions that **remove** with atomic control.

Spontaneous requires thermochemically favorable.

Self-limiting requires saturation of surface reaction.

Removal requires volatility of reaction product.

Al_2O_3 ALE Using HF-Pyridine & $Sn(acac)_2$ as Reactants

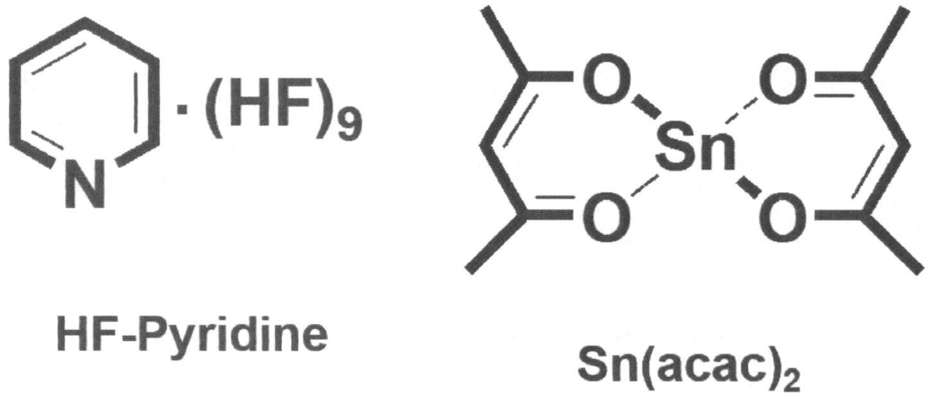

Y. Lee & S.M. George, *ACS Nano* **9**, 2061 (2015).

Al_2O_3 ALE Using HF & $Sn(acac)_2$

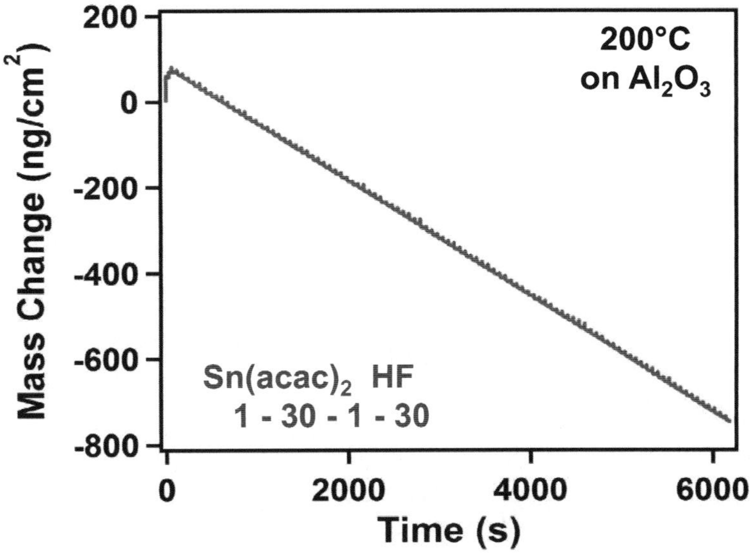

100 ALE cycles

Mass change per cycle = -8.4 ng/cm^2

Etch rate = 0.28 Å/cycle

Y. Lee & S.M. George, *ACS Nano* **9**, 2061 (2015).

Linear Decrease of Al_2O_3 Film Thickness vs Number of Al_2O_3 ALE Cycles

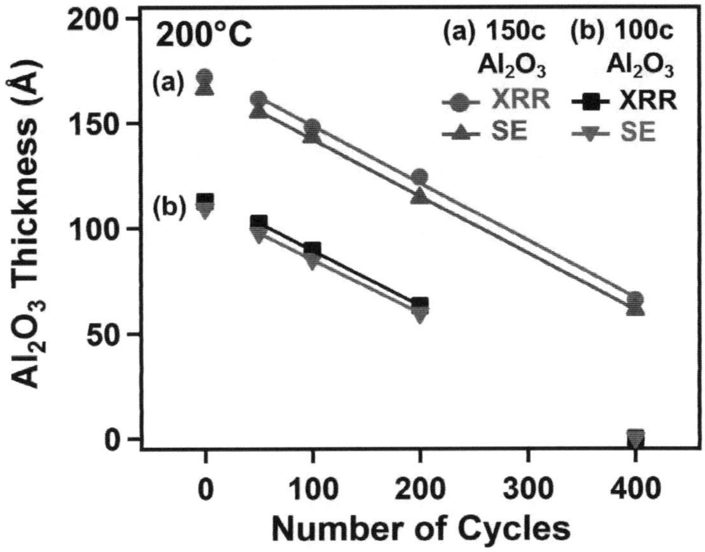

XRR measurements yield etch rate = 0.27 Å/cycle

Confirm with spectroscopic ellipsometry (SE)

Y. Lee & S.M. George, *ACS Nano* **9**, 2061 (2015).

Al_2O_3 ALE via Fluorination & Ligand Exchange

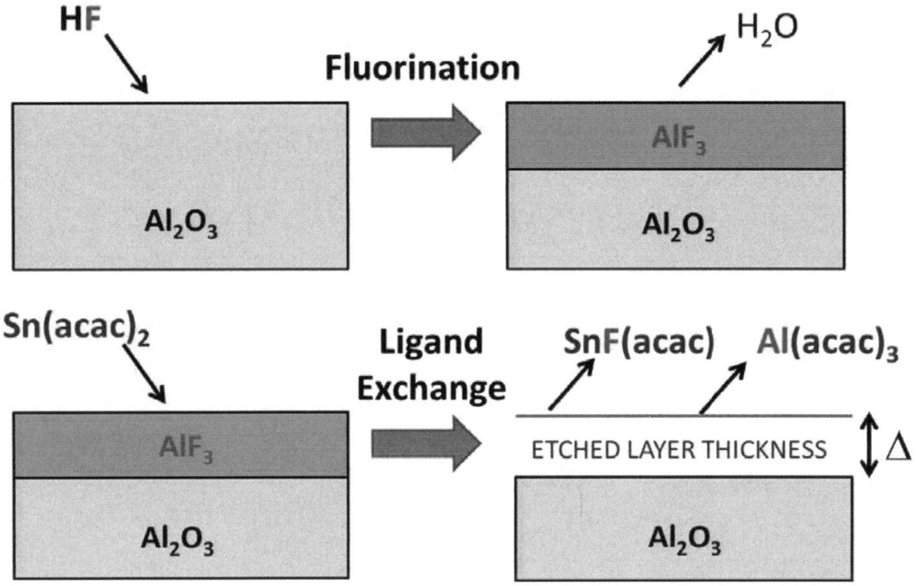

Metal Precursors for ALE

Requirements for Metal Precursor:

1. Accept fluorine from metal fluoride
2. Donate ligand to metal in metal fluoride
3. Metal reaction product is stable & volatile

Possible Metal Precursor:

Same precursor as used for ALD of etched material

e.g. $Al(CH_3)_3$ for Al_2O_3 ALE

Al_2O_3 ALE Using HF-Pyridine & $Al(CH_3)_3$ as Reactants

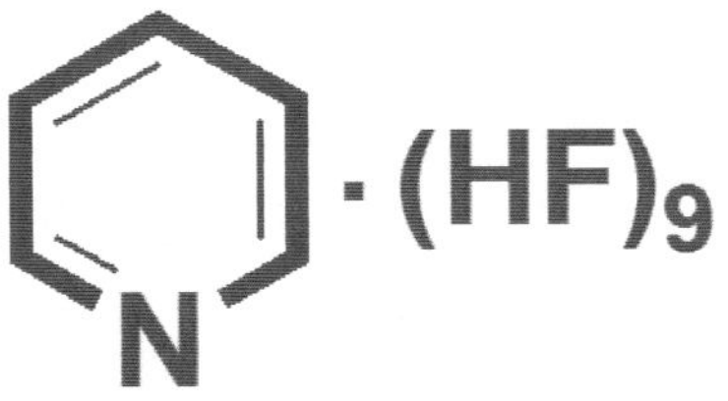

HF-Pyridine

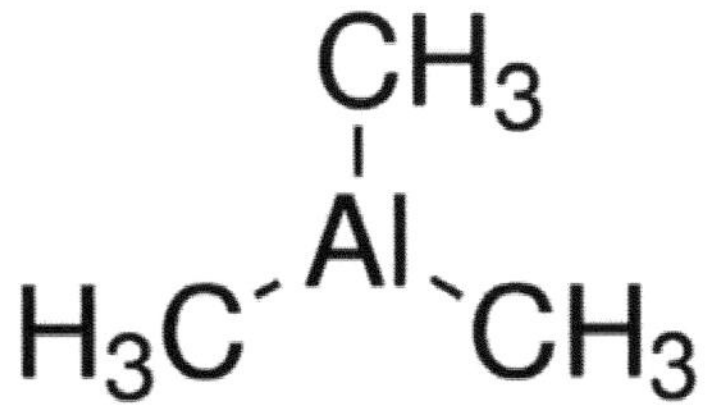

$Al(CH_3)_3$

Y. Lee, J.M. DuMont & S.M. George, *Chem. Mater. (In Press).*

Al_2O_3 ALE Using HF & $Al(CH_3)_3$

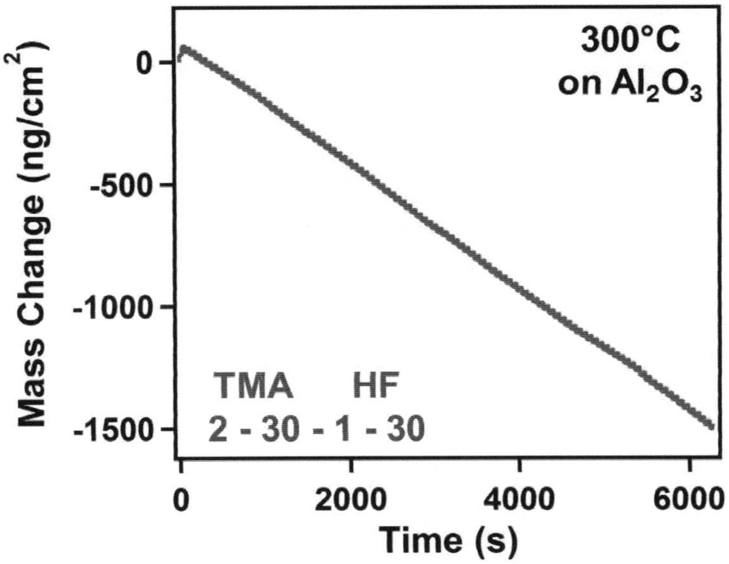

100 ALE Cycles

Mass change per cycle = -15.9 ng/cm^2

Etch rate = 0.51 Å/cycle

Y. Lee, J.M. DuMont & S.M. George, *Chem. Mater. (In Press).*

Linear Decrease of Al_2O_3 Film Thickness vs Number of Al_2O_3 ALE Cycles

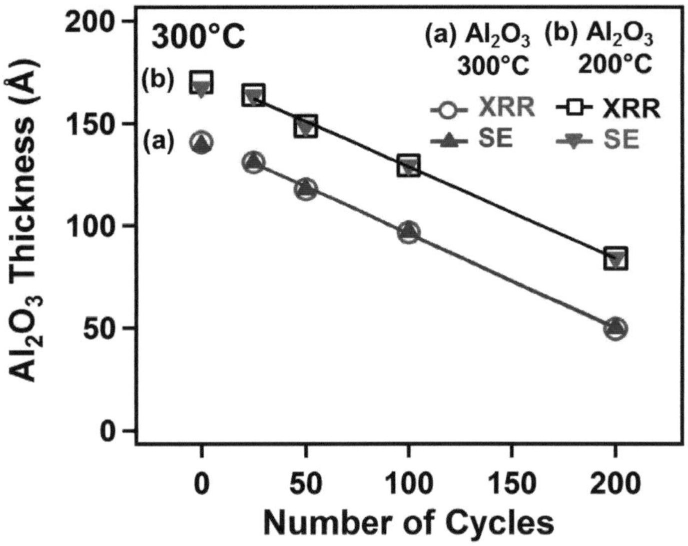

XRR measurements yield etch rate = 0.46 Å/cycle

Confirm with spectroscopic ellipsometry (SE)

Y. Lee, J.M. DuMont & S.M. George, *Chem. Mater. (In Press).*

Al_2O_3 ALE via Fluorination & Ligand Exchange

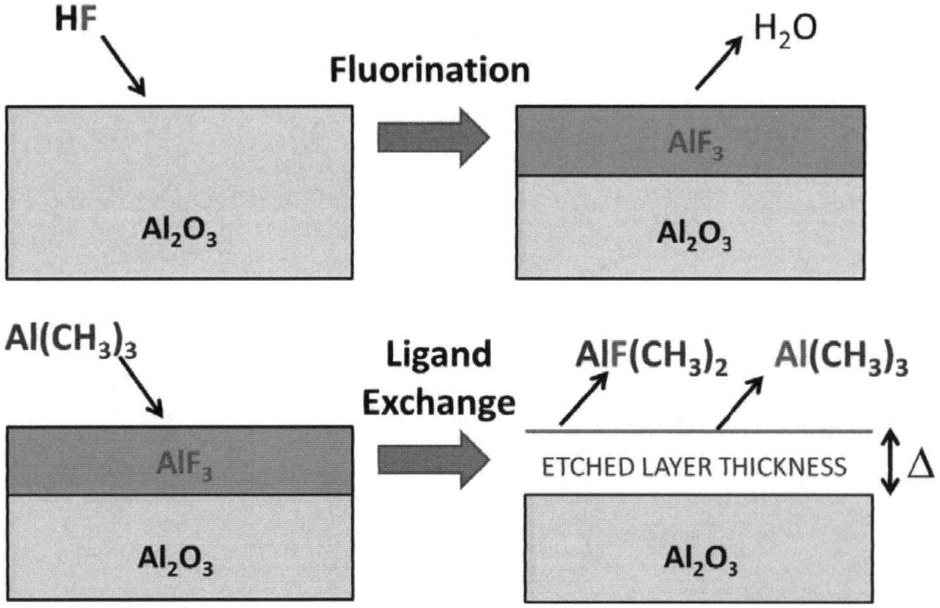

Outline

1. Al_2O_3 ALE using HF and $Sn(acac)_2$ or $Al(CH_3)_3$ as metal precursors.

2. Selectivity using $Sn(acac)_2$, $Al(CH_3)_3$, $AlCl(CH_3)_2$ and $SiCl_4$ as metal precursors.

3. Selectivity in ALE based on temperature.

Selective ALE for Different Materials

Different materials represented by various colors*

Goal to etch just one material in a background of other materials

Selectivity determined by stability & volatility of reaction products

*Adapted from C.T. Carver et al., *ECS J. Solid State Sci. Technol.* **4**, N5005 (2015).

Selectivity During ALE

Requirements for Metal Precursor:

1. Accept fluorine from metal fluoride
2. Donate ligand to metal in metal fluoride
3. Metal reaction product is stable & volatile

Strategy for Selectivity:

Use metal precursors with ligands that yield stable & volatile reaction products with target metals

ALE Using HF & $Sn(acac)_2$

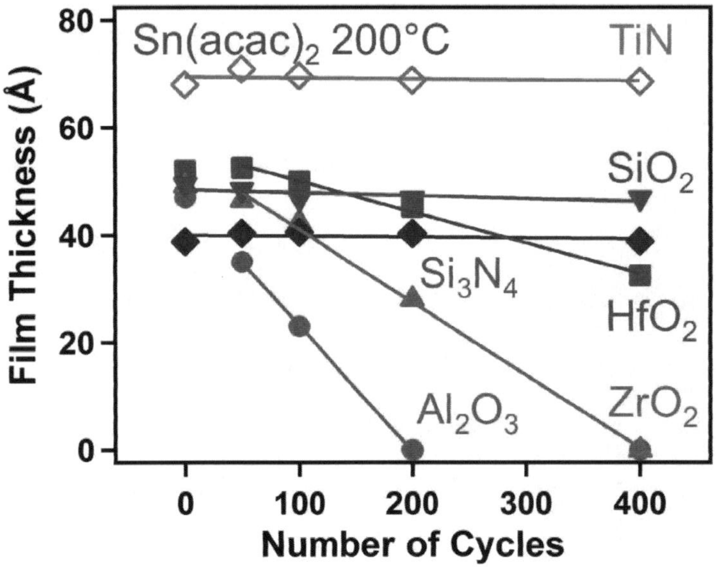

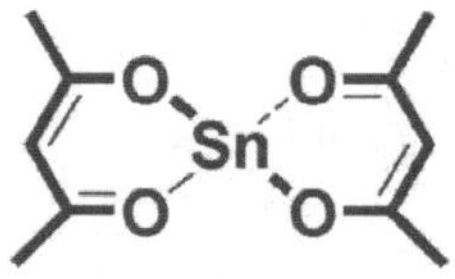

Selective etching of Al_2O_3, HfO_2 & ZrO_2.

Al, Hf & Zr form stable & volatile acac complexes.

Al_2O_3, HfO_2, ZrO_2, SiO_2, Si_3N_4, TiN

Al_2O_3, HfO_2 & ZrO_2 ALE Using HF & $Sn(acac)_2$

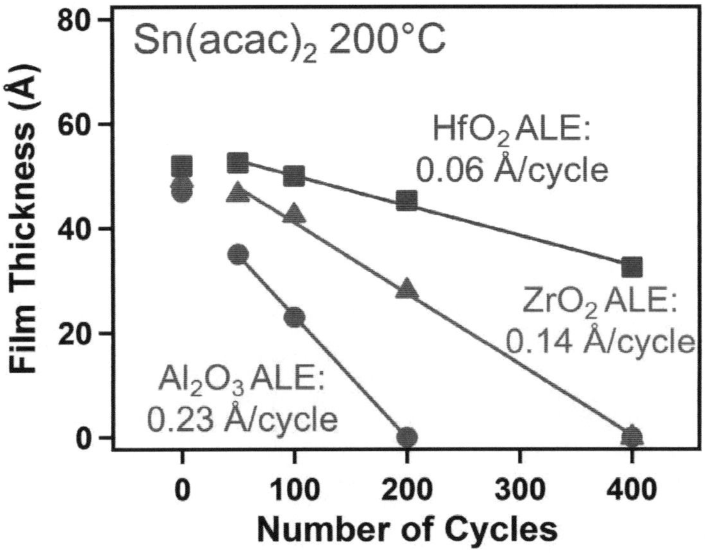

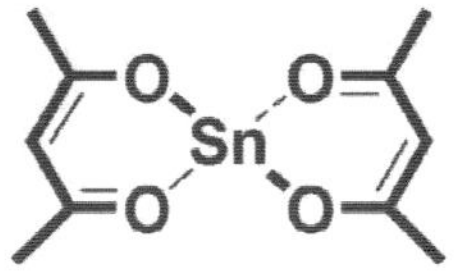

Al_2O_3, HfO_2, ZrO_2

ALE Using HF & $Al(CH_3)_3$ (TMA)

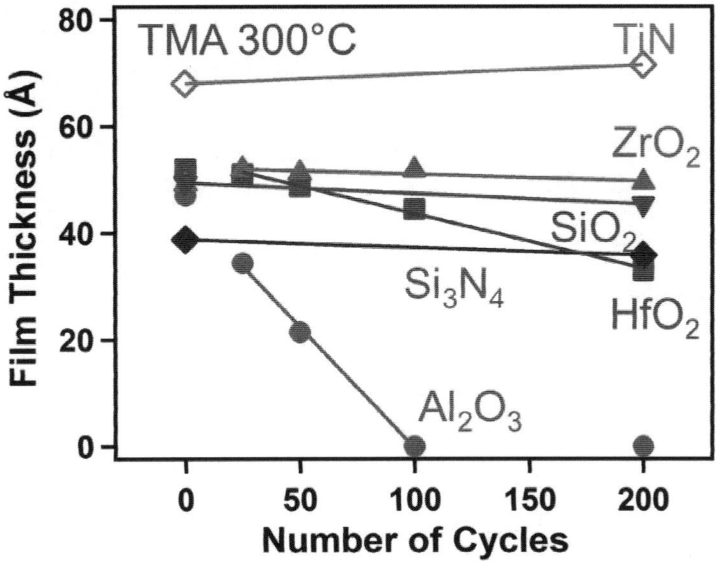

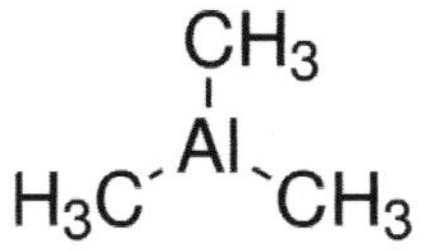

Selective etching of Al_2O_3 & HfO_2.

Al & Hf form stable & volatile complexes with methyl groups.

Al_2O_3, HfO_2, ZrO_2, SiO_2, Si_3N_4, TiN

Al_2O_3, HfO_2 & ZrO_2 ALE Using HF & $Al(CH_3)_3$ (TMA)

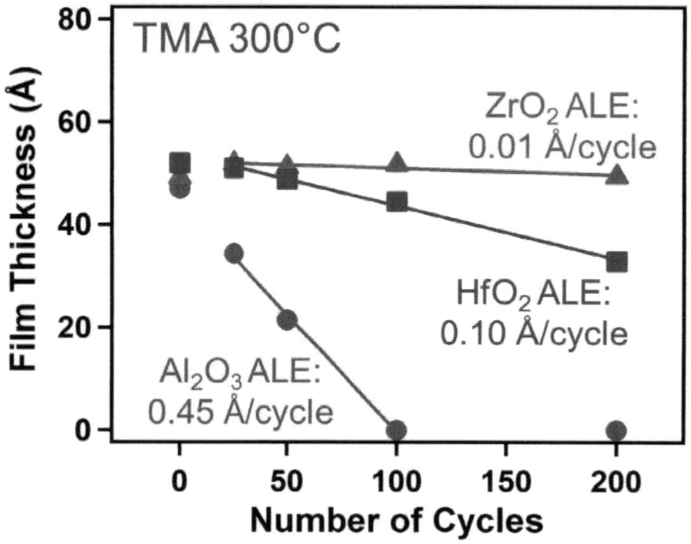

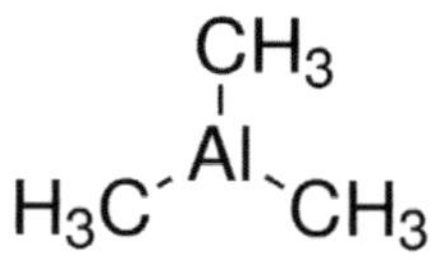

Al_2O_3, HfO_2, ZrO_2

Understanding Selectivity

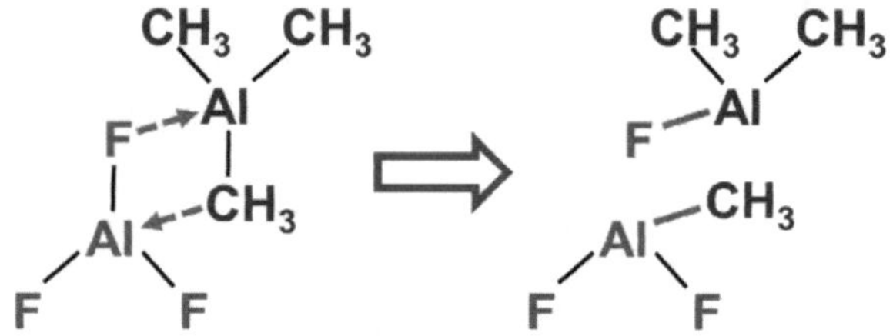

Al_2O_3 ALE with $Al(CH_3)_3$. Ligand-exchange. Stable Al-CH_3 reaction product.

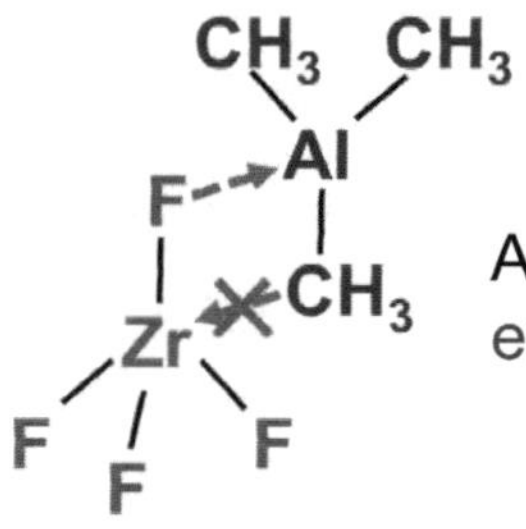

ZrO_2 ALE with $Al(CH_3)_3$. No ligand-exchange. Unstable Zr-CH_3 reaction product.

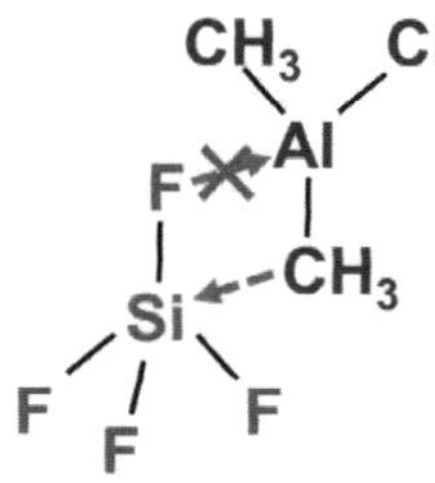

SiO_2 ALE with $Al(CH_3)_3$. No ligand-exchange. Si-F bond too stable.

ALE Using HF & $AlCl(CH_3)_2$ (DMAC)

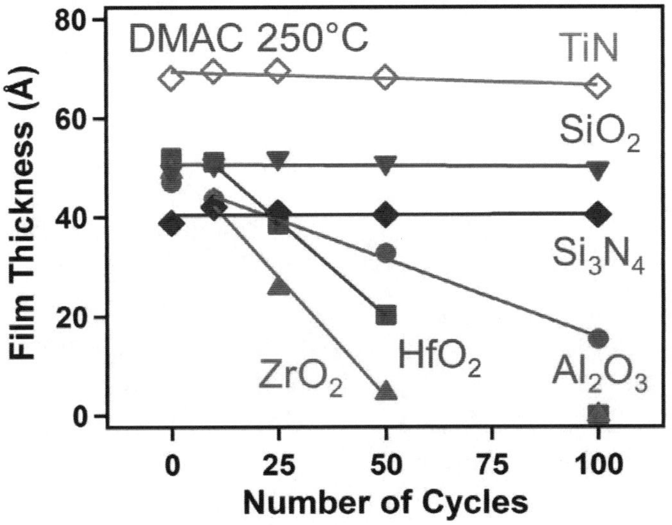

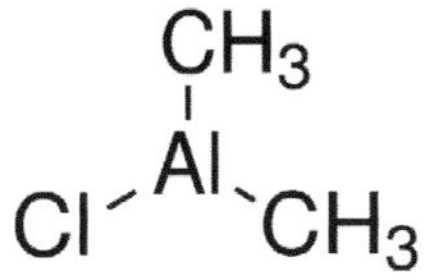

Selective etching of Al_2O_3, ZrO_2 & HfO_2.

Al, Zr & Hf form stable & volatile complexes with chloride or methyl groups.

Al_2O_3, HfO_2, ZrO_2, SiO_2, Si_3N_4, TiN

Al_2O_3, HfO_2 & ZrO_2 ALE Using HF & $AlCl(CH_3)_2$ (DMAC)

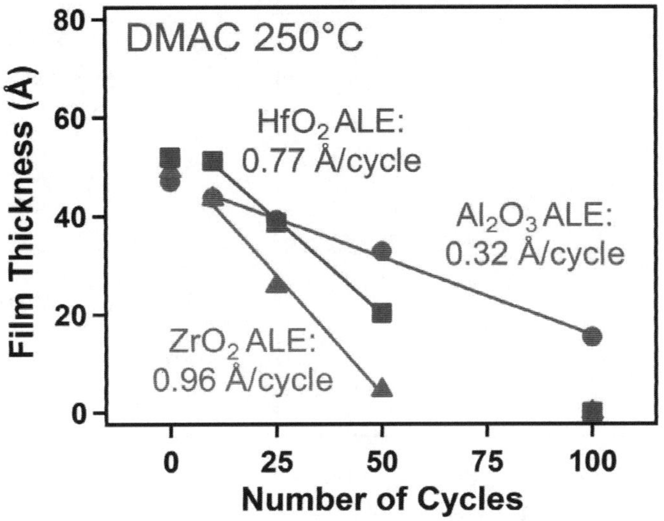

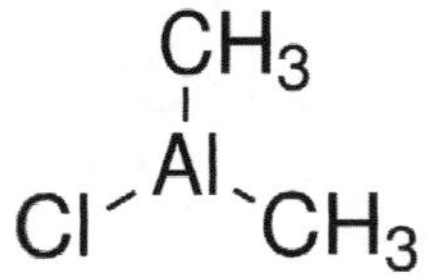

Al_2O_3, HfO_2, ZrO_2

ALE Using HF & $SiCl_4$

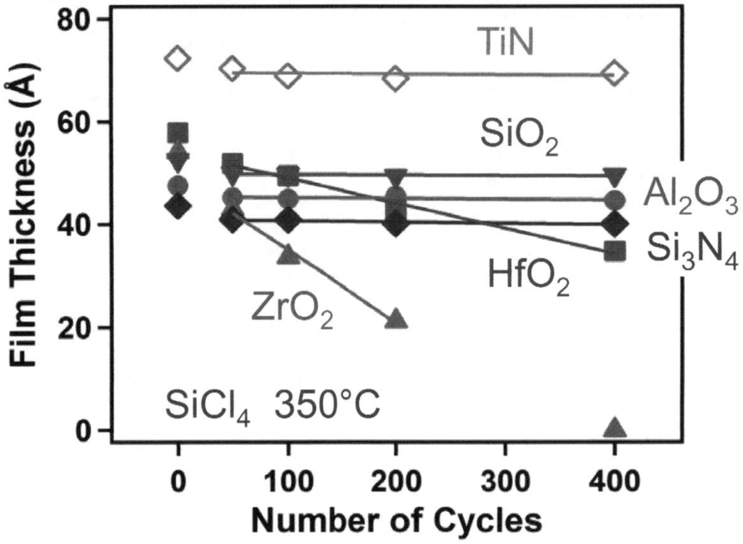

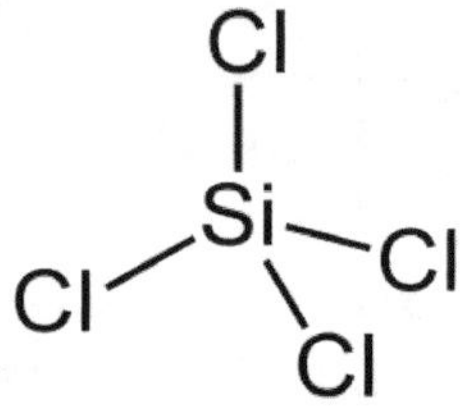

Selective etching of ZrO_2 & HfO_2.

Zr & Hf form stable & volatile complexes with chloride groups.

Al_2O_3, HfO_2, ZrO_2, SiO_2, Si_3N_4, TiN

Al_2O_3, HfO_2 & ZrO_2 ALE Using HF & $SiCl_4$

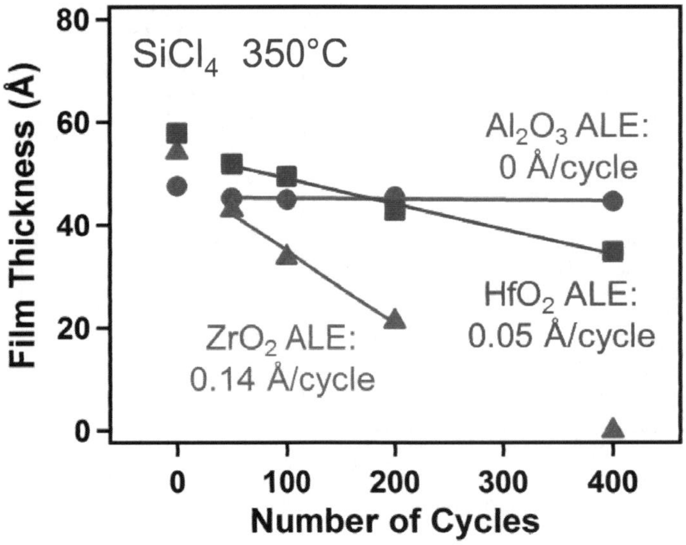

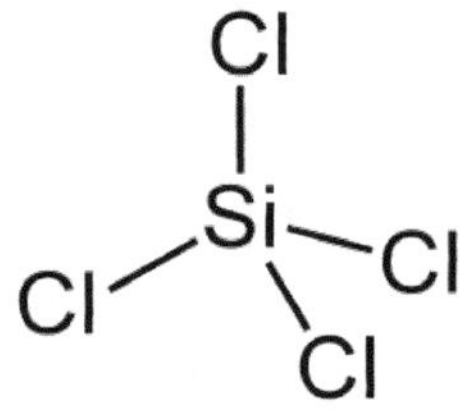

Why no etching of Al_2O_3 with $SiCl_4$?

Al_2O_3, HfO_2, ZrO_2

Ligand-Exchange Thermochemistry Explains No Al_2O_3 ALE

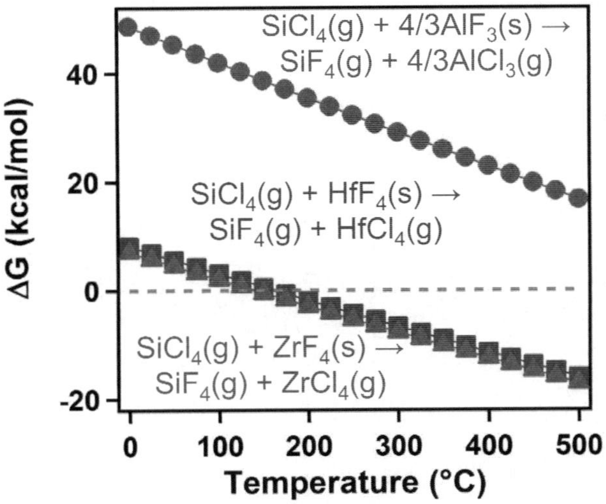

Positive ΔG for $SiCl_4$ ligand-exchange for Al_2O_3 ALE.

ΔG = 0

Negative ΔG for $SiCl_4$ ligand-exchange >150°C for HfO_2 & ZrO_2 ALE.

Outline

1. Al_2O_3 ALE using HF and $Sn(acac)_2$ or $Al(CH_3)_3$ as metal precursors.

2. Selectivity using $Sn(acac)_2$, $Al(CH_3)_3$, $AlCl(CH_3)_2$ and $SiCl_4$ as metal precursors.

3. **Selectivity in ALE based on temperature.**

Selectivity Based on Temperature for Al_2O_3 ALE Using Different Metal Precursors

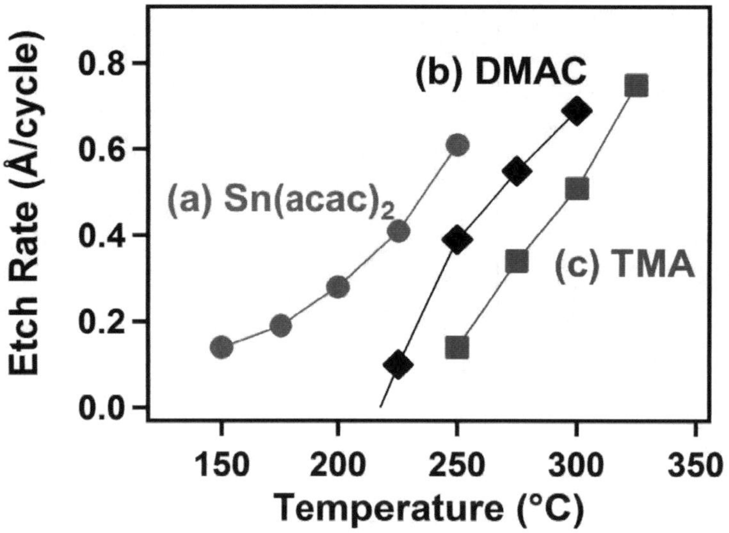

Different etch rates at various temperatures for different metal precursors.

Selectivity Based on Temperature for ALE Using $SiCl_4$ as Metal Precursor

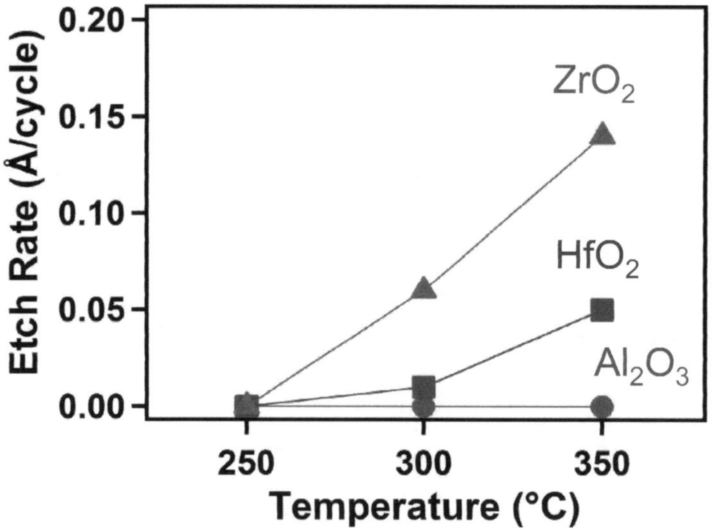

Conclusions

1. Thermal ALE possible using sequential, self-limiting fluorination & ligand-exchange reactions.
2. Thermal ALE using HF and either $Sn(acac)_2$, $Al(CH_3)_3$, $AlCl(CH_3)_2$ or $SiCl_4$ as metal precursors.
3. Selective ALE is possible. Depends on stability and volatility of reaction products.
4. Temperature provides additional pathway for selective ALE.

SOLVAY
asking more from chemistry®

Innovative and environmental friendly Fluorine F_2 based cleaning process to replace C_2F_6, CF_4 and NF_3 as cleaning gas

SPCC Conference

Santa Clara, April 19th 2016

Fraunhofer EMFT

Solvay Special Chem

Marcello Riva

Outline

- Intro – PECVD cleaning technology
- Drivers for F_2-based chamber cleaning
- Facilitation example for F_2-gas
- Hardware and experimental
- Cleaning results $F_2/Ar/N_2$
- 500 wafer run
- Waste gas data
- Summary and conclusions

PECVD Chamber Cleaning

- Currently used cleaning gases:

  Oxide / nitride films (PECVD): NF_3, CF_4, C_2F_6, C_3F_8

  Poly-Si CVD, MOCVD of W and TiN: NF_3

- CVD chambers need frequent cleaning of all process-relevant reactor surfaces to sustain particle- and defect free film deposition

3 | Fraunhofer EMFT | SOLVAY asking more from chemistry®

Drivers for F_2-based Plasma-Cleaning

- Proposed gases: Low GWP* Gas Mixtures
- F_2 (GWP=0) as NF_3, PFCs (perfluorocarbon) alternative

	GWP	Life time			GWP	Life time
CF_4	7390	5000		NF_3	17200	740
CHF_3	11700	264		SF_6	22800	3200
CH_2F_2	650	5,6		c-C_4F_8	8700	3200
C_2F_6	12200	10000		C_4F_6	~0	~0
C_3F_8	7000	2600		c-C_5F_8	90	~1

*GWP (Global Warming Potential); CO_2 = reference = 1

4 | Fraunhofer EMFT | SOLVAY asking more from chemistry®

Properties of F_2-based cleaning gas mixture

- F_2 diluted in N_2 and Ar (compressed gas)
- Transportable in standard gas cylinder
- Corrosive, **not inflammable, not explosive**
- A diluted F_2-gas mixture is best for an efficient chamber clean
- F_2 mixtures require a passivation of an existing stainless steel gas delivery system (one time procedure)

5 | Fraunhofer EMFT SOLVAY asking more from chemistry

F_2 Gas Mixture – Gas Cabinet Example

- Gas mixture 1: $F_2/Ar/N_2$ → 13,56 MHz plasma
- Gas mixture 2: $F_2/Ar/N_2$ → 400 kHz RPS-unit

- Safety gas cabinet
- 3 way valve, purge gas N_2
- Gas cylinder: 10L, 100 bar

$F_2/Ar/N_2$ cylinder

6 | Fraunhofer EMFT SOLVAY asking more from chemistry

$F_2/Ar/N_2$ Gas Mixture – Facilitation Example

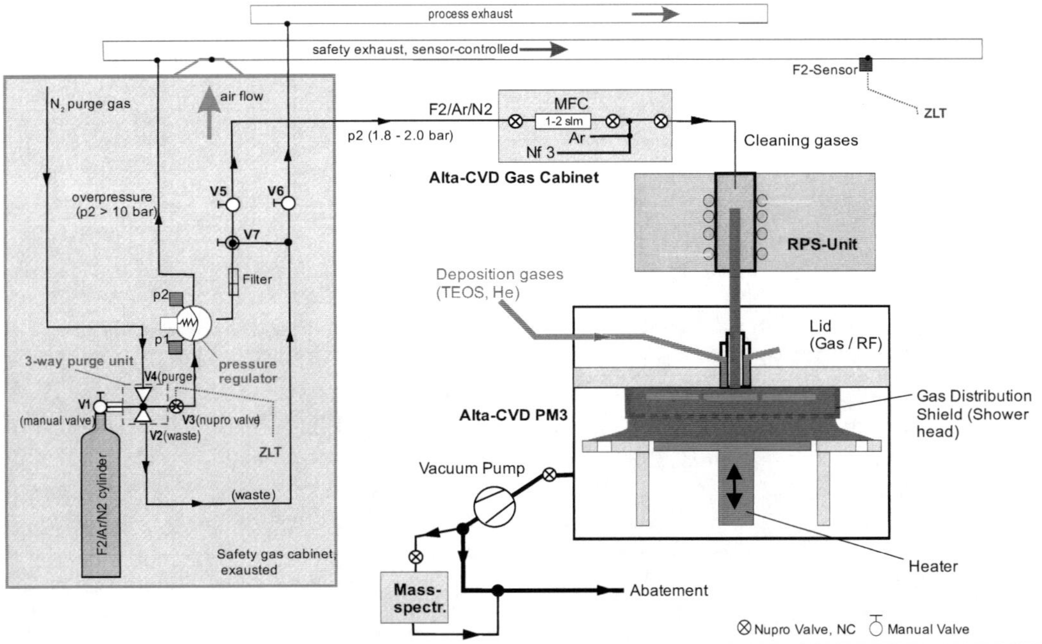

7

CVD Tools at Fraunhofer EMFT (200mm)

- Alta-CVD , Brooks VX400
  PECVD for SiO_2 films - PETEOS, BPSG
  - 13,56 MHz plasma clean, C_2F_6 / NF_3 based
  - RPS (remote plasma source) clean, NF_3 based

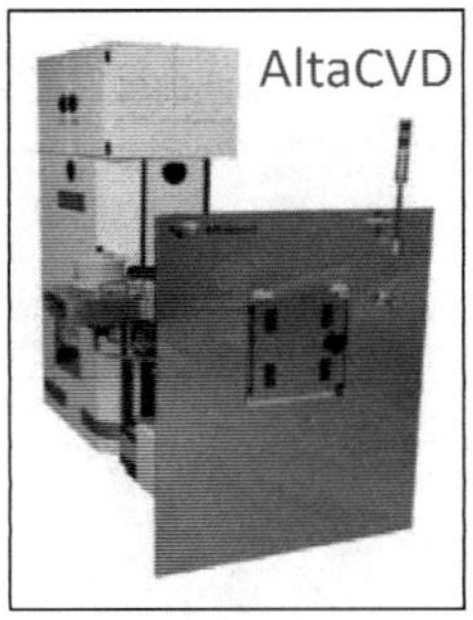

- AMAT PECVD, P 5000
  PECVD for SiO_2 films, lamp-heated
  - 13,56 MHz plasma clean, C_2F_6 /NF_3 or CF_4 based

8

RPS Unit on Top of Alta-CVD Chamber

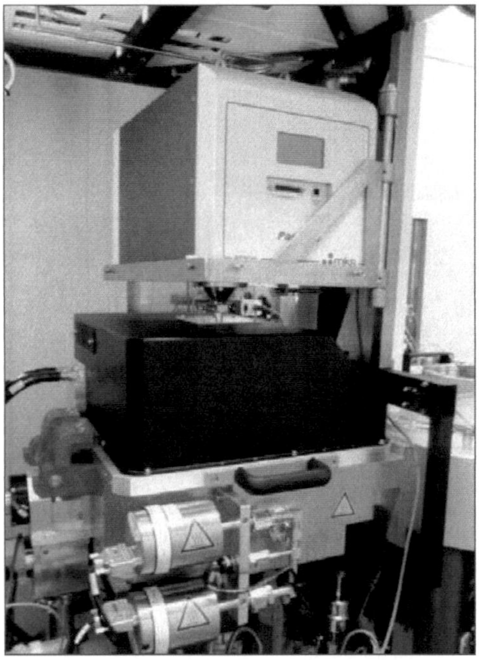

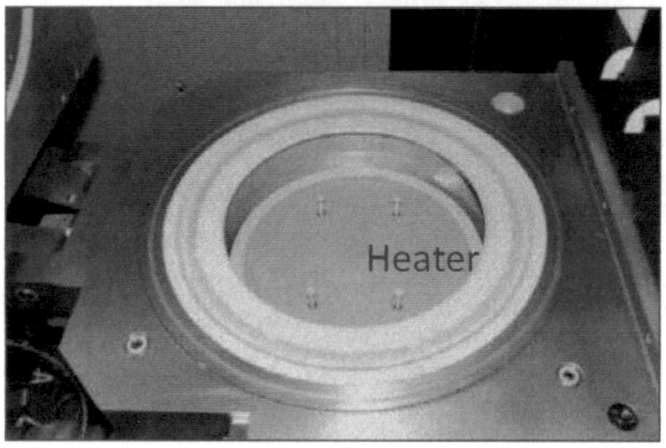

9 | Fraunhofer EMFT SOLVAY asking more from chemistry®

AMAT P5000 PECVD - Details

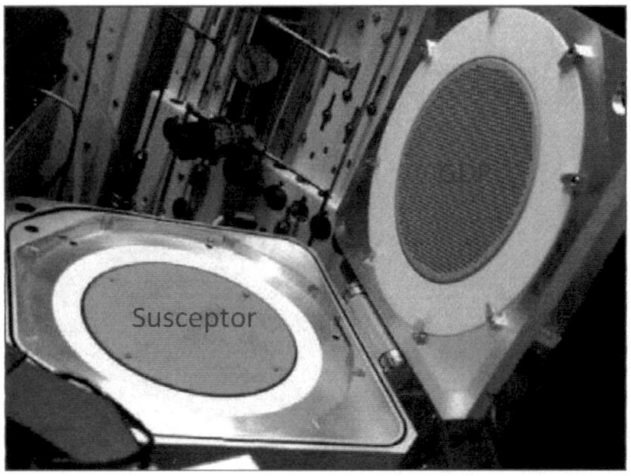

"RF isolation block"

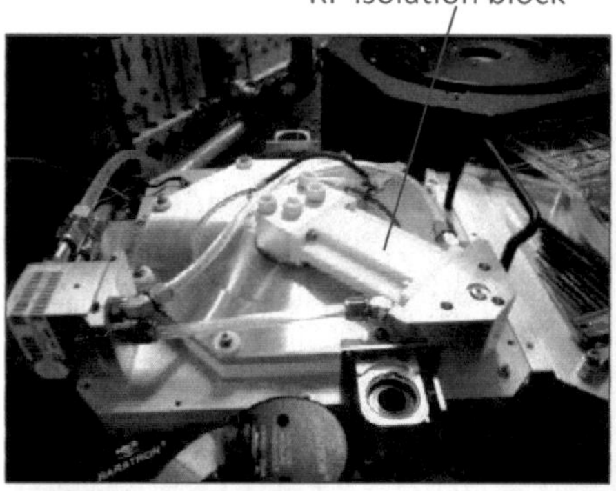

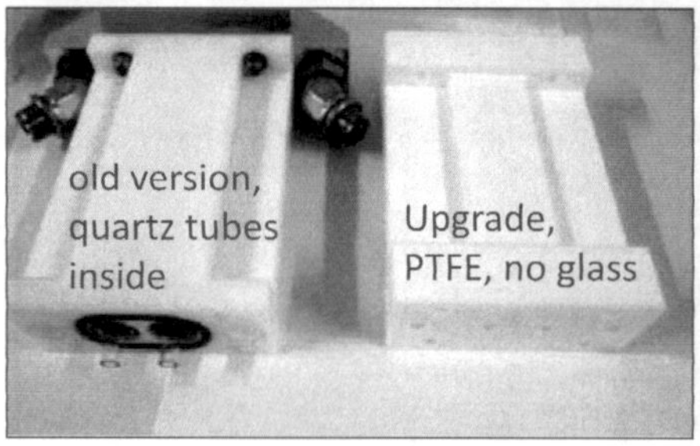

Commercially available upgrade

10 | Fraunhofer EMFT SOLVAY asking more from chemistry®

SiO_2 Cleaning Rate – Experimental

- Deposition of 2 µm PETEOS-film, 200mm Si-substrates
- Thickness measured with reflectometer / ellipsometer
- SiO_2-Etch rate calculated from post-etch TEOS film thickness
- Constant parameters:
  55 deg. C wall / lid temp., 400 deg. C heater temperature
- Parameter Variations:
  Gas flows, chamber pressure, RF-power, Spacing
- 30% higher SiO_2 cleaning rate compared to C_2F_6

Fraunhofer EMFT SOLVAY asking more from chemistry®

13,56 MHz Plasma Cleaning Process (2)

- Repeatability of SiO_2 deposition, 3 x 25 wafer

Run #	# of adders (particle size ≥ 0,25µm)			
	Slot	Slot	Slot	Average
	1	12	25	
1	38	7	21	22
2	5	5	32	14
3	11	2	6	6
Mean	18	5	20	**14**

AMAT P5000
lamp-heated
CVD, 200mm

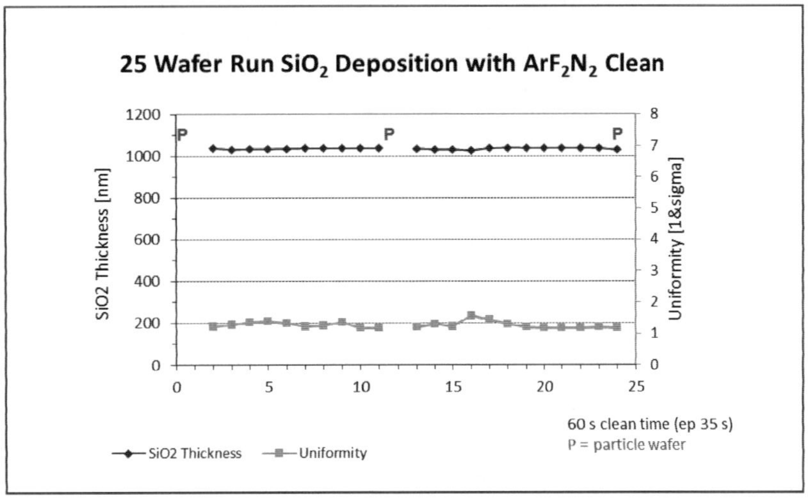

Fraunhofer EMFT SOLVAY asking more from chemistry®

13,56 MHz Plasma Cleaning Process (4)

Amount of cleaning gas needed for 25 Wafer Lot (1µm SiO2 depo / wafer):				
C2F6/O2:	tot. flow/wafer [slm]	tot. flow /lot [slm]	**tot. gas/lot** [g]	F2/lot [g]
C2F6-Flow	0,9	21,6	**132,7**	**114,4**
O2-Flow	0,9	22,2	31,7	
NF3-Flow	0,1	1,9	5,9	
Mean SiO2-rate	~1100nm/min			
CF4/O2:	tot. flow/wafer [slm]	tot. flow /lot [slm]	**tot. gas/lot** [g]	F2/lot [g]
CF4-Flow	0,8	19,3	**75,7**	**65,3**
N2O-Flow	0,3	7,0	18,8	
Mean SiO2-rate	~1200nm/min			
F2/Ar/N2:	tot. flow/wafer [slm]	tot. flow /lot [slm]	tot. gas/lot [g]	**F2/lot** [g]
F2/Ar/N2-Flow	0,8	19,5	27,1	**6,6**
Mean SiO2-rate	~1500nm/min			
Ratio C2F6 : F2	**17 : 1**			
Ratio CF4 : F2	**10 : 1**			

13 | Fraunhofer EMFT | SOLVAY asking more from chemistry®

13,56 MHz Plasma Cleaning Process (5)

Bond Dissociation energies			
Molecule			Diss. Energy [kJ/mol @ 298K]
N_2	-->	N + N	945
F_2	-->	F + F	155
NF_3	-->	NF_2 +F	243
NF_2	-->	NF + F	318
NF	-->	N + F	301
NF_3	-->	N + F +F + F	862
CF_4	-->	CF_3 + F	506
C_2	-->	C + C	607
O_2	-->	O + O	498
N_2O	-->	NO + N	115
source:	united states department of commerce,		
	national bureau od standards, Lewis M. Branscomb		

14 | Fraunhofer EMFT | SOLVAY asking more from chemistry®

400 kHz RPS Cleaning Process

- Gas mixture 2
- Monitored: SiO_2-deposition rate, particles in SiO_2-film25-wafer runs in auto-mode
- 500 wafers deposited, clean after every wafer
- Optical inspection of process kit parts
- F_2 gas mixture has identical cleaning rates of $SiO_2 \geq 1{,}0$ µm/min compared to a BKM NF_3-based recipe
- Drop-In recipe to substitute Ar/NF_3 chemistry (same pressure, total flow, spacing, step times)
- Potential for optimization still given
- Cleaning efficiency of F_2 gas mixture ≥ factor 1,2 compared to NF_3

15 | Fraunhofer EMFT SOLVAY asking more from chemistry®

400 kHz RPS Cleaning Process (2)

- 4 Step process:
  - Ar-Flow and Ignition
  - F_2-addition to Ar
  - Inner clean
  - Outer clean
- Etch behavior: Center fast

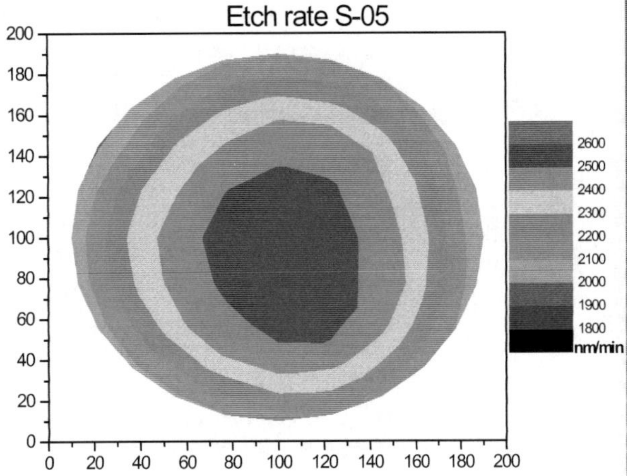

SiO_2-etch rate of complete NF_3-cleaning recipe with RPS-Unit

16 | Fraunhofer EMFT SOLVAY asking more from chemistry®

400 kHz RPS – 500 Wafer Run - Results (2)

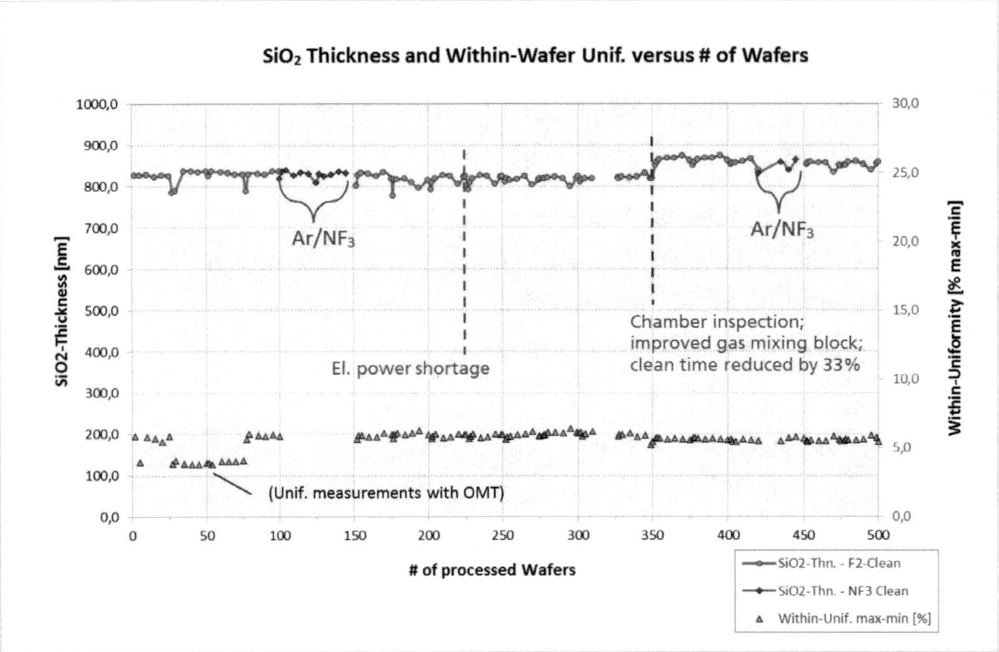

17

Fraunhofer EMFT
SOLVAY asking more from chemistry

400 kHz RPS – 500 Wafer Run - Results (3)

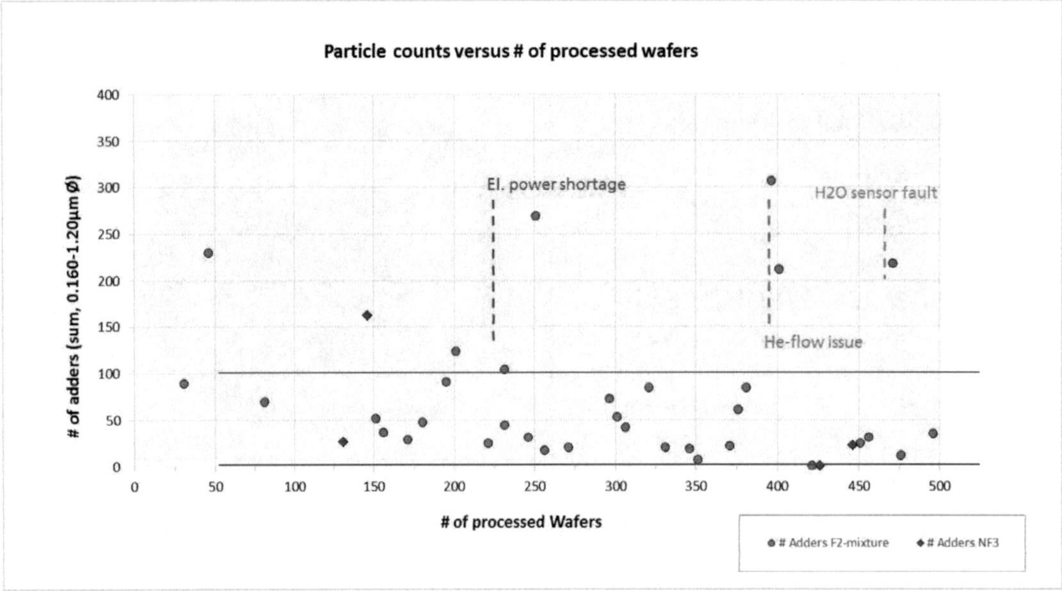

18

Fraunhofer EMFT
SOLVAY asking more from chemistry

400 kHz RPS – 500 Wafer Run - Results (5)

- Optical inspection showed no noticeable effects of the cleaning plasma on process kit parts
- PTFE O-rings fully intact, no signs of wear after 500 processed wafers

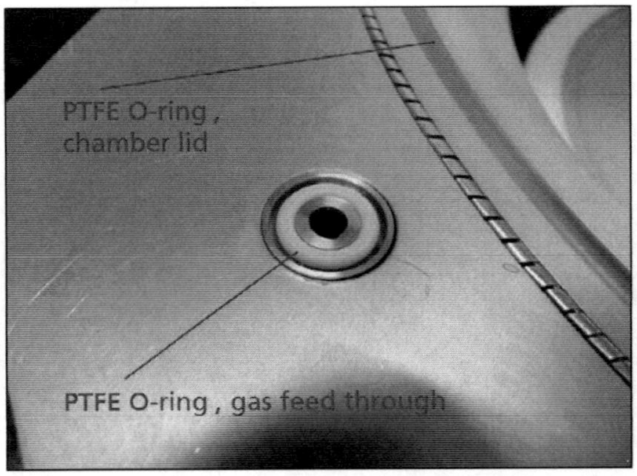

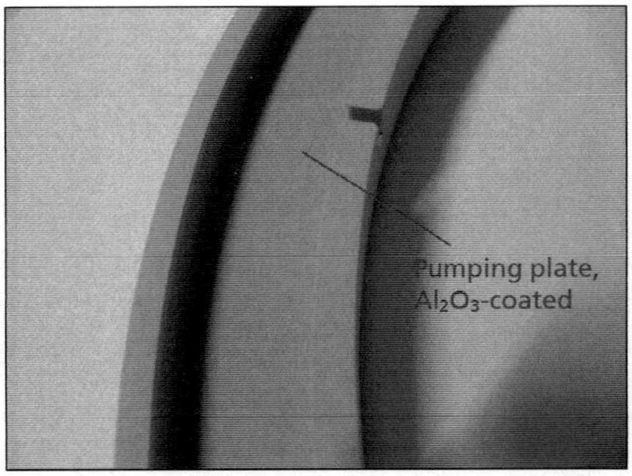

19 Fraunhofer EMFT SOLVAY asking more from chemistry®

400 kHz RPS – 500 Wafer Run - Results (6)

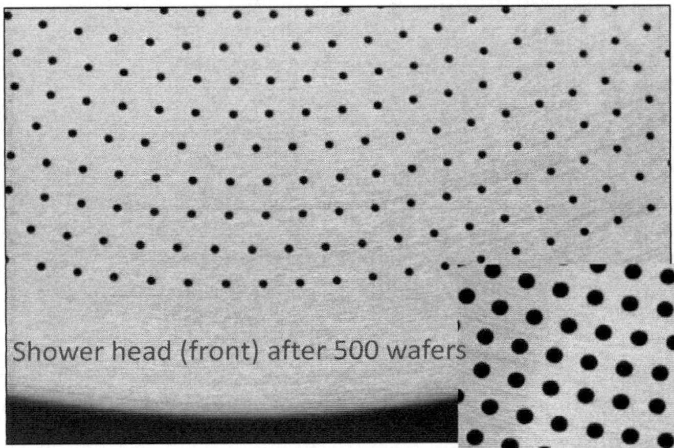

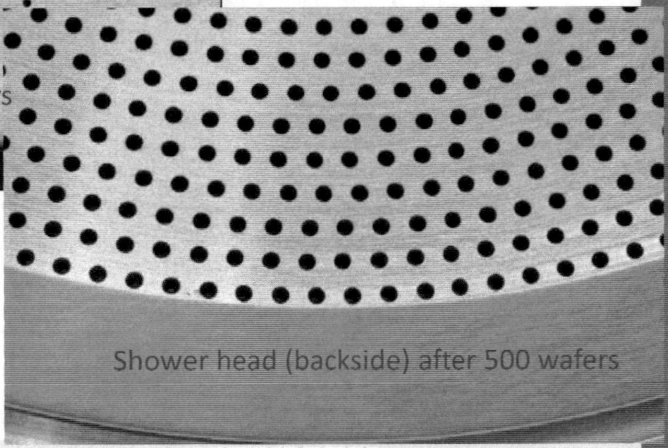

20 Fraunhofer EMFT SOLVAY asking more from chemistry®

Waste Gas Analysis

- Atmospheric mass spectrometer was mounted behind process pump exhaust
- In case of C_2F_6 compared to F_2-mixture, the absence of the big CF_3 peek confirms our early estimation of a cleaning efficiency factor >17 for F_2-mixture compared to C_2F_6

Waste Gas Analysis (2)

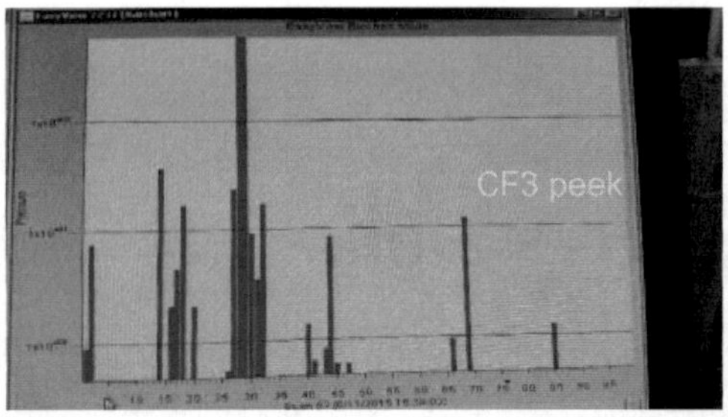

C_2F_6 Clean, 13,56 MHz parallel-plate

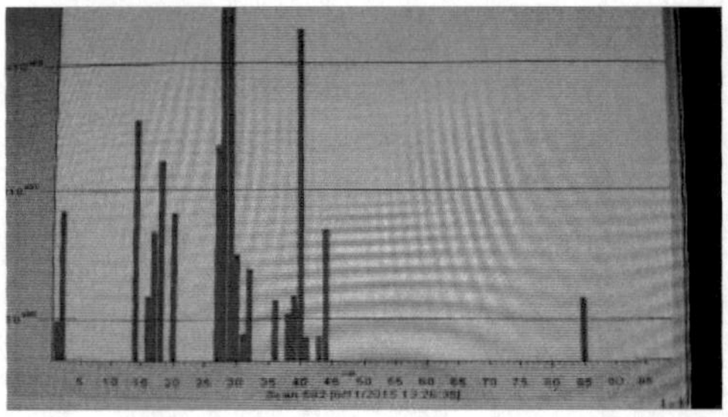

F_2 - gas mixture 2, 440kHz RPS

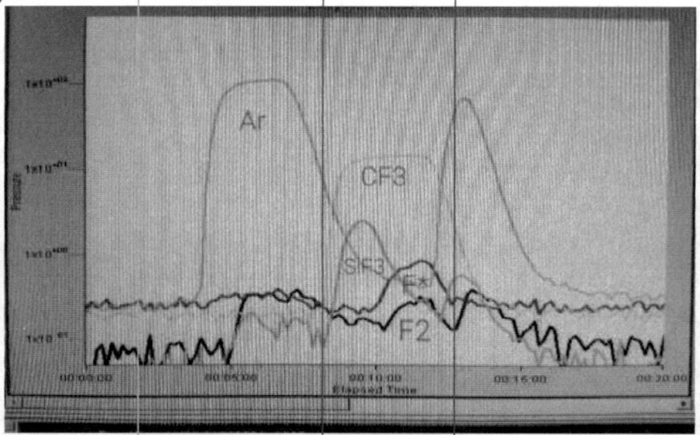

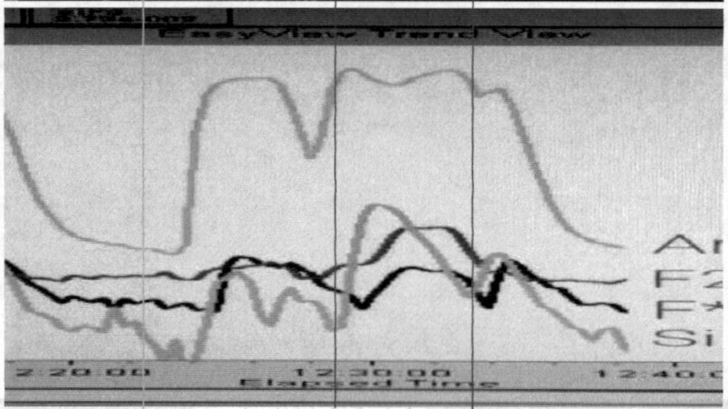

Summary and Conclusions

- Higher cleaning efficiency for F_2 gas mixtures
- F_2 gas mixtures showed similar performance for particle density and process kit degradation compared to NF_3 / PFCs
- Extended life time of process kit parts expected, especially for processes using PFCs as cleaning gases (no CF_3 radicals)
- Environmental-friendly processing
- Drop-in processes for NF_3, C_2F_6 and CF_4

23

Fraunhofer EMFT SOLVAY asking more from chemistry®

Marcello Riva
marcello.riva@solvay.com

Authors:
Michael Pittroff (michael.pittroff@solvay.com
Robert Wieland (Robert.Wieland@emft.fraunhofer.de)

www.solvay.com

SOLVAY
asking more from chemistry®

Roadmap Past, Present and Future

Paolo Gargini

Chairman ITRS2.0

Fellow IEEE, Fellow I-JSAP

Intel Fellow (1995-2012)

April 2016 SPCC P.Gargini 1

Multiple Stories

1. **Introduction**
2. 1998. ITRS 1.0
    1. Equivalent Scaling
3. 2000. NNI.
4. 2005. Nanoelectronics Research Initiative (NRI)
5. 2005-6. MPU power limits
6. 2006. More than Moore
    1. Heterogeneous Integration
7. 2010.First Selection of post CMOS devices
8. 2011. Equivalent scaling fully in production
9. 2014. Scaling acceleration
10. 2014-15.ITRS 2.0
11. 2015. Post CMOS map of devices
12. 2016. IRDS
13. 2017-2021.3D POWER Scaling

April 2016 SPCC P.Gargini 2

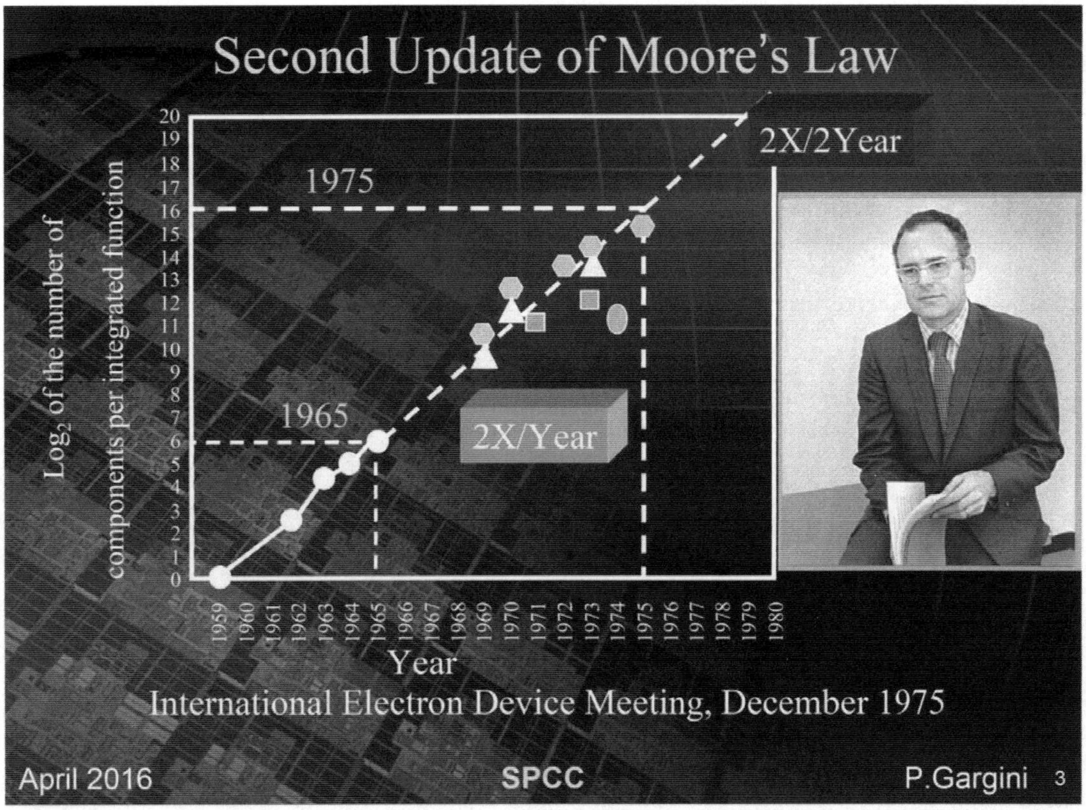

Dennard Scaling

Device or Circuit Parameter	Scaling Factor
Device dimension tox, L, W	1/K
Doping concentration Na	K
Voltage V	1/K
Current I	1/K
Capacitance eA/t	1/K
Delay time per circuit VC/I	1/K
Power dissipation per circuit VI	$1/K^2$
Power density VI/A	1

Dennard's 1974 paper summarizes transistor or circuit parameter changes under ideal MOSFET device scaling conditions, where K is the unitless scaling constant.

The benefits of scaling : as transistors get smaller, they can switch faster and use less power. Each new generation of process technology was expected to reduce minimum feature size by approximately 0.7x (K ~1.4). A 0.7x reduction in linear features size provided roughly a 2x increase in transistor density.

Dennard scaling broke down around 2004 with unscaled interconnect delays and our inability to scale the voltage and the current due to reliability concerns.

But our the ability to etch smaller transistors has continued spawning multicore designs.

April 2016 SPCC P.Gargini 4

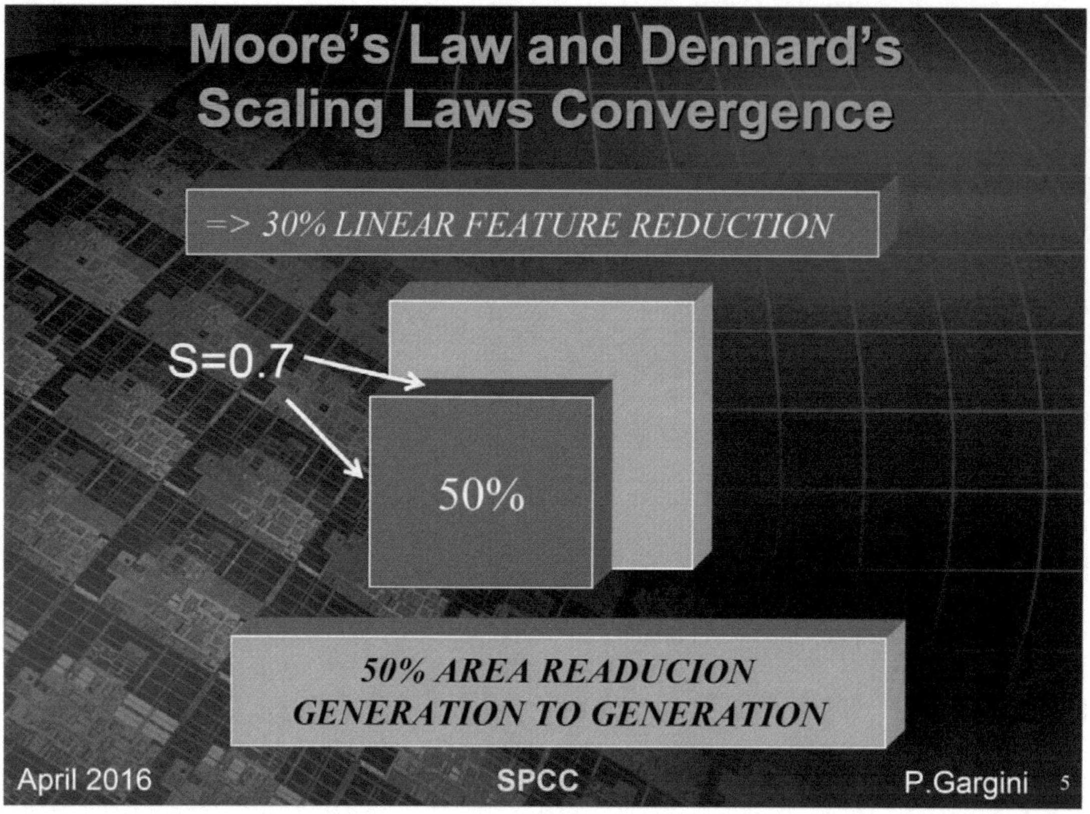
Moore's Law and Dennard's Scaling Laws Convergence
=> 30% LINEAR FEATURE REDUCTION
S=0.7
50%
50% AREA READUCION GENERATION TO GENERATION
April 2016
SPCC
P.Gargini
5

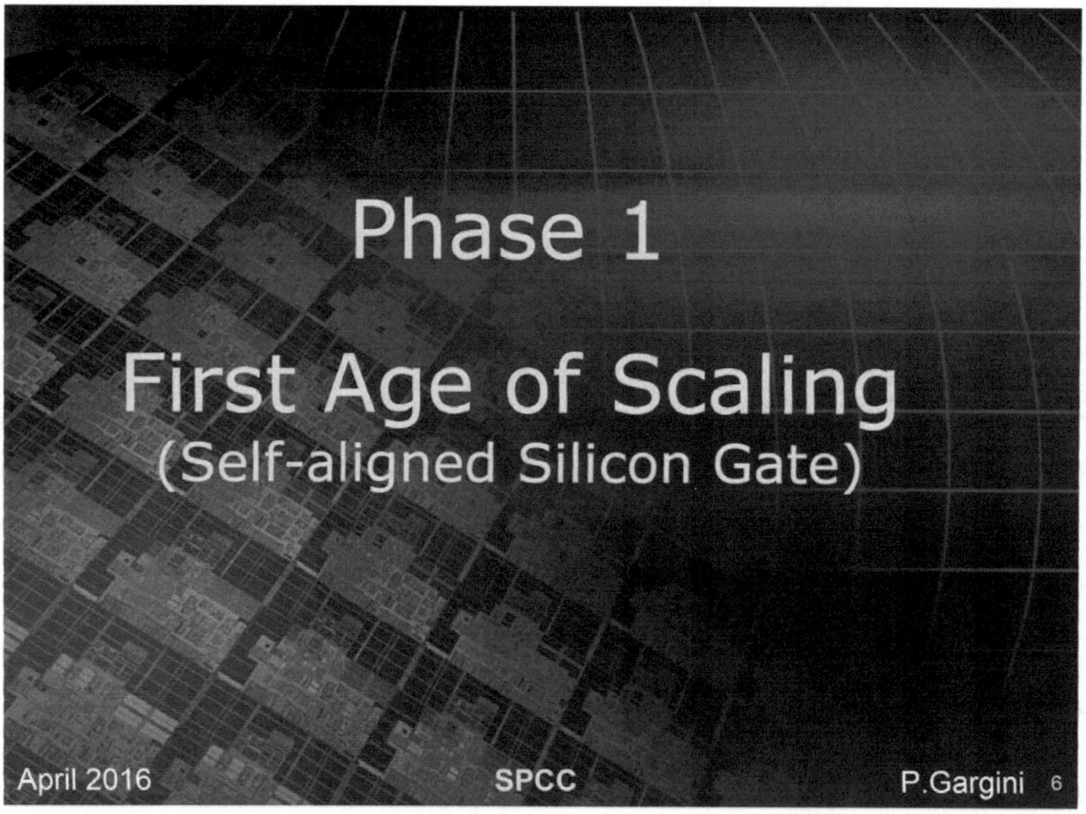
Phase 1
First Age of Scaling
(Self-aligned Silicon Gate)
April 2016
SPCC
P.Gargini
6

IC Industry at a Glance (1975-2003)

Driver ➡ Cost/transistor -> 50% Reduction

How ➡ 2x Density/2 years (Moore)

Method ➡ Geometrical Scaling (Dennard)

April 2016 SPCC P.Gargini 7

The Incredible Shrinking Silicon Technology of the 90's

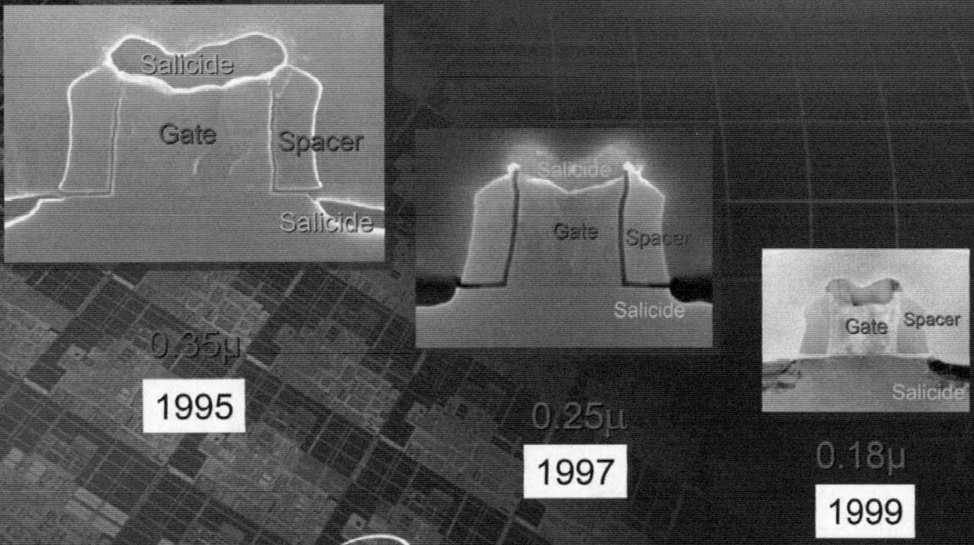

April 2016 intel SPCC P.Gargini 8

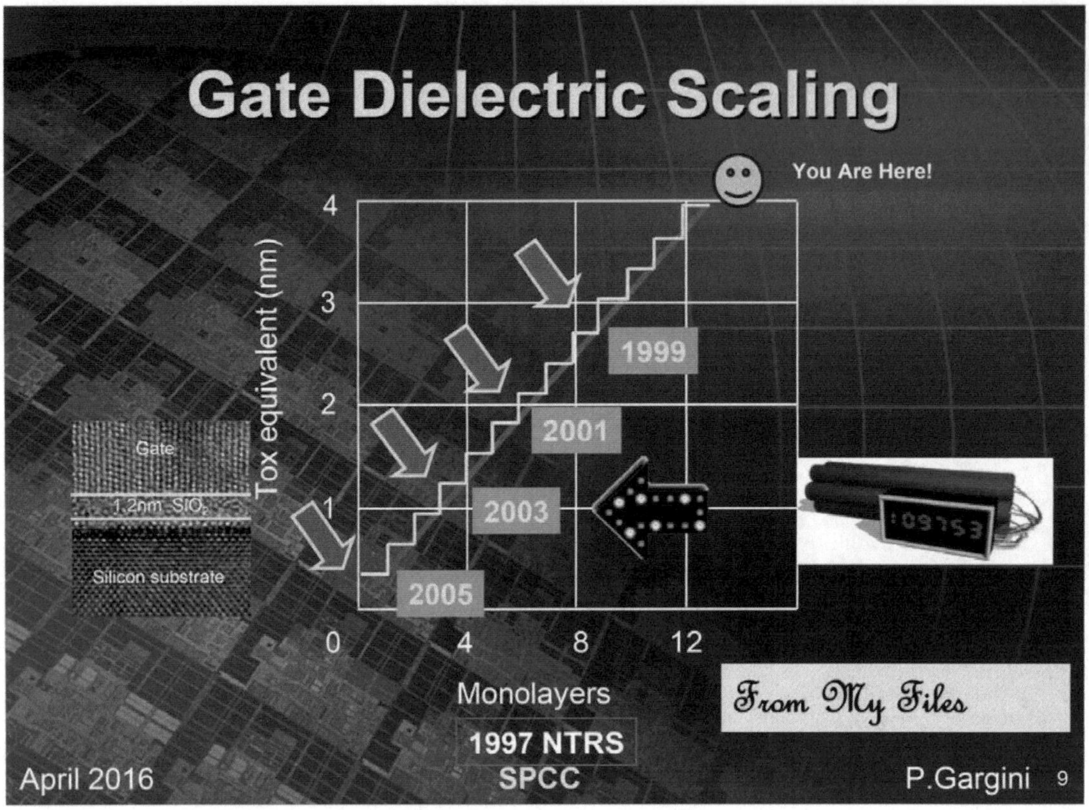
Gate Dielectric Scaling
You Are Here!
Tox equivalent (nm)
4
3
2
1
0
4
8
12
Monolayers
1999
2001
2003
2005
Gate
1.2nm SiO2
Silicon substrate
From My Files
1997 NTRS
SPCC
April 2016
P.Gargini 9

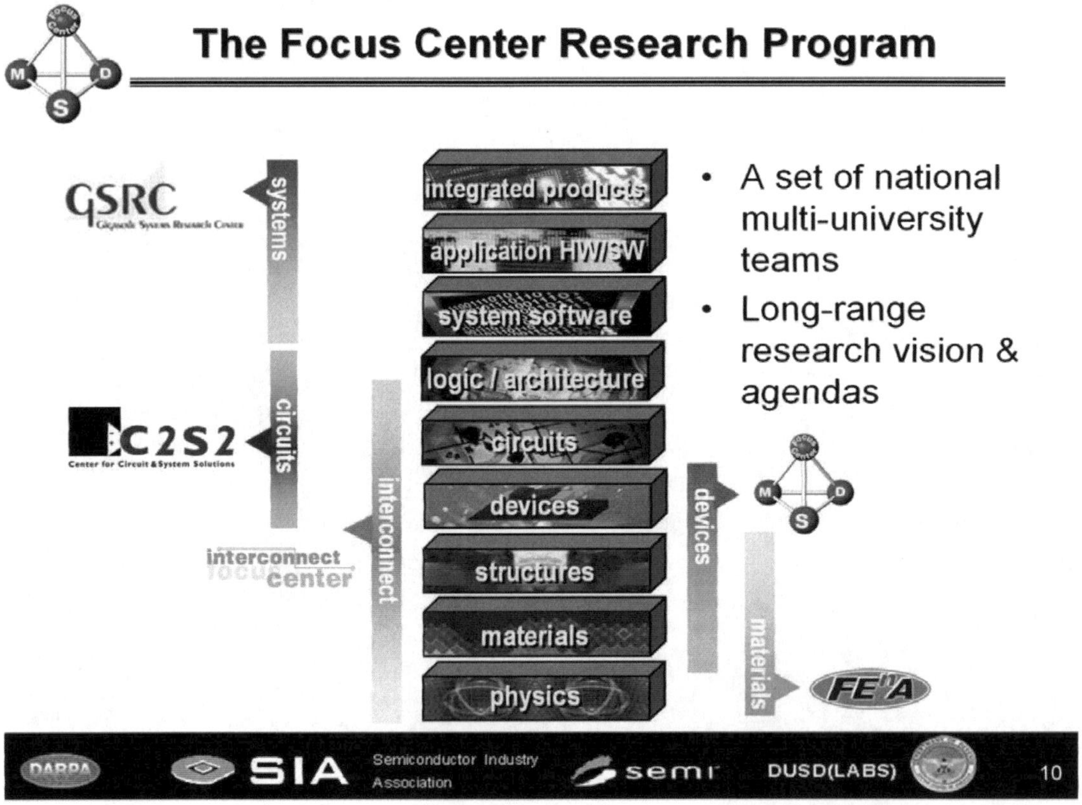
The Focus Center Research Program
GSRC
C2S2
Center for Circuit & System Solutions
interconnect focus center
systems
circuits
interconnect
integrated products
application HW/SW
system software
logic / architecture
circuits
devices
structures
materials
physics
devices
materials
A set of national multi-university teams
Long-range research vision & agendas
FENA
DARPA
SIA Semiconductor Industry Association
semi
DUSD(LABS)
10

Multiple Stories

1. Introduction
2. **1998. ITRS 1.0**
   1. **Equivalent Scaling**
3. 2000. NNI.
4. 2005. Nanoelectronics Research Initiative (NRI)
5. **2005-6. MPU power limits**
6. 2006. More than Moore
   1. Heterogeneous Integration
7. 2010.First Selection of post CMOS devices
8. 2011. **Equivalent scaling fully in production**
9. 2014. Scaling acceleration
10. 2014-15.ITRS 2.0
11. 2015. Post CMOS map of devices
12. 2016. IRDS
13. 2017-2021.3D POWER Scaling

April 2016 SPCC P.Gargini 11

1998 ITRS Update

- Participation extended to: EECA, EIAJ, KSIA, TSIA at WSC on April 23,1998
- 1st Meeting held on July 10/11,1998 in San Francisco
- 2nd meeting held on December 10/11,1998 at SFO
- 50% of tables in 1997 NTRS required some changes
- 1998 ITRS Update posted on web in April 1999

ITRS

P.Gargini

April 2016 SPCC P.Gargini 12

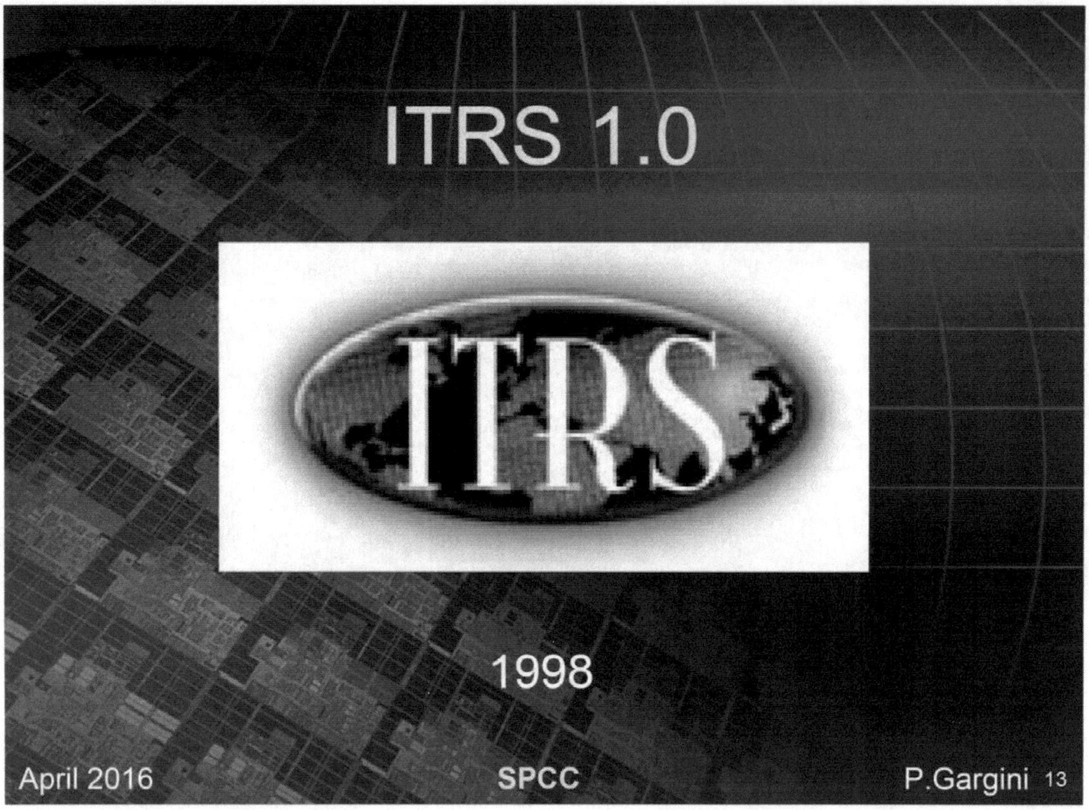
ITRS 1.0
ITRS
1998
April 2016
SPCC
P.Gargini 13

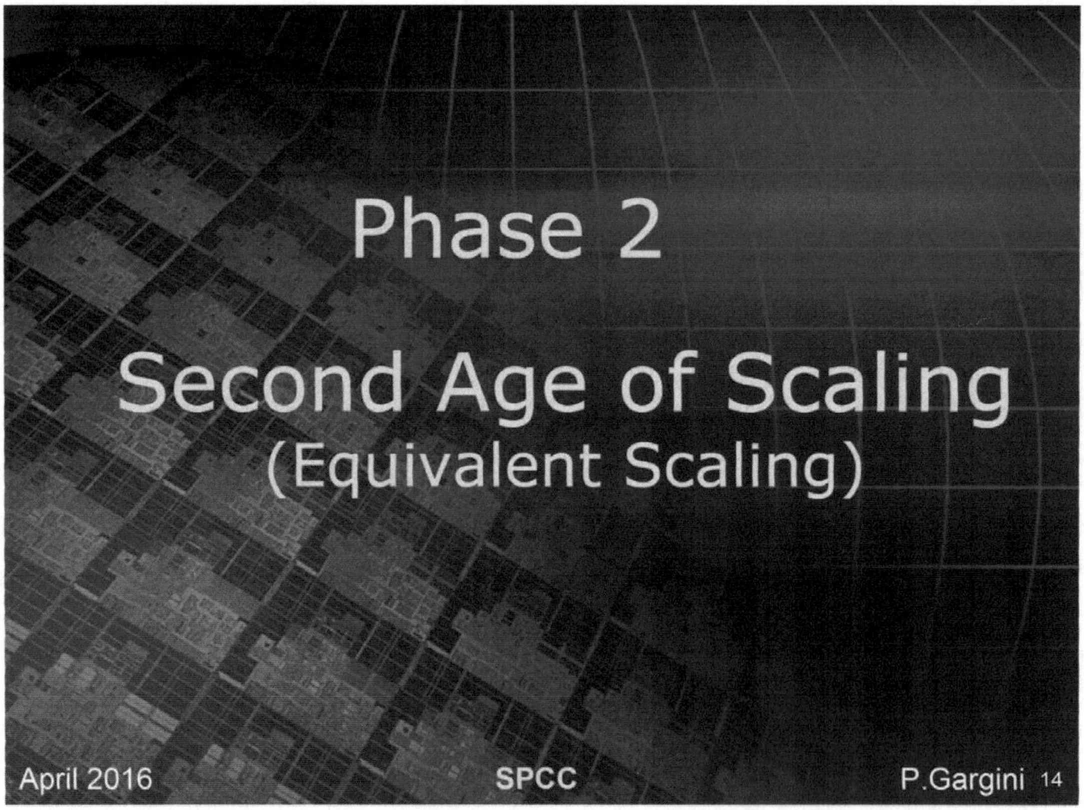
Phase 2
Second Age of Scaling
(Equivalent Scaling)
April 2016
SPCC
P.Gargini 14

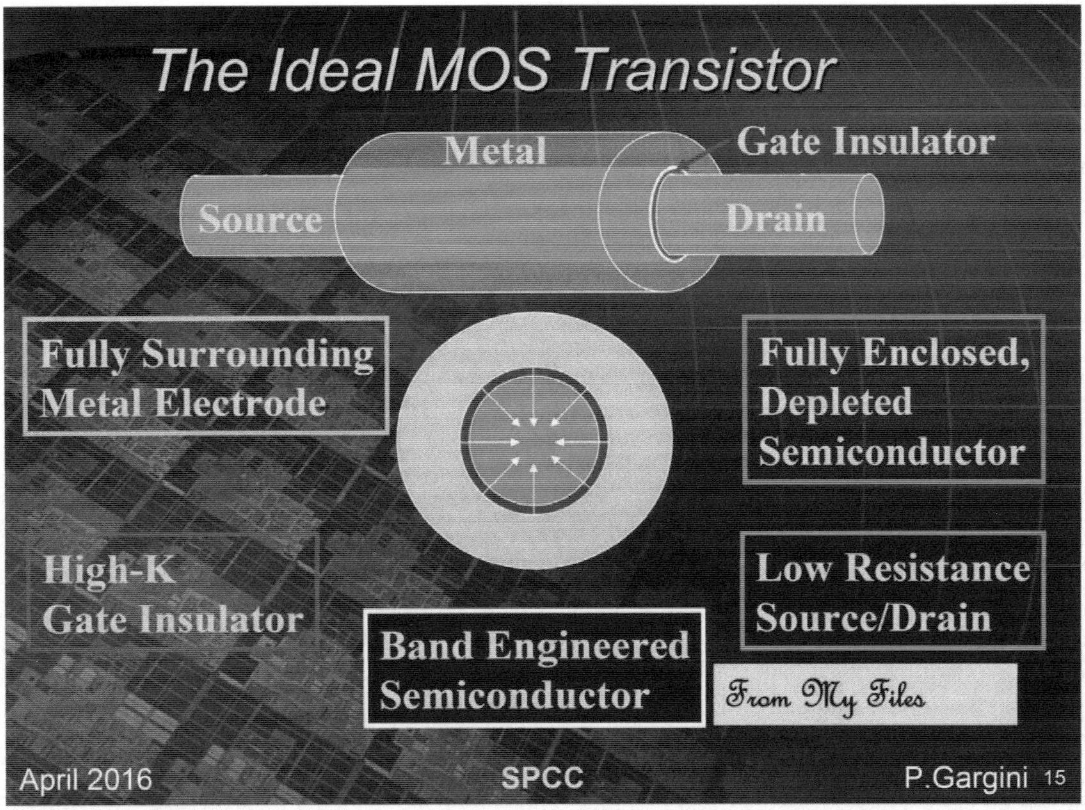
The Ideal MOS Transistor
Metal
Gate Insulator
Source
Drain
Fully Surrounding Metal Electrode
Fully Enclosed, Depleted Semiconductor
High-K Gate Insulator
Low Resistance Source/Drain
Band Engineered Semiconductor
From My Files
April 2016
SPCC
P.Gargini 15

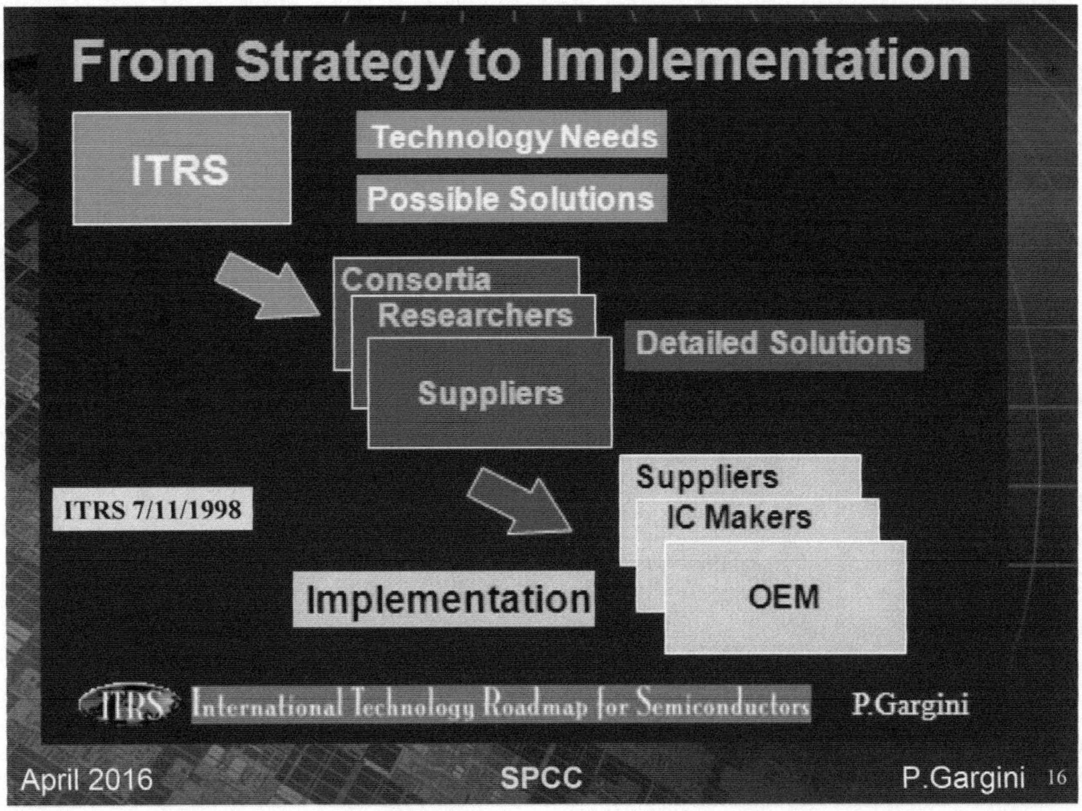
From Strategy to Implementation
ITRS
Technology Needs
Possible Solutions
Consortia
Researchers
Suppliers
Detailed Solutions
ITRS 7/11/1998
Suppliers
IC Makers
OEM
Implementation
ITRS International Technology Roadmap for Semiconductors
P.Gargini
April 2016
SPCC
P.Gargini 16

IC Industry at a Glance
(2003->2021)

Driver ➡ Cost/transistor-> 50% Reduction

How ➡ 2x Density/2 years (Moore)

Method ➡ Equivalent Scaling (ITRS1.0)

April 2016 SPCC P.Gargini 17

The Start of the ITRS

http://www.itrs2.net

1991 Micro Tech 2000 Workshop Report	1992NTRS	1994NTRS	1997NTRS

Europe Japan Korea Taiwan USA

1998 ITRS Update	1999 ITRS	2000 ITRS Update	2001 ITRS	2002 ITRS Update

April 2016 SPCC P.Gargini 18

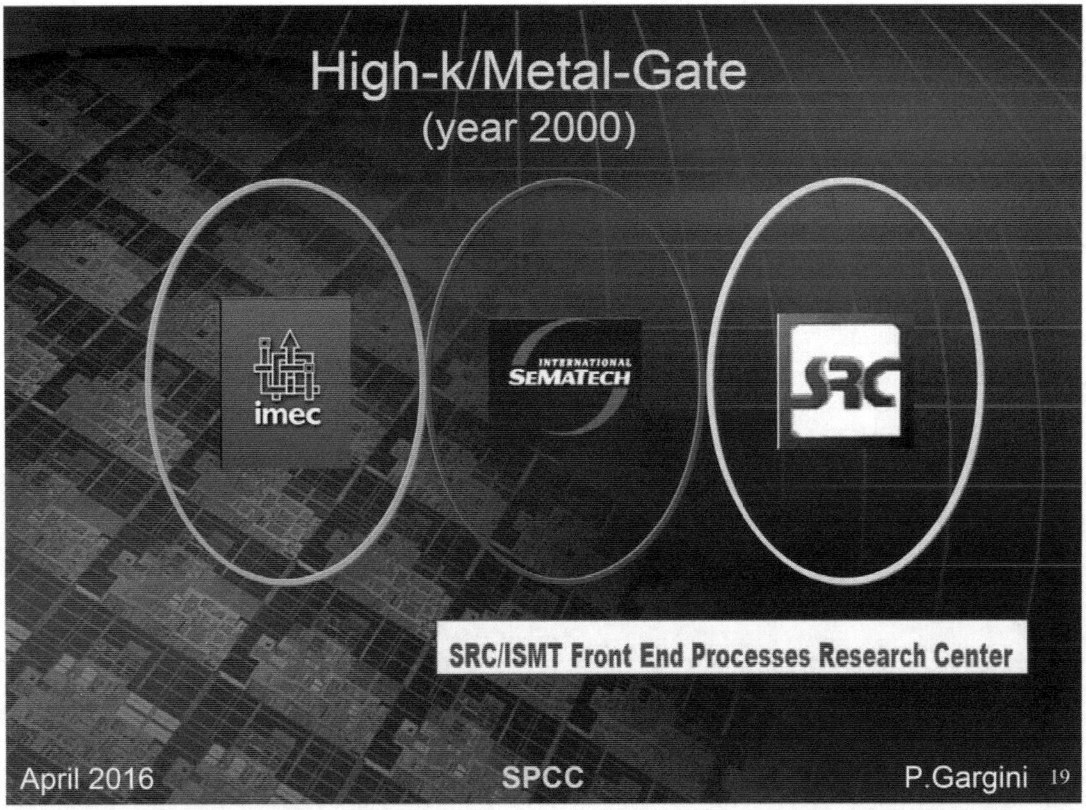
High-k/Metal-Gate
(year 2000)
imec
INTERNATIONAL SEMATECH
SRC
SRC/ISMT Front End Processes Research Center
April 2016
SPCC
P.Gargini 19

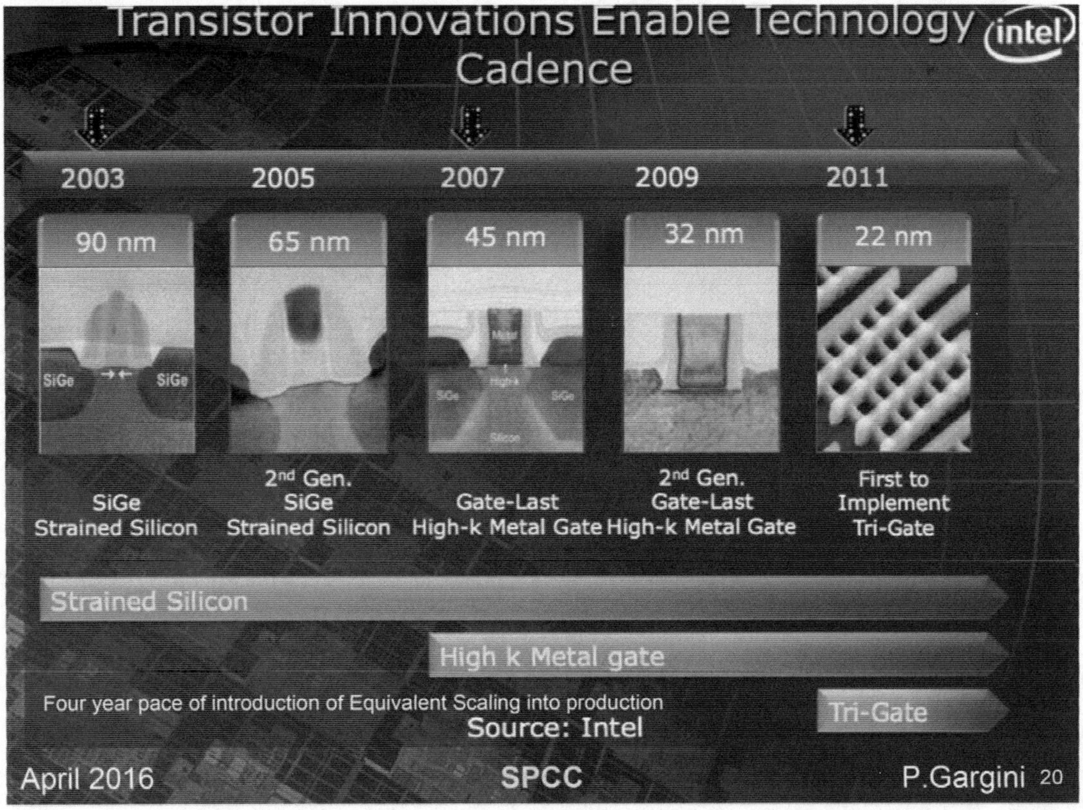
Transistor Innovations Enable Technology Cadence
intel
2003
2005
2007
2009
2011
90 nm
65 nm
45 nm
32 nm
22 nm
SiGe
SiGe
SiGe Strained Silicon
2nd Gen. SiGe Strained Silicon
Gate-Last High-k Metal Gate
2nd Gen. Gate-Last High-k Metal Gate
First to Implement Tri-Gate
Strained Silicon
High k Metal gate
Tri-Gate
Four year pace of introduction of Equivalent Scaling into production
Source: Intel
April 2016
SPCC
P.Gargini 20

Incubation Time

	Early Invention	Focused Research	Introduction Manufacturing
Strained Silicon	1992	1998	2003
HKMG	1996	1998	2007
Raised S/D	1993	1998	2009
MultiGates	1988	2000	2011

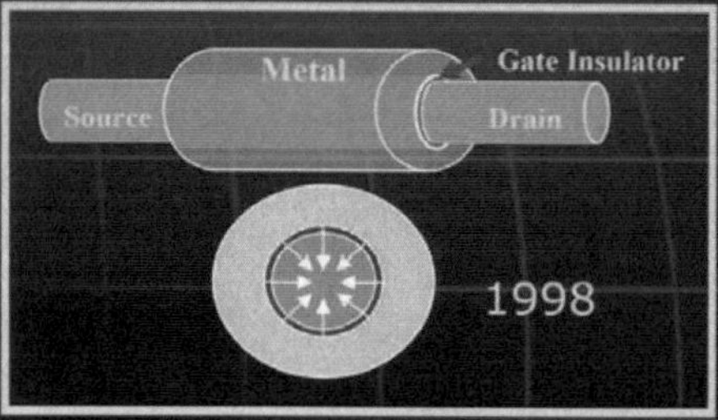

<11 years

April 2016 SPCC P.Gargini 21

Sortable table

Model	Core	Clock Speed	Thermal Design Power
Pentium 4 1.3	Willamette (180 nm)	1.3 GHz	51.6 W
Pentium 4 1.4 (Socket 423)	Willamette (180 nm)	1.4 GHz	54.7 W
Pentium 4 1.4 (Socket 478)	Willamette (180 nm)	1.4 GHz	55.3 W
Pentium 4 1.5 (Socket 423)	Willamette (180 nm)	1.5 GHz	57.8 W
Pentium 4 1.5 (Socket 478)	Willamette (180 nm)	1.5 GHz	57.9 W
Pentium 4 1.6 (Socket 423)	Willamette (180 nm)	1.6 GHz	61 W
Pentium 4 1.6 (Socket 478)	Willamette (180 nm)	1.6 GHz	60.8 W
Pentium 4 1.7 (Socket 423)	Willamette (180 nm)	1.7 GHz	64 W
Pentium 4 1.7 (Socket 478)	Willamette (180 nm)	1.7 GHz	63.5 W
Pentium 4 1.8 (Socket 423)	Willamette (180 nm)	1.8 GHz	66.7 W
Pentium 4 1.8 (Socket 478)	Willamette (180 nm)	1.8 GHz	66.1 W
Pentium 4 1.9 (Socket 423)	Willamette (180 nm)	1.9 GHz	69.2 W
Pentium 4 1.9 (Socket 478)	Willamette (180 nm)	1.9 GHz	72.8 W
Pentium 4 HT 560	Prescott (90 nm)	3.6 GHz	115 W
Pentium 4 HT 560J	Prescott (90 nm)	3.6 GHz	115 W
Pentium 4 HT 561	Prescott (90 nm)	3.6 GHz	115 W

April 2016 SPCC P.Gargini 22

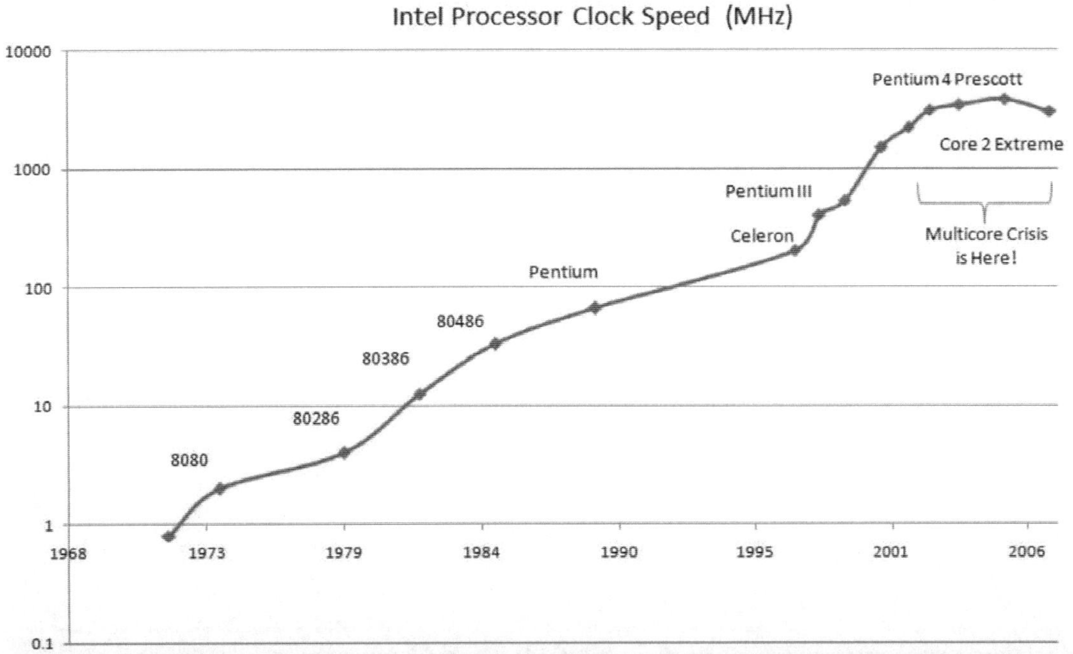

April 2016 SPCC P.Gargini 23

Sortable table

Model	Core	Clock Speed	Thermal Design Power
Core 2 Duo E4200	Conroe (65 nm)	1.6 GHz	65 W
Core 2 Duo E4300	Conroe (65 nm)	1.8 GHz	65 W
Core 2 Duo E4400	Conroe (65 nm)	2 GHz	65 W
Core 2 Duo E4500	Conroe (65 nm)	2.2 GHz	65 W
Core 2 Duo E4600	Conroe (65 nm)	2.4 GHz	65 W
Core 2 Duo E4700	Conroe (65 nm)	2.6 GHz	65 W
Core 2 Duo E8190	Wolfdale (45 nm)	2.66 GHz	65 W
Core 2 Duo E8200	Wolfdale (45 nm)	2.66 GHz	65 W
Core 2 Duo E8300	Wolfdale (45 nm)	2.83 GHz	65 W
Core 2 Duo E8400	Wolfdale (45 nm)	3 GHz	65 W
Core 2 Duo E8500	Wolfdale (45 nm)	3.16 GHz	65 W
Core 2 Duo E8600	Wolfdale (45 nm)	3.33 GHz	65 W

April 2016 SPCC P.Gargini 24

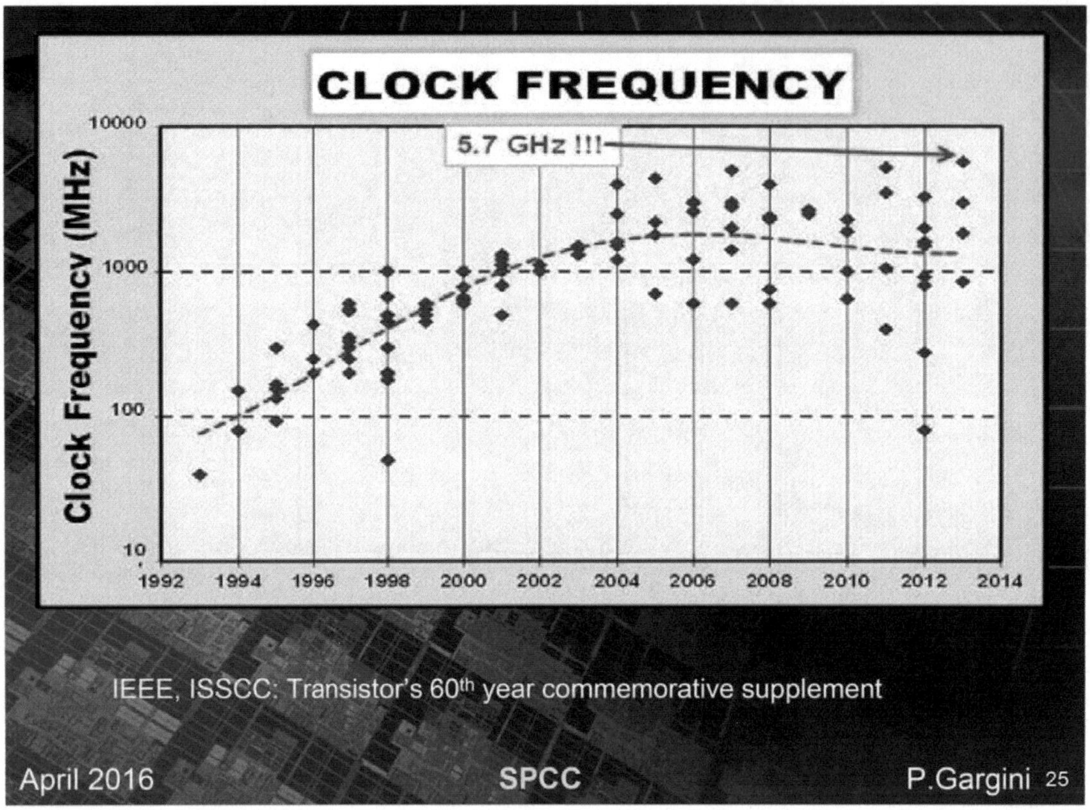
CLOCK FREQUENCY
5.7 GHz !!!
Clock Frequency (MHz)
10000
1000
100
10
1992 1994 1996 1998 2000 2002 2004 2006 2008 2010 2012 2014
IEEE, ISSCC: Transistor's 60th year commemorative supplement
April 2016
SPCC
P.Gargini 25

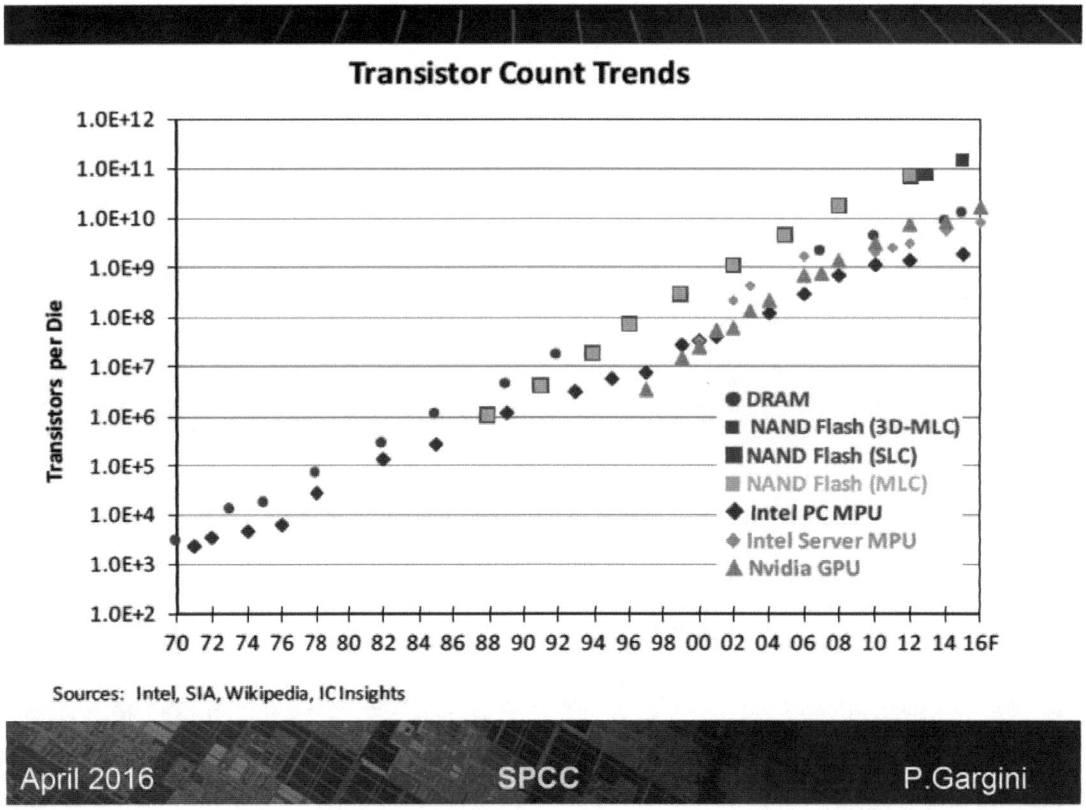
Transistor Count Trends
Transistors per Die
1.0E+12
1.0E+11
1.0E+10
1.0E+9
1.0E+8
1.0E+7
1.0E+6
1.0E+5
1.0E+4
1.0E+3
1.0E+2
70 72 74 76 78 80 82 84 86 88 90 92 94 96 98 00 02 04 06 08 10 12 14 16F
DRAM
NAND Flash (3D-MLC)
NAND Flash (SLC)
NAND Flash (MLC)
Intel PC MPU
Intel Server MPU
Nvidia GPU
Sources: Intel, SIA, Wikipedia, IC Insights
April 2016
SPCC
P.Gargini

Multiple Stories

1. Introduction
2. 1998. ITRS 1.0
   1. Equivalent Scaling
3. **2000. NNI.**
4. **2005. Nanoelectronics Research Initiative (NRI)**
5. 2005-6. MPU power limits
6. 2006. More than Moore
   1. Heterogeneous Integration
7. **2010.First Selection of post CMOS devices**
8. 2011. Equivalent scaling fully in production
9. 2014. Scaling acceleration
10. 2014-15.ITRS 2.0
11. **2015. Post CMOS map of devices**
12. 2016. IRDS
13. 2017-2021.3D POWER Scaling

April 2016 SPCC P.Gargini 27

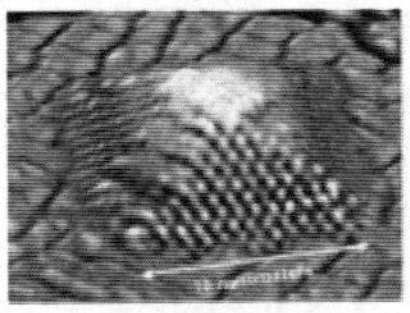

National Nanotechnology Initiative (NNI)

"My budget supports a major new National Nanotechnology Initiative, worth $500 million.... the ability to manipulate matter at the atomic and molecular level. Imagine the possibilities: materials with ten times the strength of steel and only a small fraction of the weight -- shrinking all the information housed at the Library of Congress into a device the size of a sugar cube -- detecting cancerous tumors when they are only a few cells in size. Some of our research goals may take 20 or more years to achieve, but that is precisely why there is an important role for the federal government."

--President William J. Clinton
January 21, 2000
California Institute Of Technology

28

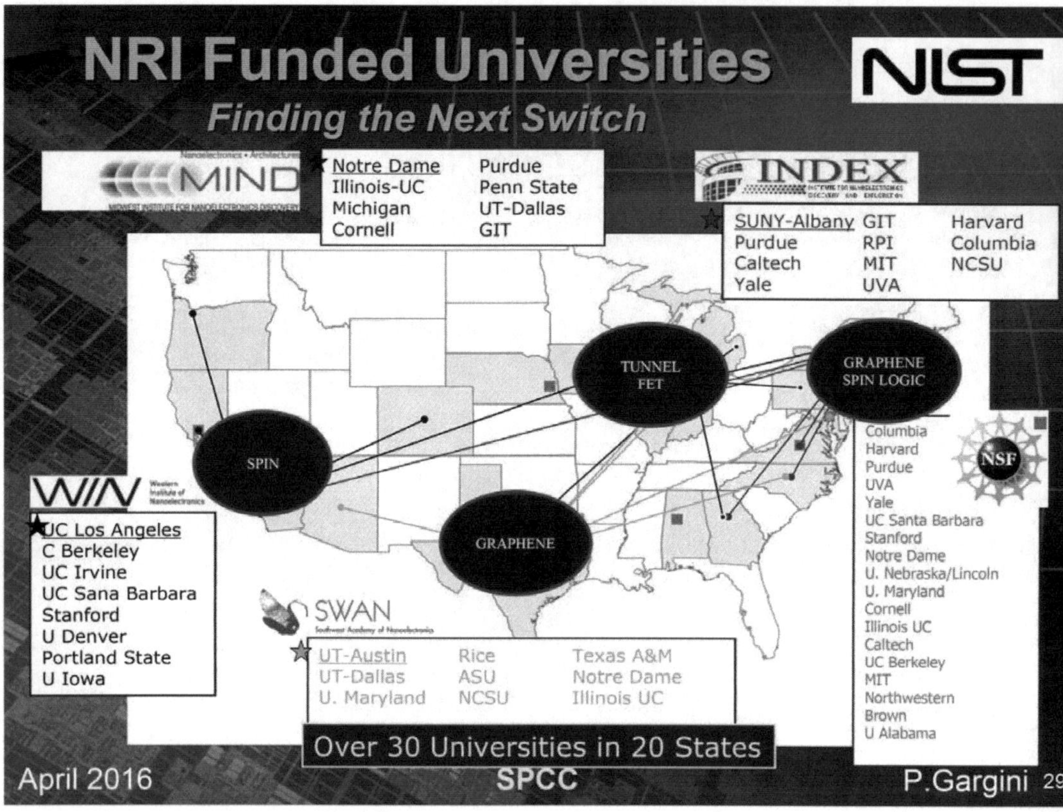

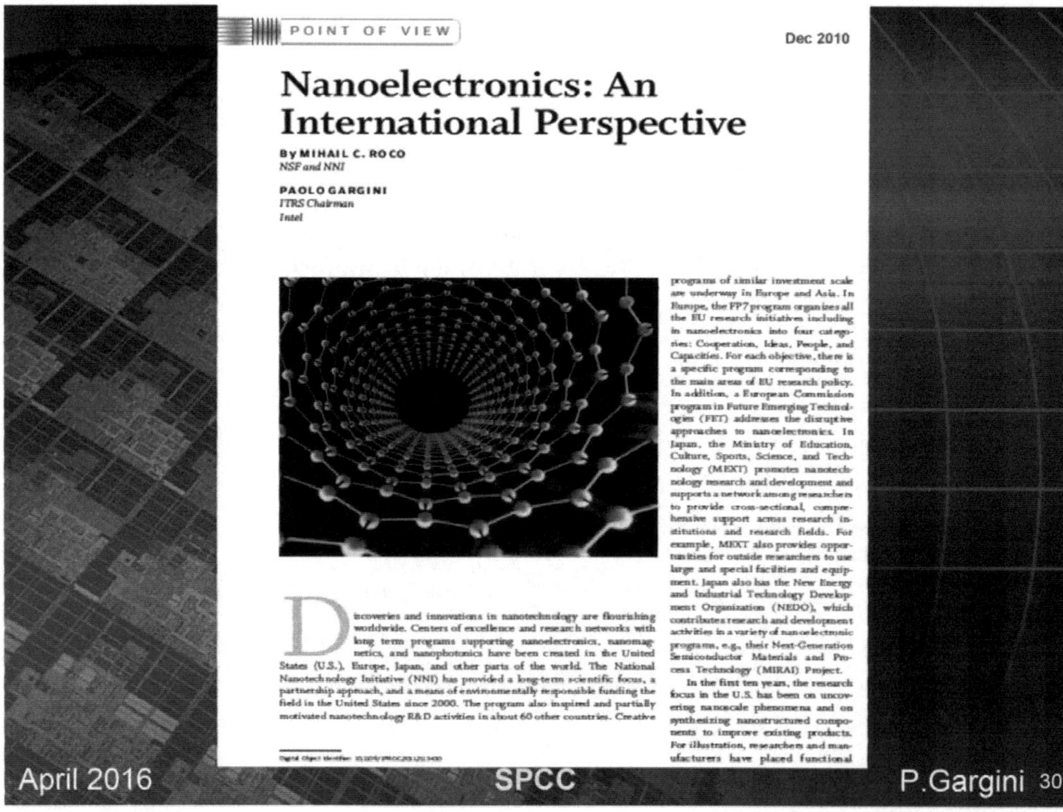
POINT OF VIEW

Dec 2010

Nanoelectronics: An International Perspective

By MIHAIL C. ROCO
NSF and NNI

PAOLO GARGINI
ITRS Chairman
Intel

Discoveries and innovations in nanotechnology are flourishing worldwide. Centers of excellence and research networks with long term programs supporting nanoelectronics, nanomagnetics, and nanophotonics have been created in the United States (U.S.), Europe, Japan, and other parts of the world. The National Nanotechnology Initiative (NNI) has provided a long-term scientific focus, a partnership approach, and a means of environmentally responsible funding the field in the United States since 2000. The program also inspired and partially motivated nanotechnology R&D activities in about 60 other countries. Creative programs of similar investment scale are underway in Europe and Asia. In Europe, the FP7 program organizes all the EU research initiatives including in nanoelectronics into four categories: Cooperation, Ideas, People, and Capacities. For each objective, there is a specific program corresponding to the main areas of EU research policy. In addition, a European Commission program in Future Emerging Technologies (FET) addresses the disruptive approaches to nanoelectronics. In Japan, the Ministry of Education, Culture, Sports, Science, and Technology (MEXT) promotes nanotechnology research and development and supports a network among researchers to provide cross-sectional, comprehensive support across research institutions and research fields. For example, MEXT also provides opportunities for outside researchers to use large and special facilities and equipment. Japan also has the New Energy and Industrial Technology Development Organization (NEDO), which contributes research and development activities in a variety of nanoelectronic programs, e.g., their Next-Generation Semiconductor Materials and Process Technology (MIRAI) Project.

In the first ten years, the research focus in the U.S. has been on uncovering nanoscale phenomena and on synthesizing nanostructured components to improve existing products. For illustration, researchers and manufacturers have placed functional

Nanoelectronics Research Initiative

Benchmarking of Beyond-CMOS Exploratory Devices for Logic Integrated Circuits

Dmitri E. Nikonov and Ian A. Young, 2015, IEEE J. on Exploratory Solid-State Computational Devices and Circuits

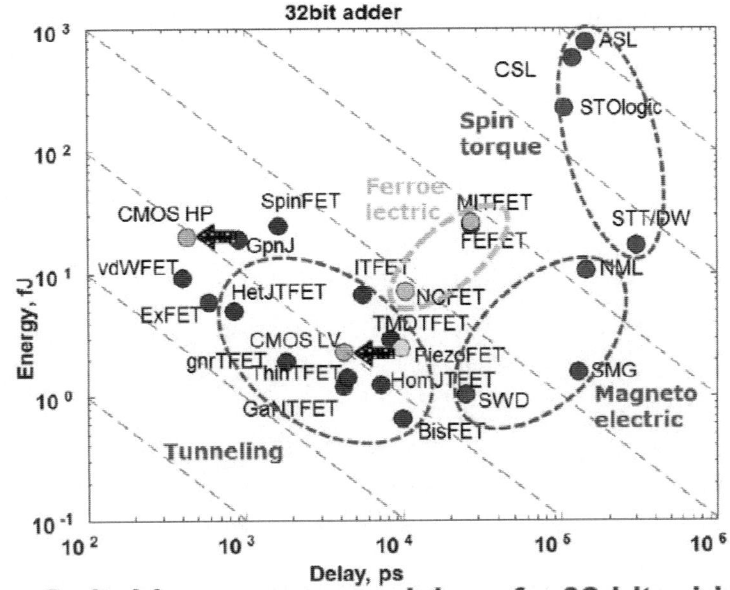

Switching energy vs. delay of a 32-bit adder

Moore’s Law: A Path Forward

Bill Holt

Executive Vice President, Intel

General Manager, Technology and Manufacturing Group

intel

The Changing Measure of Improvement

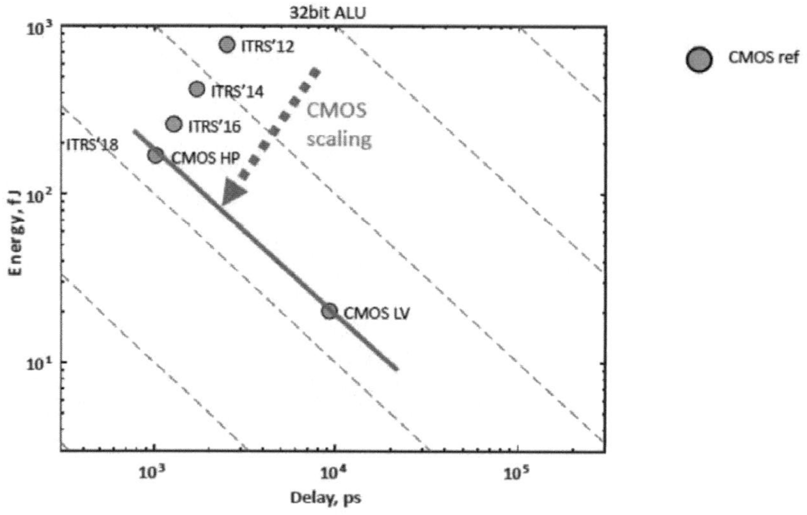

Source: ITRS 2011

The Changing Measure of Improvement

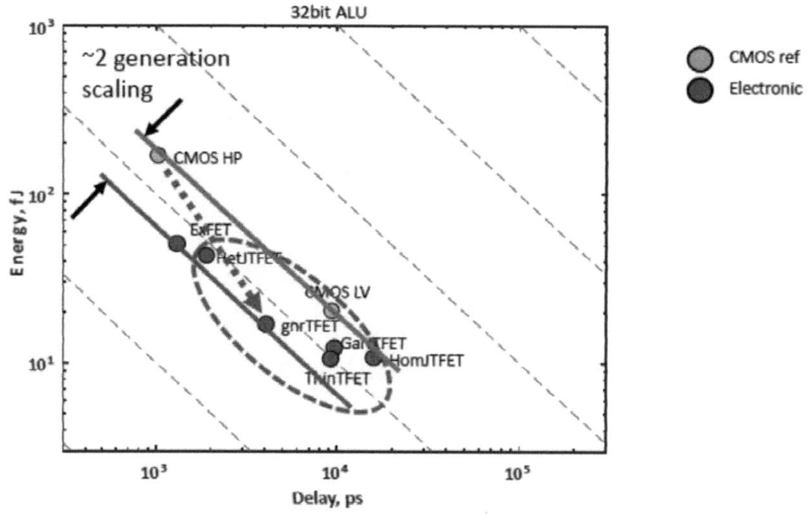

Sources: ITRS 2011, Nikonov and Young, IEEE JxCDC, 1, 3-11 (2015);
Manipatruni, Nikonov and Young, Arxiv cond-mat 1512.05428 (2015)

The Changing Measure of Improvement

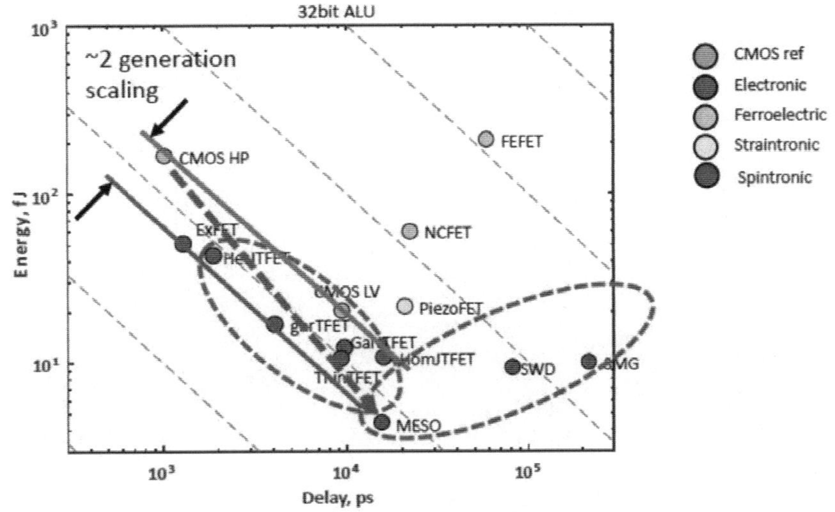

Sources: ITRS 2011, Nikonov and Young, IEEE JxCDC, 1, 3-11 (2015); Manipatruni, Nikonov and Young, Arxiv cond-mat 1512.05428 (2015)

1.1: Moore's Law: A Path Forward

Multiple Stories

1. Introduction
2. 1998. ITRS 1.0
   1. Equivalent Scaling
3. **2000. NNI.**
4. **2005. Nanoelectronics Research Initiative (NRI)**
5. 2005-6. MPU power limits
6. **2006. More than Moore**
   1. **Heterogeneous Integration**
7. **2010.First Selection of post CMOS devices**
8. 2011. Equivalent scaling fully in production
9. 2014. Scaling acceleration
10. 2014-15.ITRS 2.0
11. **2015. Post CMOS map of devices**
12. 2016. IRDS
13. 2017-2021.3D POWER Scaling

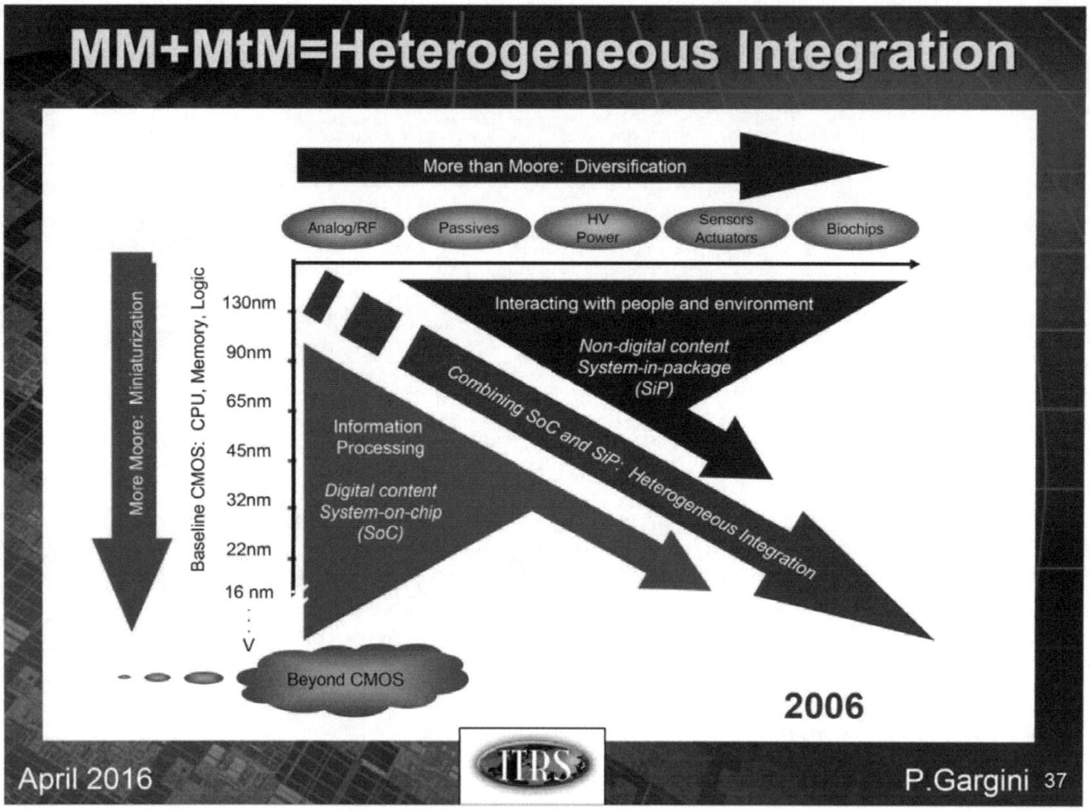
MM+MtM=Heterogeneous Integration
More than Moore: Diversification
Analog/RF
Passives
HV
Power
Sensors
Actuators
Biochips
More Moore: Miniaturization
Baseline CMOS: CPU, Memory, Logic
130nm
90nm
65nm
45nm
32nm
22nm
16 nm
V
Interacting with people and environment
Non-digital content
System-in-package
(SiP)
Combining SoC and SiP: Heterogeneous Integration
Information
Processing
Digital content
System-on-chip
(SoC)
Beyond CMOS
2006
April 2016
P.Gargini 37

iPhone
June 2007
On January 9, 2007 Steve Jobs announced the iPhone at the Macworld convention, receiving substantial media attention,[16] and that it would be released later that year. On June 29, 2007 the first iPhone was released.
April 2016
SPCC
P.Gargini 38

Tablet
April 2010
9:41
A WiFi-only model of the tablet was released in April 2010, and a WiFi+3G model was introduced about a month later
April 2016
SPCC
P.Gargini 39

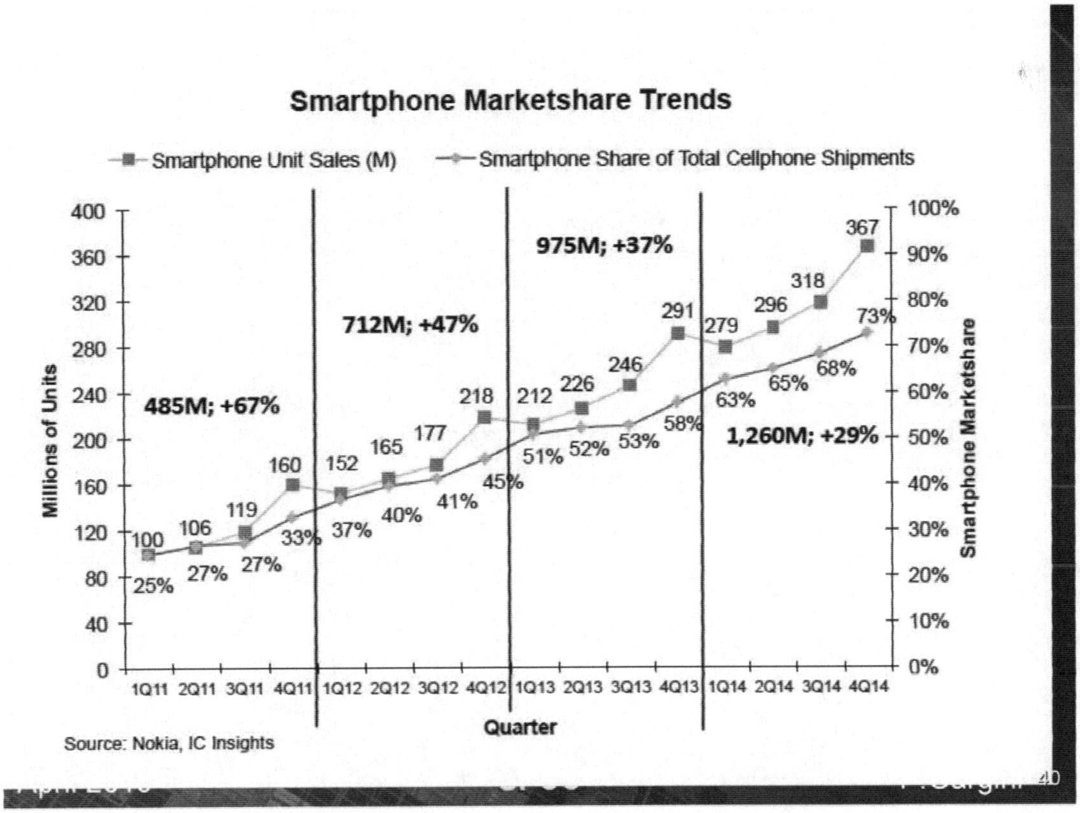
Smartphone Marketshare Trends
Smartphone Unit Sales (M)
Smartphone Share of Total Cellphone Shipments
Millions of Units
Smartphone Marketshare
485M; +67%
712M; +47%
975M; +37%
1,260M; +29%
100 106 119 160 152 165 177 218 212 226 246 291 279 296 318 367
25% 27% 27% 33% 37% 40% 41% 45% 51% 52% 53% 58% 63% 65% 68% 73%
1Q11 2Q11 3Q11 4Q11 1Q12 2Q12 3Q12 4Q12 1Q13 2Q13 3Q13 4Q13 1Q14 2Q14 3Q14 4Q14
Quarter
Source: Nokia, IC Insights
40

Multiple Stories

1. Introduction
2. 1998. ITRS 1.0
   1. Equivalent Scaling
3. 2000. NNI.
4. 2005. Nanoelectronics Research Initiative (NRI)
5. 2005-6. MPU power limits
6. 2006. More than Moore
   1. Heterogeneous Integration
7. 2010.First Selection of post CMOS devices
8. 2011. Equivalent scaling fully in production
9. **2014. Scaling acceleration**
10. 2014-15.ITRS 2.0
11. 2015. Post CMOS map of devices
12. 2016. IRDS
13. 2017-2021.3D POWER Scaling

April 2016 SPCC P.Gargini 41

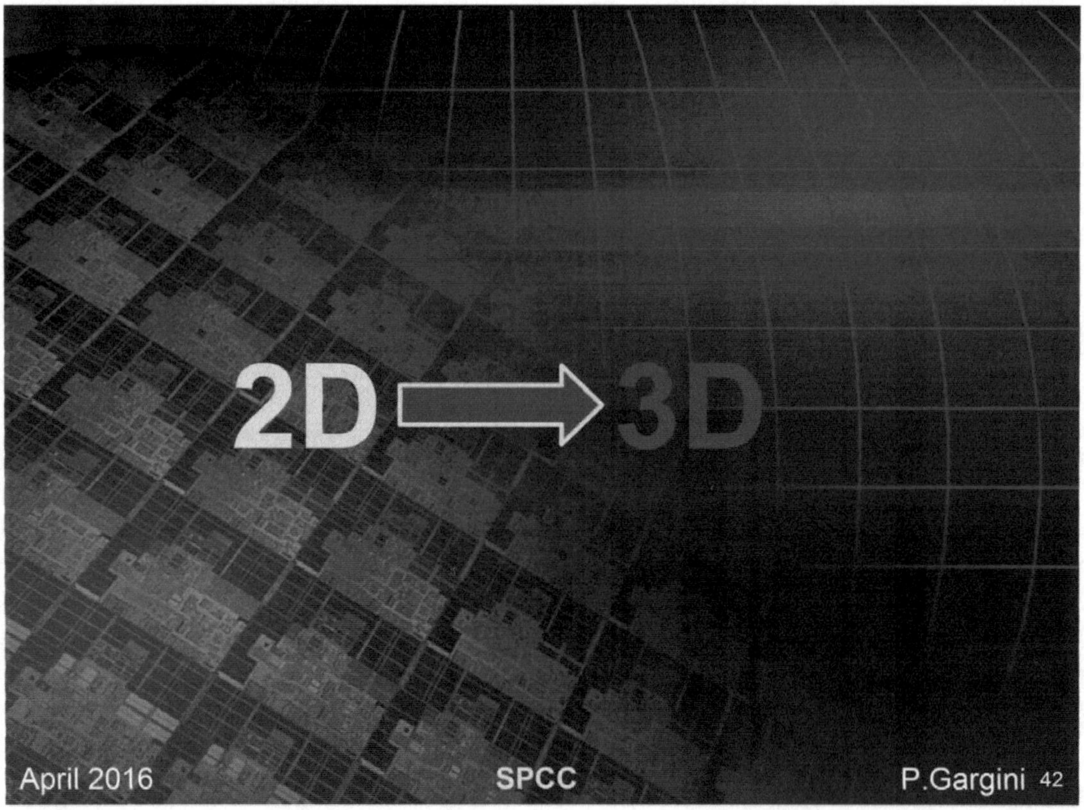

22 nm Tri-Gate Transistor

intel

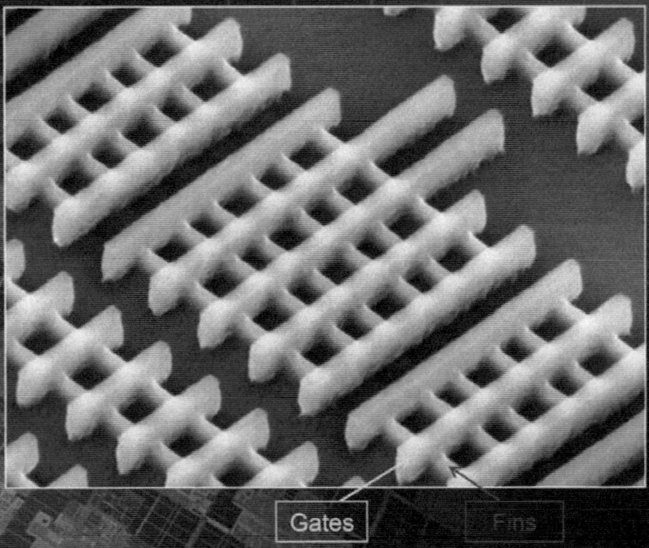

Mark Bohr, Kaizad Mistry, May 2011

Question

How many more technology generations can Equivalent Scaling be extended for ?

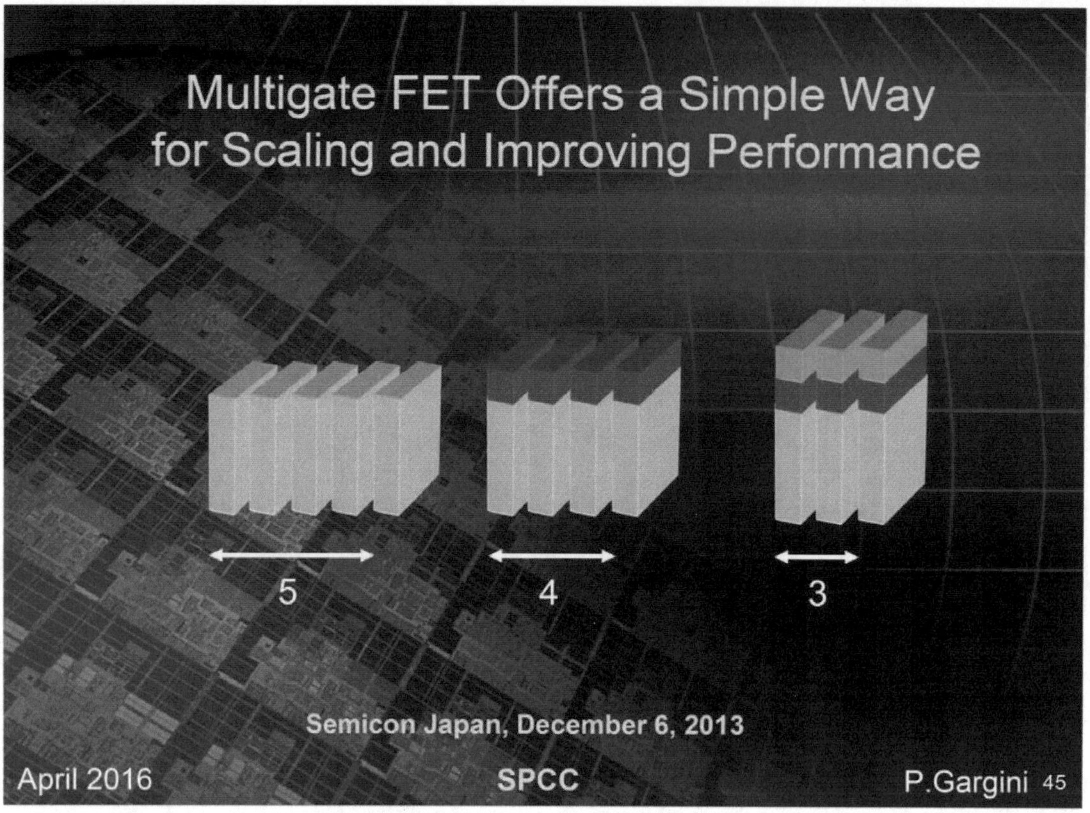
Multigate FET Offers a Simple Way for Scaling and Improving Performance
5
4
3
Semicon Japan, December 6, 2013
April 2016
SPCC
P.Gargini 45

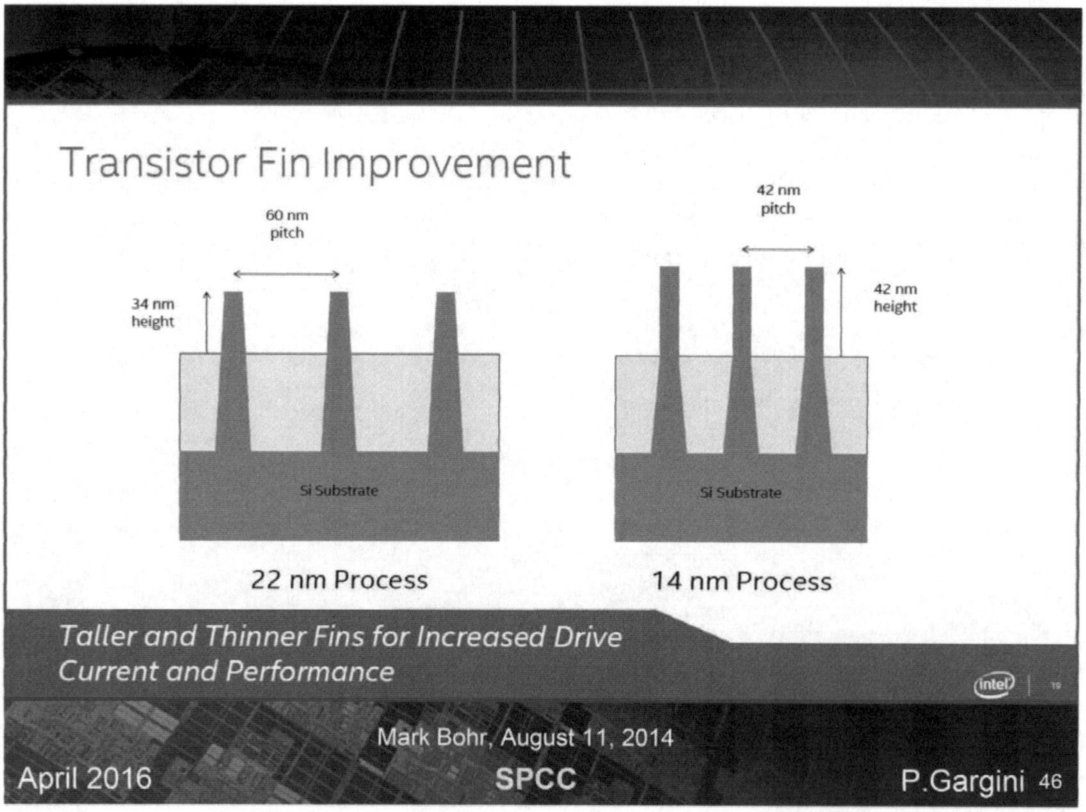
Transistor Fin Improvement
60 nm pitch
34 nm height
Si Substrate
22 nm Process
42 nm pitch
42 nm height
Si Substrate
14 nm Process
Taller and Thinner Fins for Increased Drive Current and Performance
Mark Bohr, August 11, 2014
April 2016
SPCC
P.Gargini 46

Transistor Fin Improvement

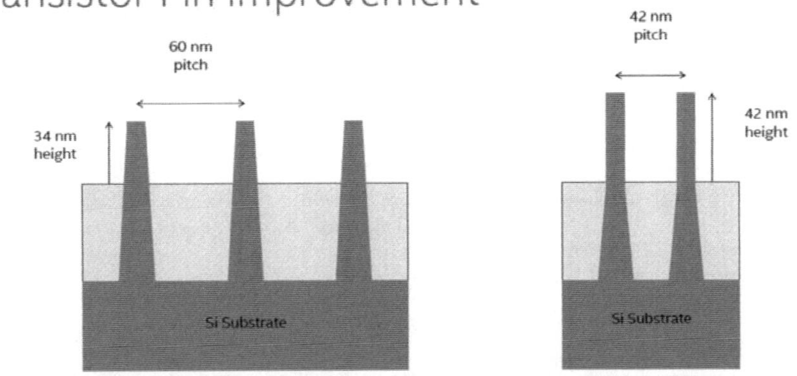

22 nm Process

14 nm Process

Reduced Number of Fins for Improved Density and Lower Capacitance

intel | 20

Mark Bohr, August 11, 2014

April 2016 SPCC P.Gargini 47

Fin FET Moore's Law Acceleration

April 2016 SPCC P.Gargini 48

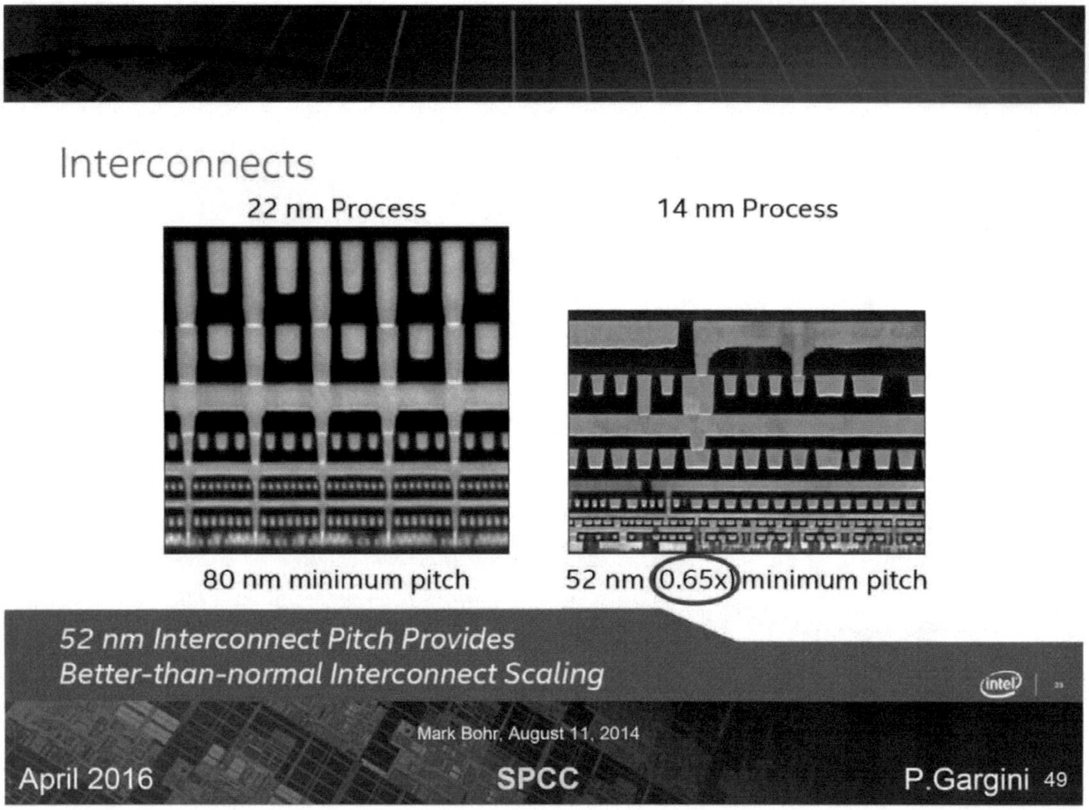
Interconnects
22 nm Process
14 nm Process
80 nm minimum pitch
52 nm 0.65x minimum pitch
52 nm Interconnect Pitch Provides
Better-than-normal Interconnect Scaling
Mark Bohr, August 11, 2014
April 2016
SPCC
P.Gargini 49

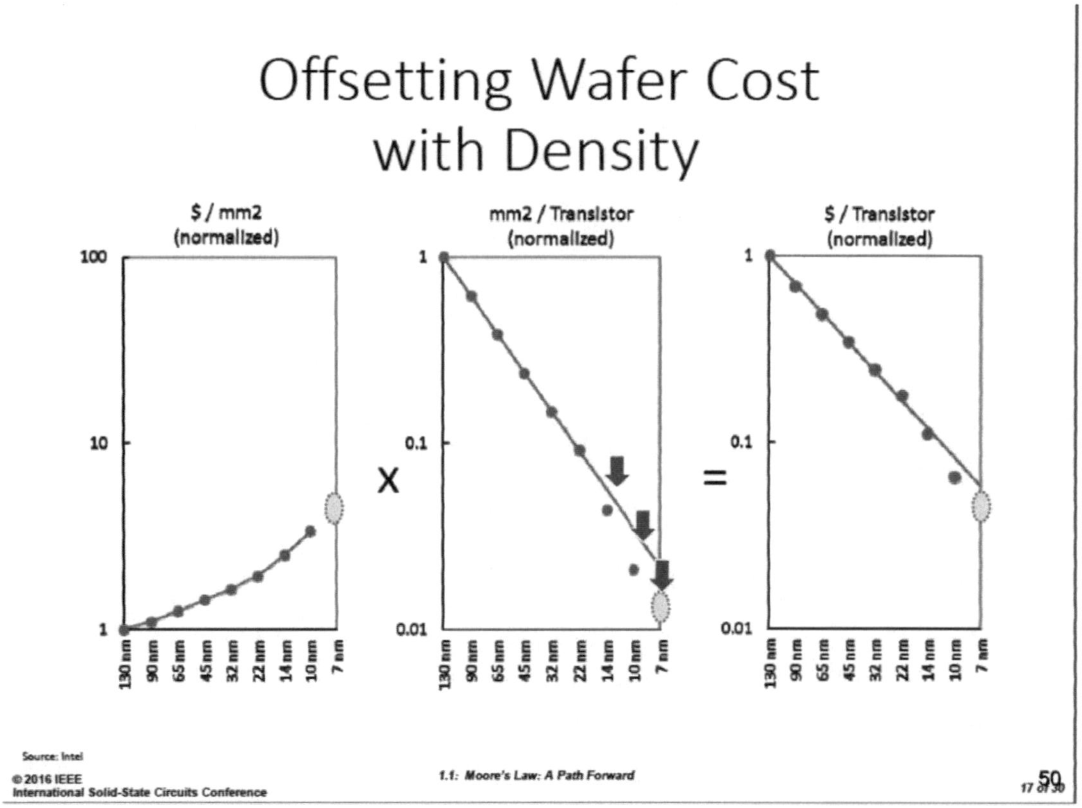
Offsetting Wafer Cost
with Density
$ / mm2
(normalized)
mm2 / Transistor
(normalized)
$ / Transistor
(normalized)
100
10
1
0.1
0.01
x
=
130 nm
90 nm
65 nm
45 nm
32 nm
22 nm
14 nm
10 nm
7 nm
Source: Intel
© 2016 IEEE
International Solid-State Circuits Conference
1.1: Moore's Law: A Path Forward
17 of 30
50

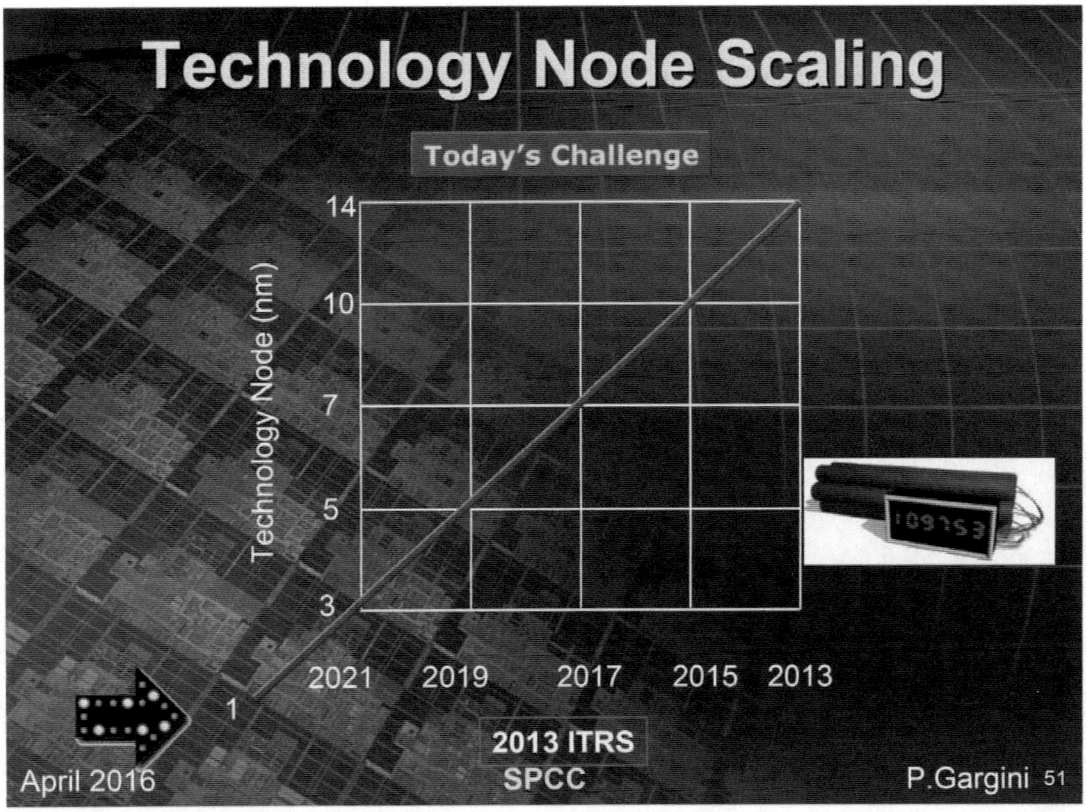

Technology Node Scaling
Today's Challenge
Technology Node (nm)
14
10
7
5
3
1
2021 2019 2017 2015 2013
2013 ITRS
April 2016
SPCC
P.Gargini 51

STOP
WRONG WAY
DO NOT ENTER

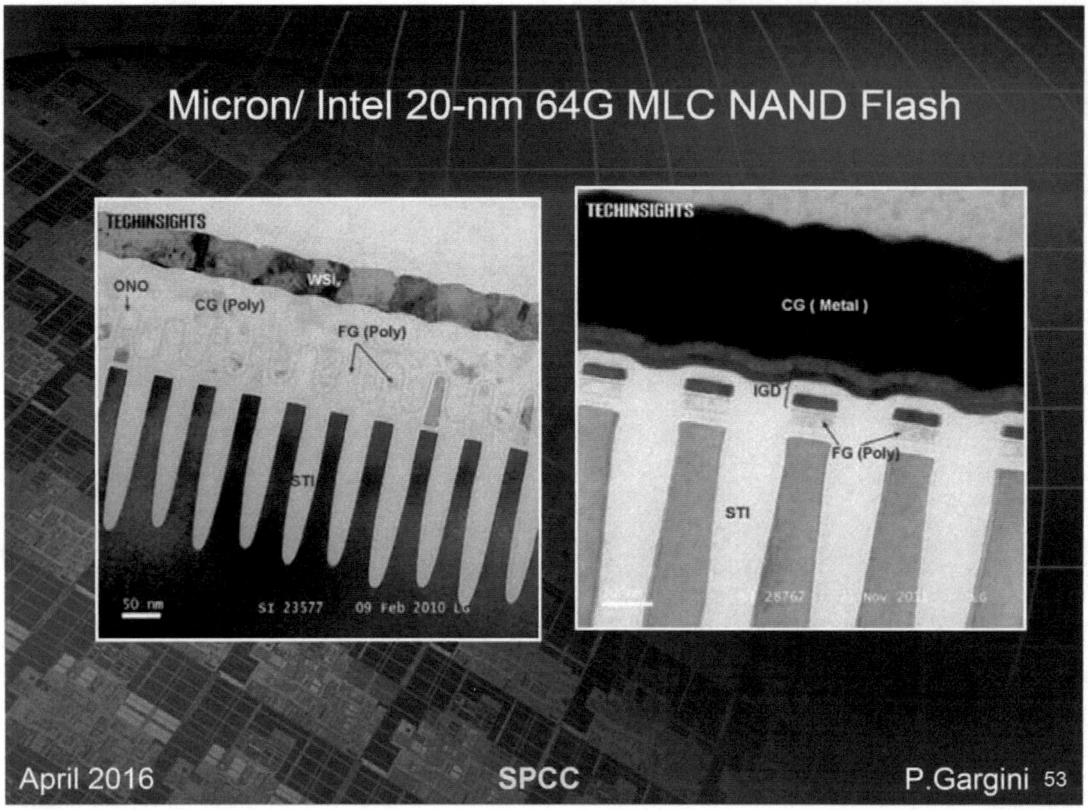
Micron/ Intel 20-nm 64G MLC NAND Flash
TECHINSIGHTS
ONO
WSi$_x$
CG (Poly)
FG (Poly)
STI
50 nm
SI 23577 09 Feb 2010 LG
TECHINSIGHTS
CG (Metal)
IGD
FG (Poly)
STI
April 2016
SPCC
P.Gargini 53

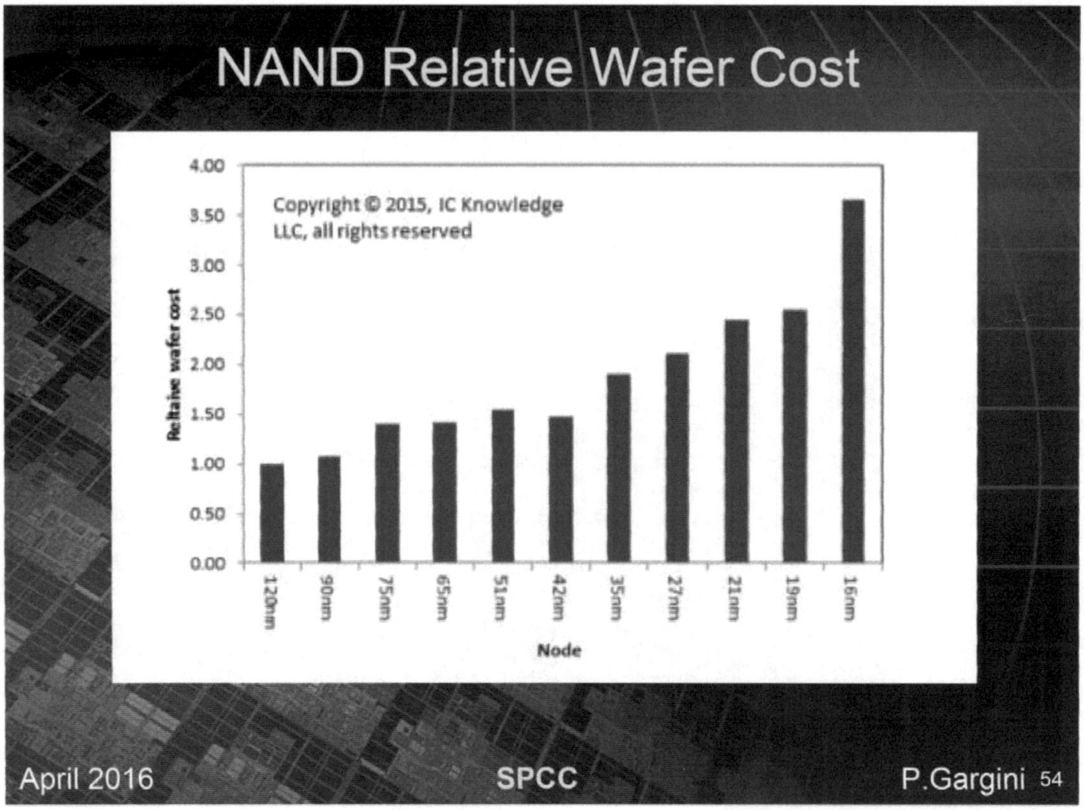
NAND Relative Wafer Cost
Copyright © 2015, IC Knowledge LLC, all rights reserved
Reltaive wafer cost
4.00
3.50
3.00
2.50
2.00
1.50
1.00
0.50
0.00
120nm
90nm
75nm
65nm
51nm
42nm
35nm
27nm
21nm
19nm
16nm
Node
April 2016
SPCC
P.Gargini 54

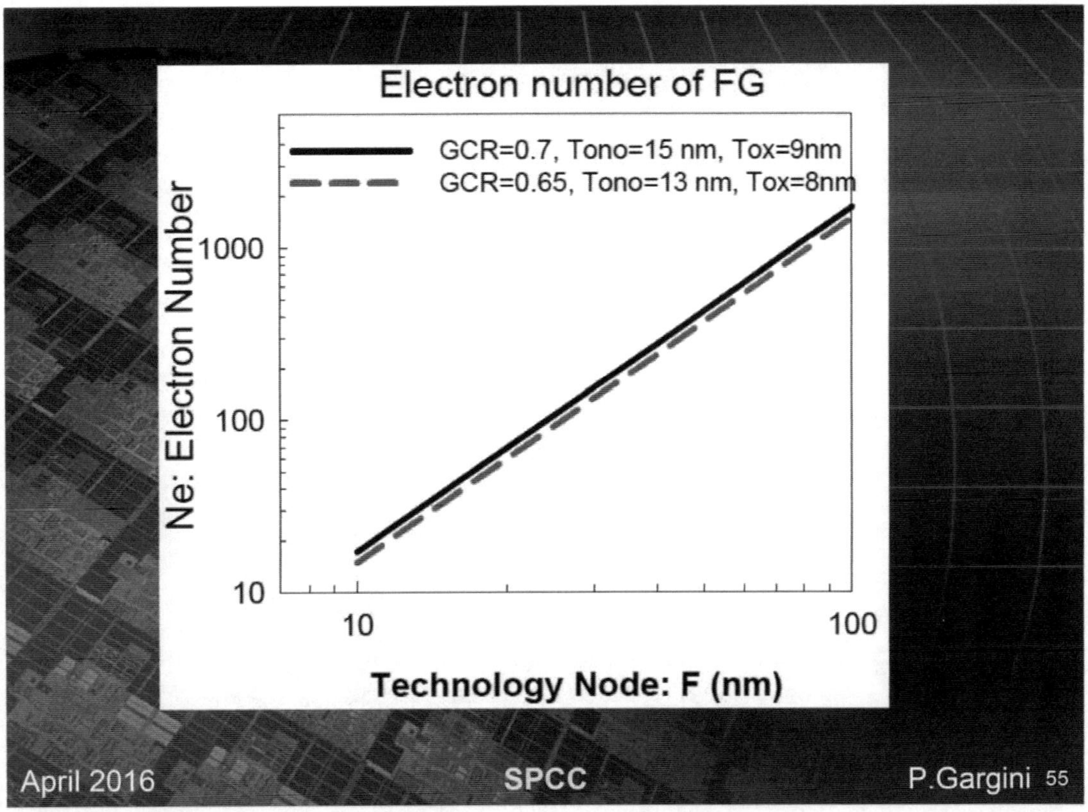
Electron number of FG
GCR=0.7, Tono=15 nm, Tox=9nm
GCR=0.65, Tono=13 nm, Tox=8nm
Ne: Electron Number
1000
100
10
10
100
Technology Node: F (nm)
April 2016
SPCC
P.Gargini 55

Samsung's 32-Layer 3D V-NAND Memory Chip
Source: PC Perspective, Samsung
April 2016
SPCC
P.Gargini 56

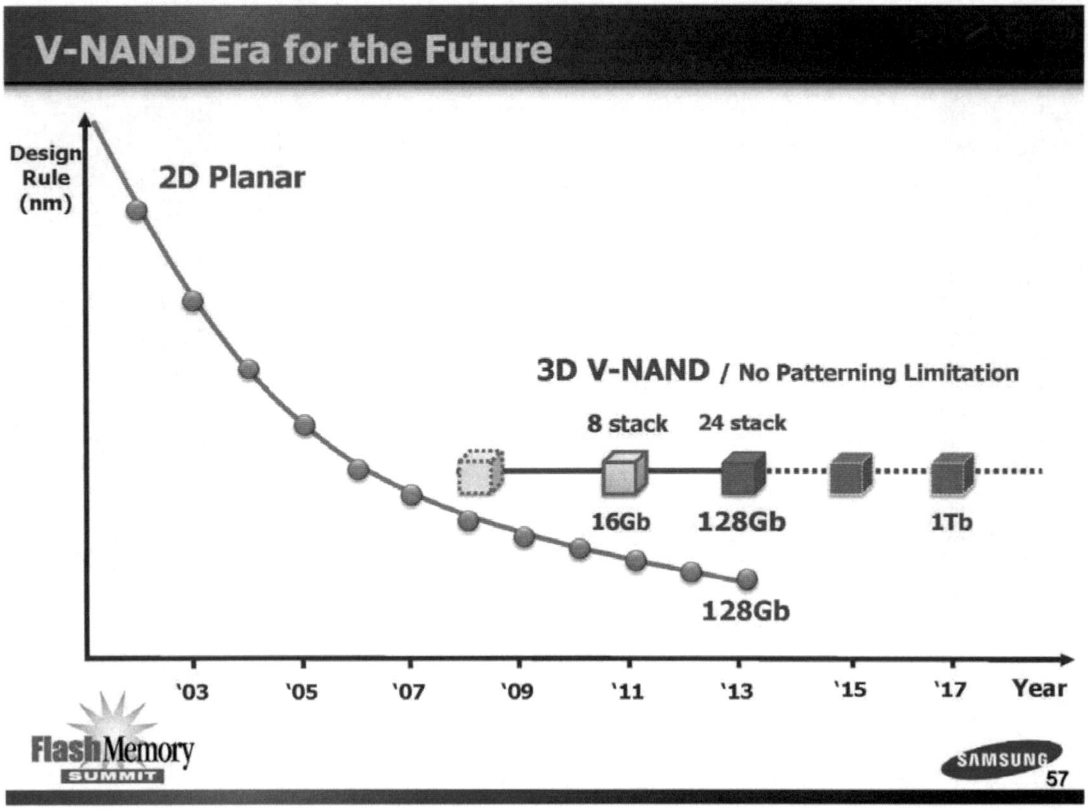
V-NAND Era for the Future
Design Rule (nm)
2D Planar
3D V-NAND / No Patterning Limitation
8 stack
24 stack
16Gb
128Gb
1Tb
128Gb
'03
'05
'07
'09
'11
'13
'15
'17
Year
Flash Memory SUMMIT
SAMSUNG
57

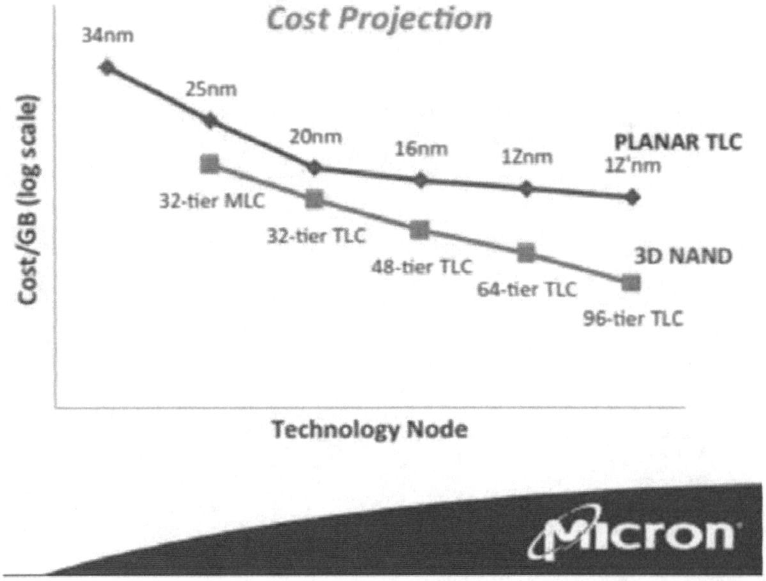
Cost Projection
Cost/GB (log scale)
34nm
25nm
20nm
16nm
12nm
12'nm
PLANAR TLC
32-tier MLC
32-tier TLC
48-tier TLC
64-tier TLC
96-tier TLC
3D NAND
Technology Node
Micron

58

Vertical Logic Architecture

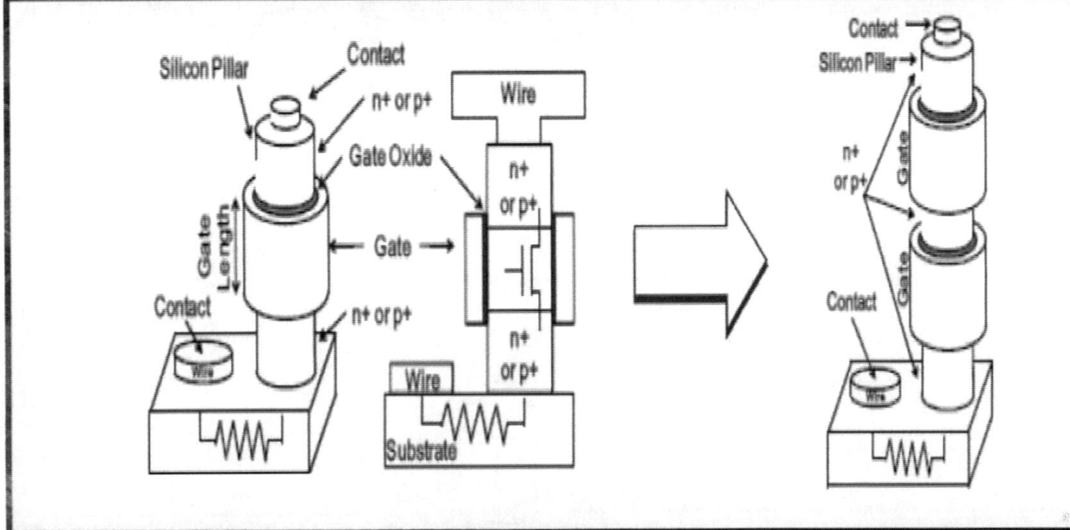

3D Moore's Law Acceleration

3D Architecture

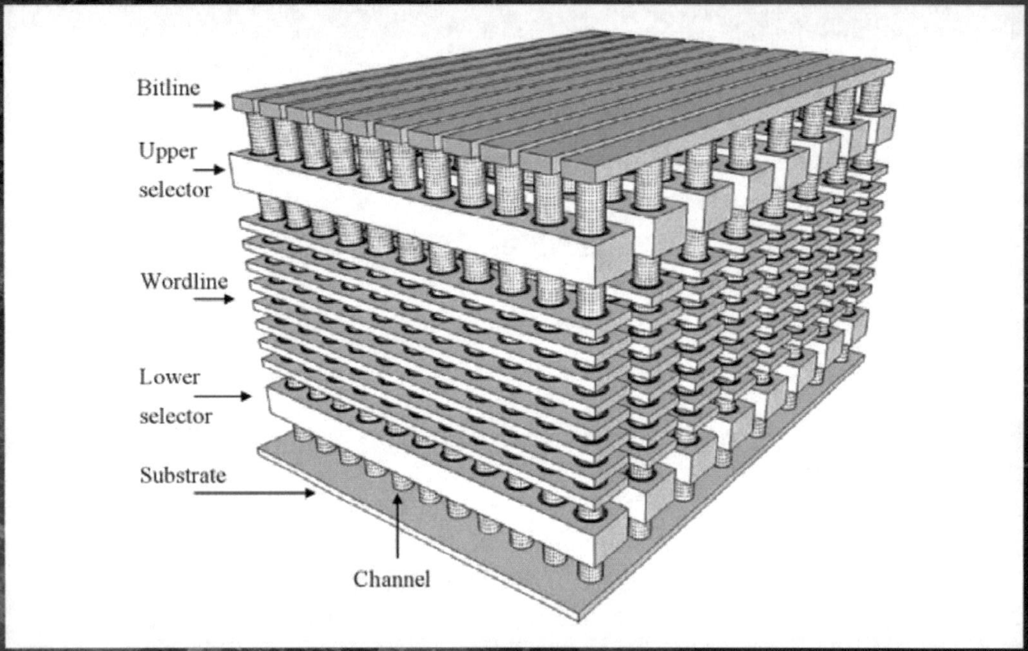

April 2016 SPCC P.Gargini 61

Multiple Stories

1. Introduction
2. 1998. ITRS 1.0
   1. Equivalent Scaling
3. 2000. NNI.
4. 2005. Nanoelectronics Research Initiative (NRI)
5. 2005-6. MPU power limits
6. 2006. More than Moore
   1. Heterogeneous Integration
7. 2010.First Selection of post CMOS devices
8. 2011. Equivalent scaling fully in production
9. 2014. Scaling acceleration
10. **2014-15.ITRS 2.0**
11. **2015. Post CMOS map of devices**
12. 2016. IRDS
13. **2017-2021.3D POWER Scaling**

April 2016 SPCC P.Gargini 62

Phase 3

Third Age of Scaling

(3D Power Scaling)

April 2016 SPCC P.Gargini 63

IC Industry at a Glance

(2021->203X)

Driver ➡ Cost/transistor & power reduction

How ➡ 2x Density/2 years (Moore)

Method ➡ 3D Power Scaling (ITRS2.0)

April 2016 SPCC P.Gargini 64

The Different Ages of Scaling
(Different methods for different times)

① Geometrical Scaling (1975-2003)

- Reduction of horizontal and vertical physical dimensions in conjunction with improved performance of planar transistors

② Equivalent Scaling (2003~2021)

- Reduction of only horizontal dimensions in conjunction with introduction of new materials and new physical effects. New vertical structures replace the planar transistor

③ 3D Power Scaling (2021~203X)

- Transition to complete vertical device structures. Heterogeneous integration in conjunction with reduced power consumption become the technology drivers

April 2016 SPCC P.Gargini 65

Beyond 2020

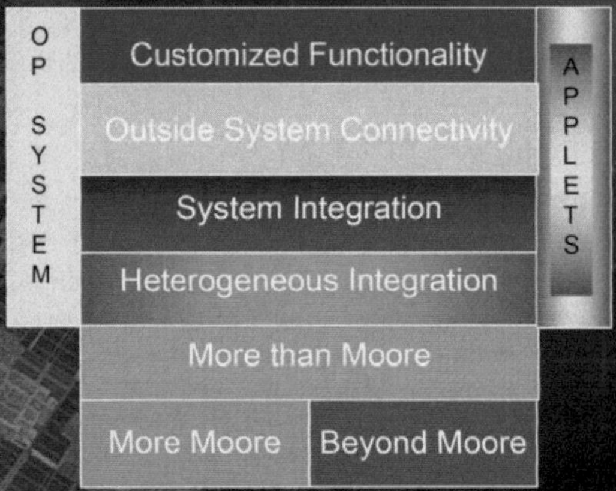

ITRS 2012

April 2016 P.Gargini 66

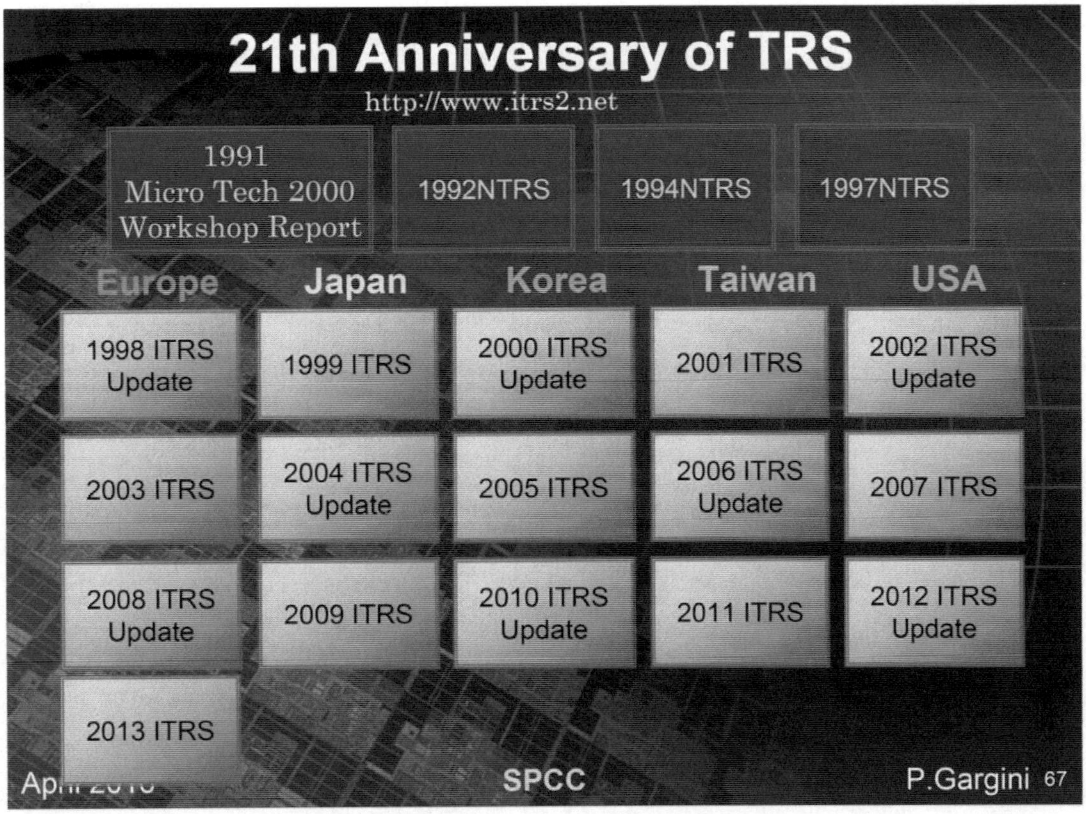
21th Anniversary of TRS
http://www.itrs2.net
1991
Micro Tech 2000
Workshop Report
1992NTRS
1994NTRS
1997NTRS
Europe
Japan
Korea
Taiwan
USA
1998 ITRS Update
1999 ITRS
2000 ITRS Update
2001 ITRS
2002 ITRS Update
2003 ITRS
2004 ITRS Update
2005 ITRS
2006 ITRS Update
2007 ITRS
2008 ITRS Update
2009 ITRS
2010 ITRS Update
2011 ITRS
2012 ITRS Update
2013 ITRS
SPCC
P.Gargini 67

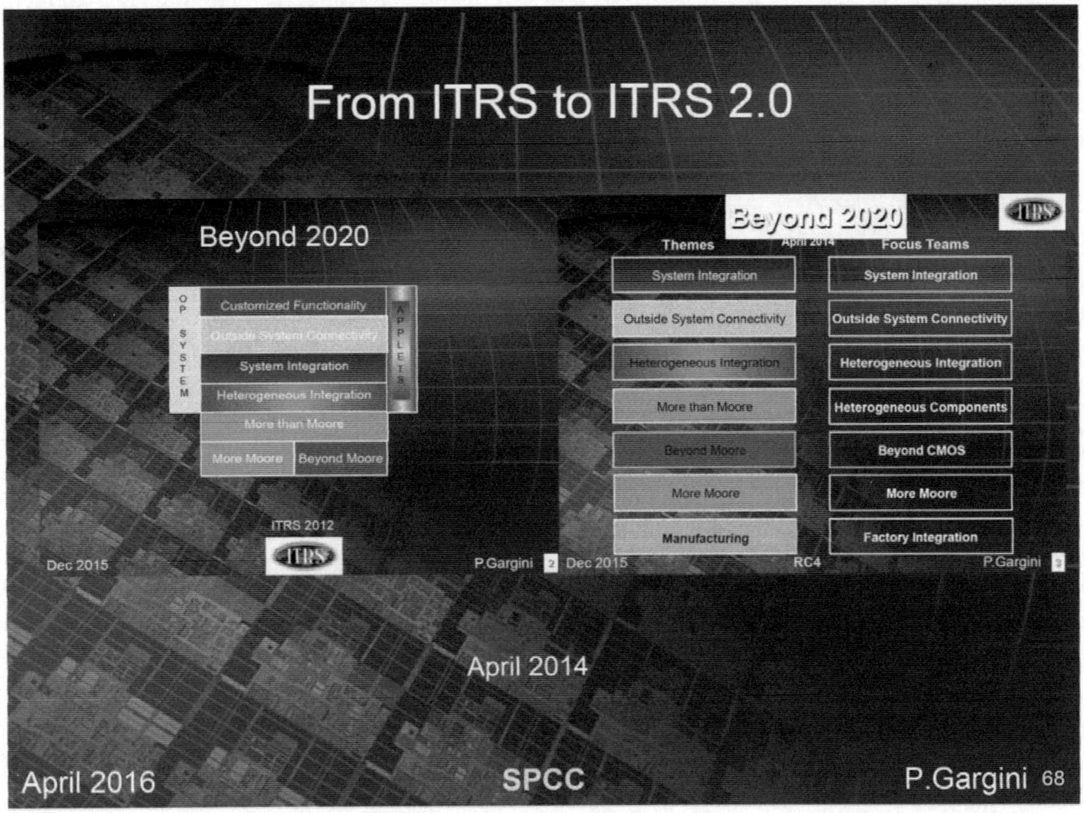
From ITRS to ITRS 2.0
Beyond 2020
Customized Functionality
Outside System Connectivity
System Integration
Heterogeneous Integration
More than Moore
More Moore
Beyond Moore
ITRS 2012
Dec 2015
P.Gargini
Beyond 2020
Themes
Focus Teams
System Integration
Outside System Connectivity
Heterogeneous Integration
More than Moore
Beyond Moore
More Moore
Manufacturing
System Integration
Outside System Connectivity
Heterogeneous Integration
Heterogeneous Components
Beyond CMOS
More Moore
Factory Integration
Dec 2015
RC4
P.Gargini
April 2014
April 2016
SPCC
P.Gargini 68

ITRS 2.0
2.0
ITRS
http://www.itrs2.net
April 2016
SPCC
P.Gargini 69

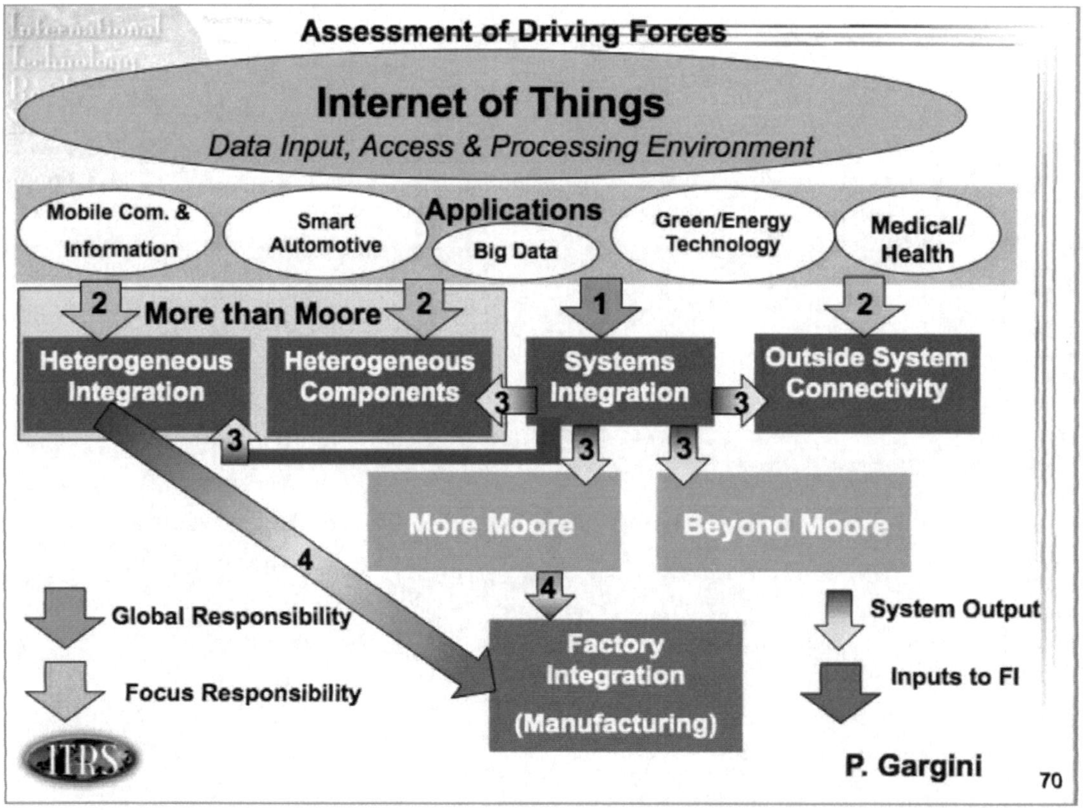
Assessment of Driving Forces
Internet of Things
Data Input, Access & Processing Environment
Applications
Mobile Com. & Information
Smart Automotive
Big Data
Green/Energy Technology
Medical/ Health
More than Moore
Heterogeneous Integration
Heterogeneous Components
Systems Integration
Outside System Connectivity
More Moore
Beyond Moore
Global Responsibility
Focus Responsibility
Factory Integration (Manufacturing)
System Output
Inputs to FI
P. Gargini
70

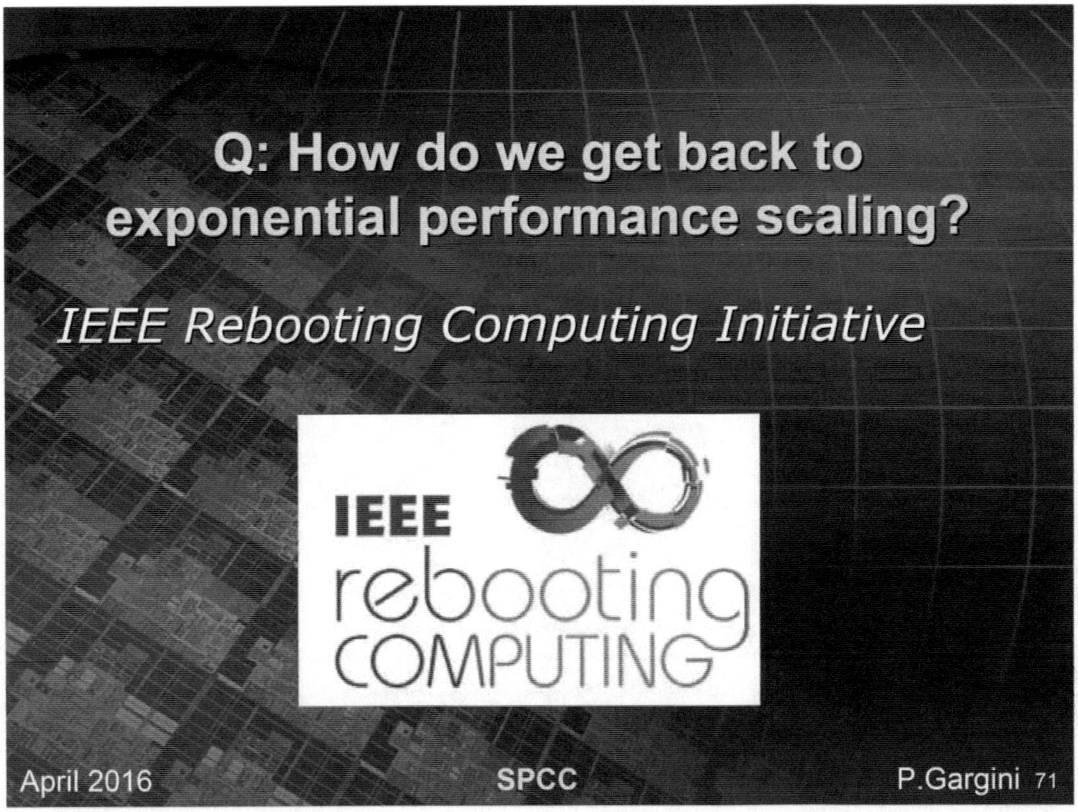
Q: How do we get back to exponential performance scaling?
IEEE Rebooting Computing Initiative
IEEE
rebooting
COMPUTING
April 2016
SPCC
P.Gargini 71

IEEE
rebooting
COMPUTING
ITRS
International Technology
Roadmap for Semiconductors
IEEE
14
April 2016
SPCC
P.Gargini 72

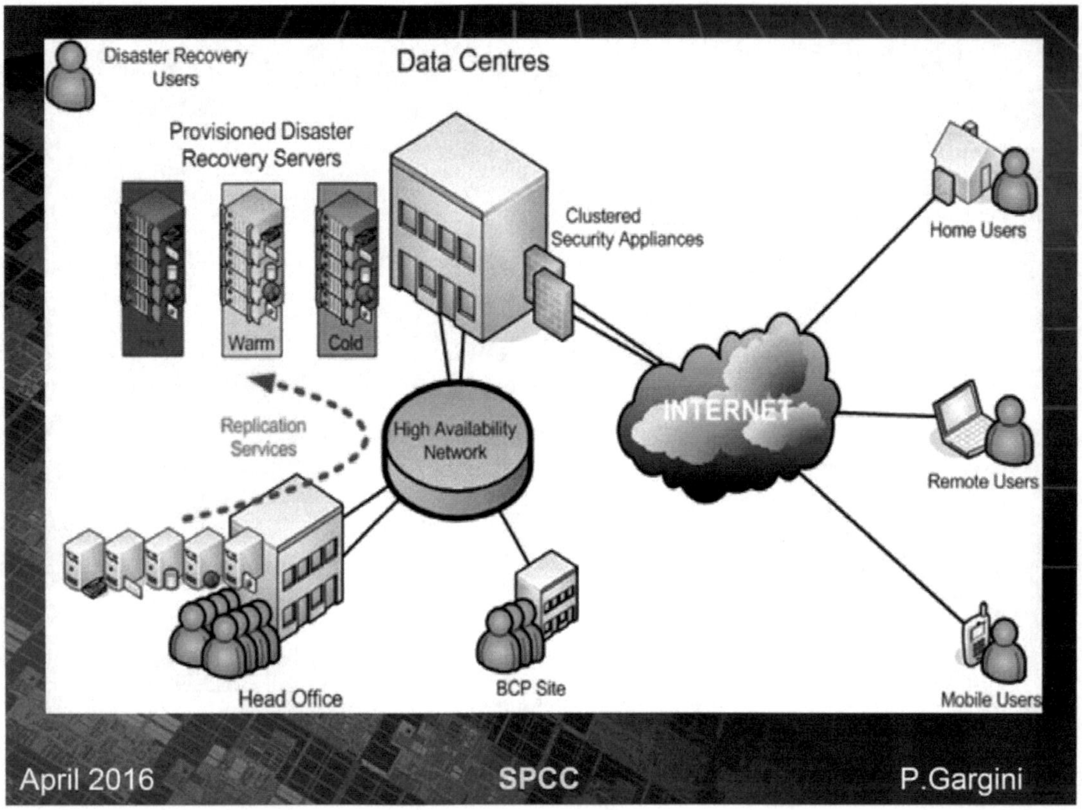

Data Centers

Categories	Year	2015	2017	2019	2021	2023	2025	2027	2029
Data Center Global Indicators	Number of Cores (K)	360	1044	3008	4935	5825	7578	8967	10602
	Total Memory Storage (PB)	300	1559	4676	14029	42088	126264	378792	1136377
	Area of One Server Building (MSF)	0.5	0.9	1.6	2.2	[illegible]	2.42	2.42	2.42
	Total Power Consumed by Datacenter (MkWh)	779.8	839	1004.7	1137.2	[illegible]	1380.7	1635.6	2044.3
	Total Switching Capability of Datacenter BW (Tb/s)	1000	2512	6309	[illegible]	15849	[illegible]	39811	63096
	Data Center Efficiency (GFLOPS/W)	2.4	4.9	10	17	24	33.9	47.9	67.8
Servers	Server Units/Rack	40	40	40	[illegible]	40	40	40	40
	Number of Cores/socket	18	29	[illegible]	59	74	93	117	147
	Power /Single Server Unit (W)	700	700	700	[illegible]	700	700	700	700
	Main Memory / Single Server Unit (GB)	32	45	64	76	91	108	129	154
	Power Consumed by Cores/single socket (W)	165	[illegible]	[illegible]	149	145	141	137	133
Switching	Network Bandwidth/Unit (Gb/s)	40	40	100	100	100	400	400	400
	Network Bandwidth/Rack (Gb/s)	1600	[illegible]	[illegible]	4000	4000	16000	16000	16000
	Rack Switch Capacity BW (Gb/s)	[illegible]	1200	3000	3000	3000	12000	12000	12000
Power	Power efficiency (Grid delivery/Data Center Use)	0.55	[illegible]	0.59	0.61	0.63	0.65	0.67	0.69
	Power Consumed by Networking and Switching (MkWh)	21.91	55.05	138.26	87.66	138.93	220.19	348.97	553.08
	Power Consumed by Storage (MkWh)	0.0657	0.259	0.449	0.778	1.347	2.334	4.043	7.004
	Power Consumed for facility cooling (MkWh)	38.99	55.02	86.13	237.31	264.26	307.03	374.89	482.56

DRAFT

April 2016 SPCC P.Gargini 74

Mobile

	Year	2015	2017	2019	2021	2023	2025	2027	2029
Input Metrics	Number of AP cores	4	4	6	8	10	12	14	18
	Number of GPU cores	14	22	34	54	86	138	220	352
	Max frequency of any Component in System (GHz)	2.7	2.9	3.2	3.4	3.7	4	4.3	4.7
	Number of Mega pixels in Display	2.1	2.1	3.7	8.8	8.8	33.2	33.2	33.2
	Band Width between AP and Main memory (Gb/s)	45.3	45.3	[illegible]	89.6	89.6	148.2	148.2	148.2
	Number of Sensors	14	16	[illegible]	20	21	21	22	22
	Number of Antennas	11	13	13	14	15	15	15	15
	Number of ICs	7-10	[illegible]	7-10	7-10	7-10	7-10	7-10	7-10
	Cellular data rate growing ~1.3x/year (MB/s)	[illegible]	[illegible]	21.63	40.75	40.75	40.75	40.75	40.75
	Wi-Fi data rate evolving with standards (Mb/s)	867	[illegible]	867	7000	7000	28000	28000	28000
Output Metrics	PCB area of main Components (cm2)	62	69	76	84	93	103	103	103
	Board power averaged at ~7%/year (mW)	4274	4706	5183	5708	6287	6926	7630	8406

NAND Flash							
Year of Production	***2015***	***2016***	***2020***	***2022***	***2024***	***2028***	***2030***
2D NAND Flash uncontacted poly 1/2 pitch – F (nm)	15	14	12	[illegible]	12	12	12
3D NAND minimum array 1/2 pitch - F(nm)	80nm	80nm	80nm	[illegible]	80nm	80nm	80nm
Number of word lines in one 3D NAND string	32	32-48	[illegible]	[illegible]	128-192	256-384	384-512
Dominant Cell type (FG, CT, 3D, etc.)	FG/CT/3D	FG/CT/3D	[illegible]	FG/CT/3D	FG/CT/3D	FG/CT/3D	FG/CT/3D
Product highest density (2D or 3D)	256G	[illegible]	[illegible]	1T	1.5T	3T	4T
3D NAND number of memory layers	32	[illegible]	64-96	96-128	128-192	256-384	384-512
Maximum number of bits per cell for 2D NAND	3	3	3	3	3	3	3
Maximum number of bits per cell for 3D NAND	3	3	3	3	3	3	3

DRAM TECHNOLOGY							
YEAR OF PRODUCTION	***2015***	***2017***	***2019***	***202[illegible]***	***2024***	***2027***	***2030***
Half Pitch (Calculated Half pitch) (nm)	24	20	17	[illegible]	[illegible]	8.4	7.7
DRAM cell size (μm^2)	0.00346	0.00240	0.00116	[illegible]	0.00048	0.00028	0.00024
DRAM cell FET structure	RCAT+Fin	RCAT+Fin	VCT	[illegible]	VCT	VCT	VCT
Cell Size Factor: a	6	6	4	[illegible]	4	4	4
Array Area Efficiency	0.55	0.55	0.5	0.5	0.5	0.5	0.5
V_{int} (support FET voltage) [V]	1.1	[illegible]	[illegible]	1.1	0.95	0.95	0.95
Support min. V_{tn} (25C, $G_{m,max}$, V_d=55mV)	0.40	0.40	0.40	0.40	0.37	0.37	0.37
Minimum DRAM retention time (ms)	64	64	64	64	64	64	64
DRAM soft error rate (fits)	1000	[illegible]	1000	1000	1000	1000	1000
Gb/1chip target	8G	8G	16G	16G	32G	32G	32G

YEAR OF PRODUCTION	2015	2017	2019	2021	2024	2027	2030
Logic device technology naming	P70M56	P48M36	P42M24	P32M20	P24M12G1	P24M12G2	P24M12G3
Logic industry "Node Range" Labeling (nm)	"16/14"	"11/10"	"8/7"	"6/5"	"4/3"	"3/2.5"	"2/1.5"
Logic device structure options	finFET FDSOI	finFET FDSOI	finFET LGAA	finFET LGAA VGAA	VGAA, M3D	VGAA, M3D	VGAA, M3D
LOGIC DEVICE GROUND RULES							
MPU/SoC Metalx ½ Pitch (nm)[1,2]	28.0	18.0	12.0	10.0	[illegible]	6.0	6.0
MPU/SoC Metal0/1 ½ Pitch (nm)	28.0	18.0	[illegible]	10.0	6.0	6.0	6.0
Contacted poly half pitch (nm)	35.0	24.0	[illegible]	[illegible]	12.0	12.0	12.0
L_g: Physical Gate Length for HP Logic (nm) [3]	24	18	14	10	10	10	10
L_g: Physical Gate Length for LP Logic (nm)	26	20	16	[illegible]	12	12	12
FinFET Fin Half-pitch (new) =0.75 or 1.0 M0/M1 (nm)	21.0	18.0	[illegible]				
FinFET Fin Width (nm)	8.0	[illegible]	[illegible]				
FinFET Fin Height (nm)	42.0	[illegible]	[illegible]				
Footprint drive efficiency - finFET	[illegible]	[illegible]	[illegible]				
Lateral GAA Lateral Half-pitch (nm)			12.0	10.0			
Lateral GAA Vertical Half-pitch (nm)			12.0	9.0			
Lateral GAA Diameter (nm)			6.0	6.0			
Footprint drive efficiency - lateral GAA, 3x NWs stacked			2.4	2.8			
Vertical GAA Lateral Half-pitch (nm)				10.0	6.0	6.0	6.0
Vertical GAA Diameter (nm)				6.0	5.0	5.0	5.0
Footprint drive efficiency - vertical GAA, 3x NWs stacked				2.8	3.9	3.9	3.9
Defice effective width - [nm]	92.0	90.0	56.5	56.5	56.5	56.5	56.5
Device lateral half pitch (nm)	21.0	18.0	12.0	10.0	6.0	6.0	6.0
Device width or diameter (nm)	8.0	6.0	6.0	6.0	5.0	5.0	5.0

DRAFT

April 2016 SPCC P.Gargini 79

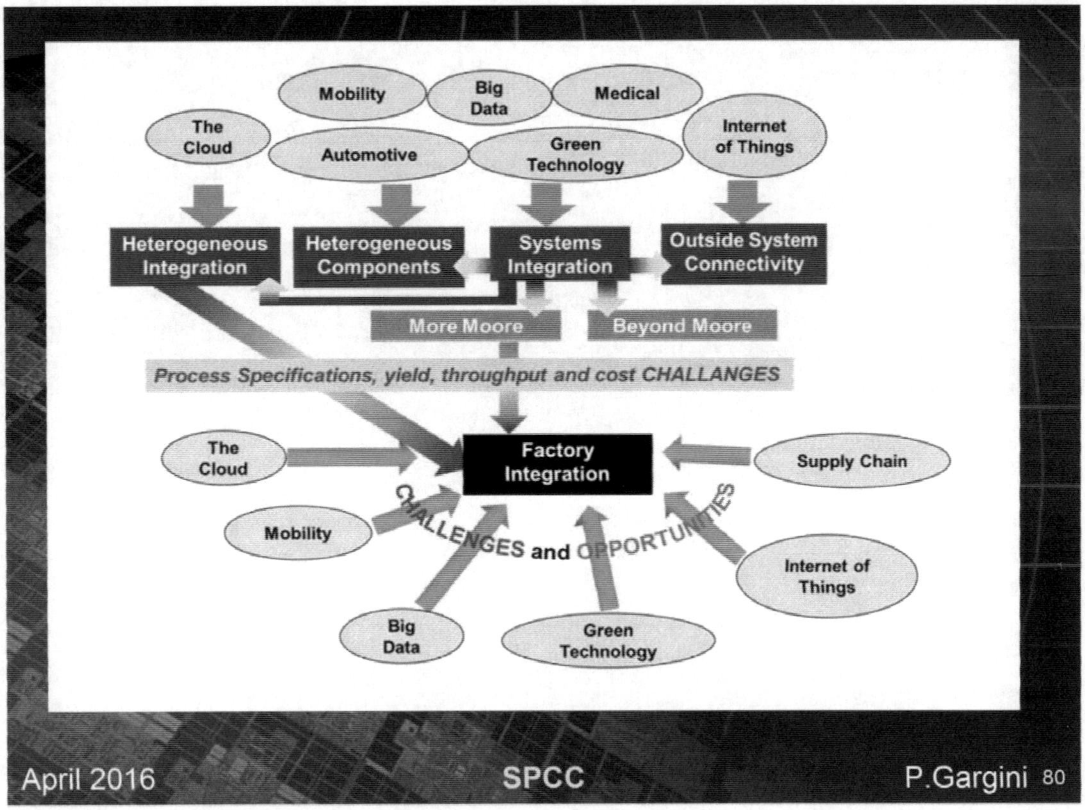

Conclusions

- **"Geometrical Scaling"** led the IC Industry for 3 decades

ITRS 1.0

- Cooperative and distributed research and manufacturing methods highlighted by **ITRS** emerged as cost effective means of reducing costs since the mid-90s
- FCRP, NRI, Sematech, IMEC and Government organizations actively cooperated in **advanced research**

- **"Equivalent Scaling"** saved the Semiconductor Industry since the beginning of the previous decade
- Preliminary evaluation of post-CMOS candidates published in 2010

ITRS 2.0

- **"3D Power Scaling" is the next phase of (accelerated) scaling**
- **Post CMOS devices and emerging architectures are being jointly evaluated->ITRS/IEEE RC->IRDS**

April 2016 SPCC P.Gargini 81

SPCC 2016
Surface Preparation and Cleaning Conference

Effect of Surface Preparation of Copper on Self-Assembly of Fullerene Molecules

Dongni Ma, Selene Sandoval, Krishna Muralidharan, Srini Raghavan

University of Arizona

Department of Materials Science and Engineering
Department of Chemical and Environmental Engineering

Objective

- Effect of <u>surface preparation of copper</u> on:
- ➢ Substrate mediated controllable self-assembly of **fullerene rods**

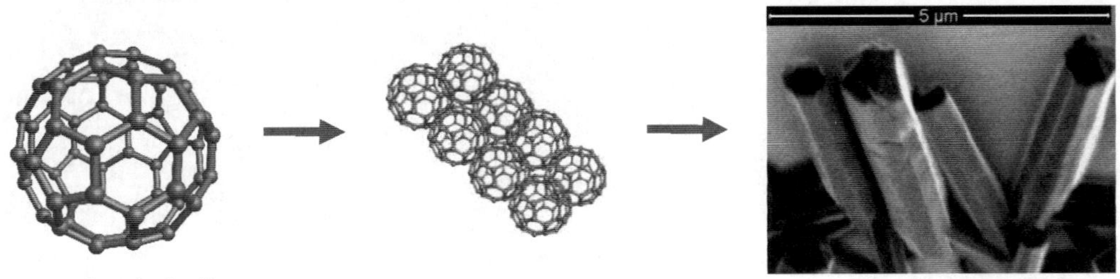

C_{60} buckyball C_{60} nano-rods

➢ Ultimately enable high-aspect ratio molecular C_{60} wires as interconnects

Background:
Conventional Methods for Fullerene Nanostructures

➢ **Conventional synthesis techniques**

- Liquid-Liquid Interfacial Precipitation (LLIP): *time consuming (a few hours - two weeks)—no substrate needed.*

- Template based self-assembly: *longer processing time, expensive and broken nanotubes obtained—Porous alumina template to be infiltrated by fullerene solution.*

THE UNIVERSITY OF ARIZONA.

Background: Surface mediated synthesis
Directed Air Stream: leads to fullerene rods

C_{60} dissolved in Toluene

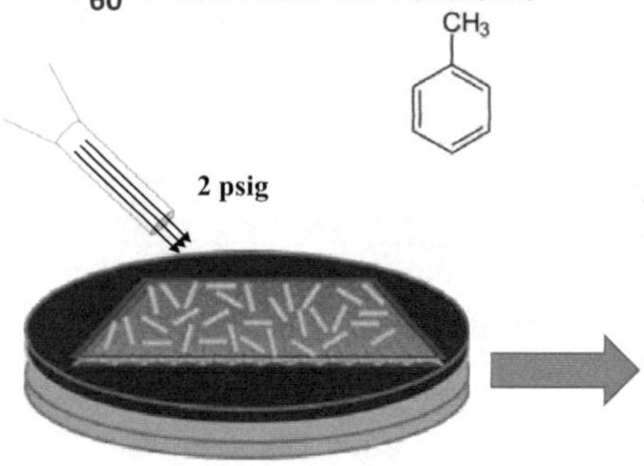

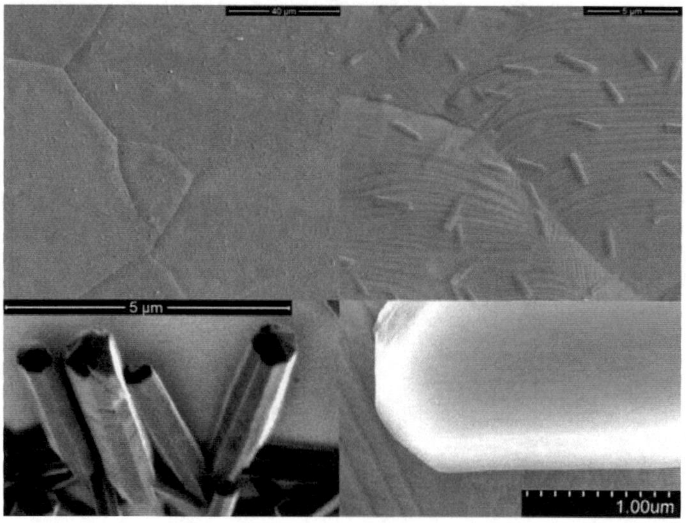

Self-assemblies size:

- **Length: ~2 µm**
- **Width: ~500-700 nm**

THE UNIVERSITY OF ARIZONA.

Current Method: **Spin coating based substrate mediated route**

Substrates	Solution	Spin coating procedure	Spin coating RPM
Cold rolled Cu	Fullerene dispersed in Toluene (2mg/ml)	**1. Dispense solution.** **2. Wait for 1 minute.** **3. Spin substrate at a predetermined RPM.** **4. Spin for 30 seconds.** Repeated 1-4 procedure for total of 4 times.	200-500 RPM
Annealed Cu			
Electropolished Cu			
Graphene coated Cu			

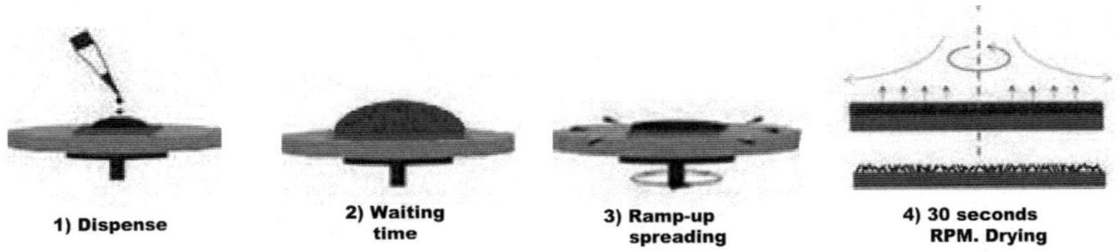

- **Overall processing time: < 10 minutes for one substrate of size 1 cm^2.**

THE UNIVERSITY OF ARIZONA.

Substrate Preparation

Copper foil (0.25 mm thick, 99.99% metals basis, Alfa Aesar Puratronic®)

Cold rolled Cu (Contact angle of 72°)	Annealed Cu (Contact angle of 64°)	Graphene on Cu (Contact angle of 80°)	Electropolished Cu (Contact angle of 72°)
Organic impurities removal: IPA rinsed, DI water rinsed, and blown dry with nitrogen.			
	Annealed in a tube furnace (Lindberg Blue M) for 2 hours at 1050°C.	Graphene on Cu via chemical vapor deposition (CVD).	Acetic acid treatment: 1) Immersed in 2M acetic acid solution at 60°C for 10 min. 2) DI water rinsed and blown dry with nitrogen. *Electropolishing procedure shown on next slide.*

THE UNIVERSITY OF ARIZONA.

Preparation of Substrates: Electropolishing Procedure

- Solution: 85% phosphoric acid.
- Applied a constant potential of 1.5 V vs. SCE for 1 hour.

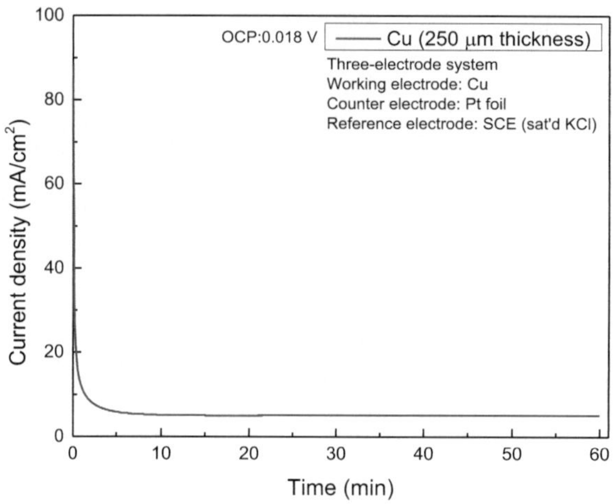

THE UNIVERSITY OF ARIZONA.

Graphene grown *via* chemical vapor deposition (CVD)

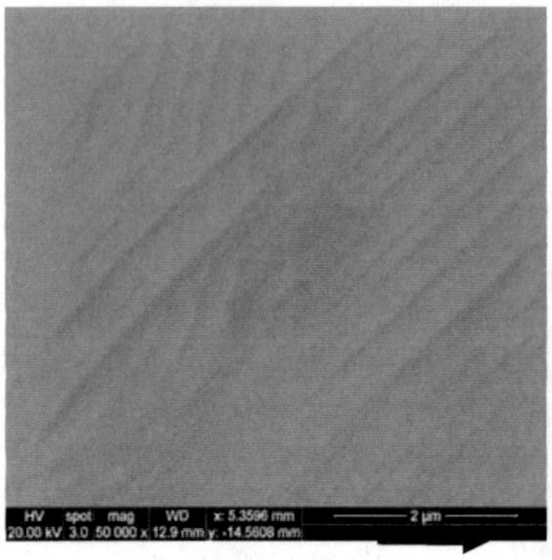

- CVD grown graphene shows characteristic ripple structure.

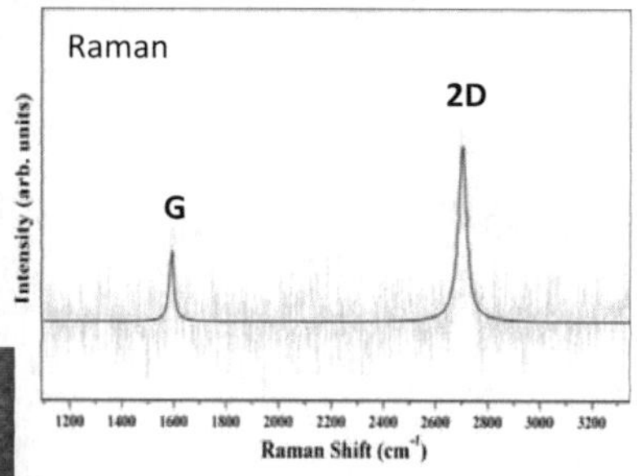

CVD conditions:
Pressure of 200 mTorr
Temperature of 1050°C
100 sccm Argon
60 sccm Hydrogen
20 sccm Methane

SURFACE CHARACTERIZATION

THE UNIVERSITY OF ARIZONA.

Surface Roughness of
Cold Rolled Cu and Electropolished Cu
an AFM Analysis

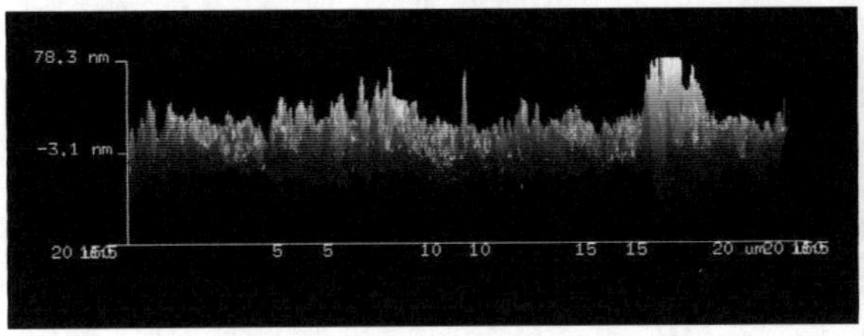

Cold rolled Cu
(**surface roughness rms : 15.5nm**)

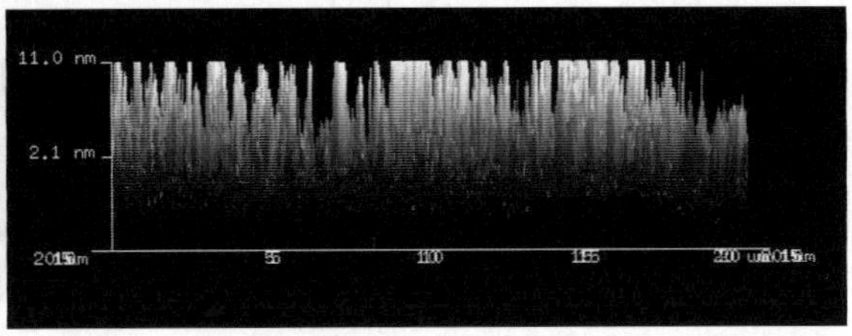

Electropolished Cu
(**surface roughness rms: 4.7nm**)

Surface Roughness of <u>Annealed Cu and Graphene Coated Cu</u> an AFM Analysis

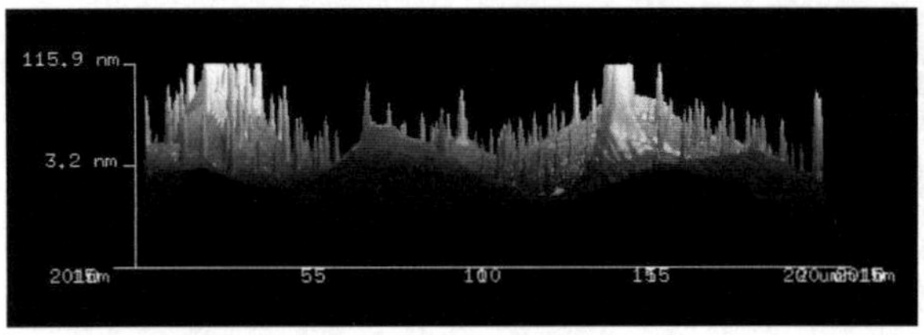

Annealed Cu
(**surface roughness rms: 53.3nm**)

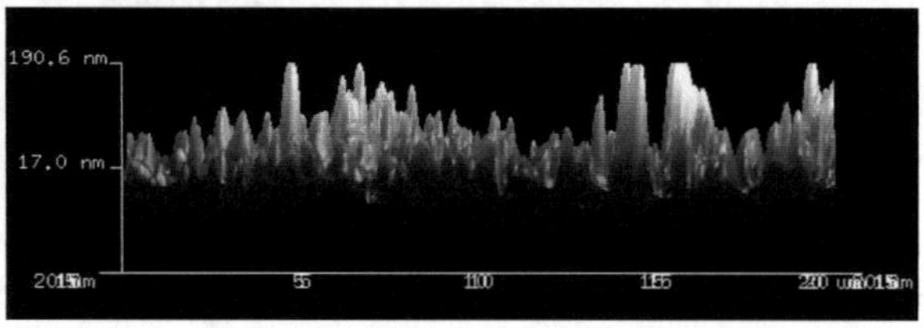

Graphene Coated Cu
(**surface roughness rms: 41.5nm**)

Results and discussion

Substrates	Solution
Cold rolled Cu	▪ Fullerene dispersed in Toluene (2mg/ml)
Annealed Cu	
Electropolished Cu	
Graphene coated Cu	

Uniform distribution of Rod-like C_{60} Self-Assemblies on Cold Rolled Cu Substrate

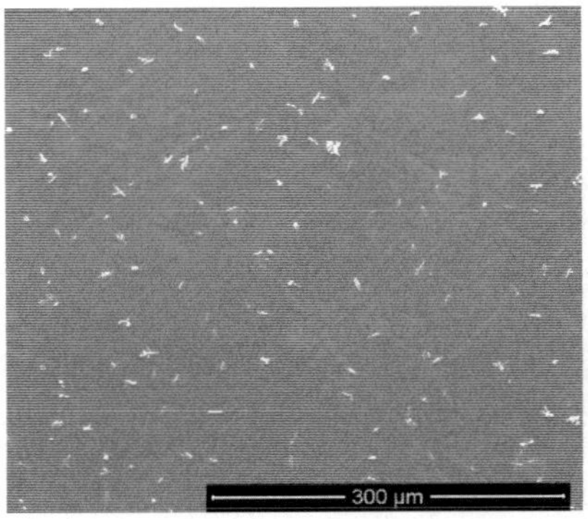

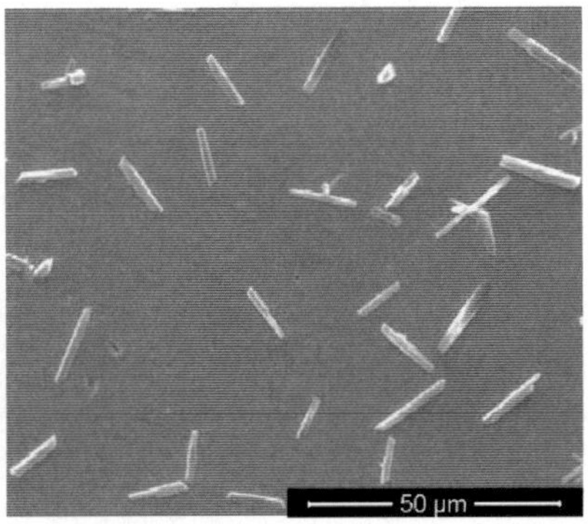

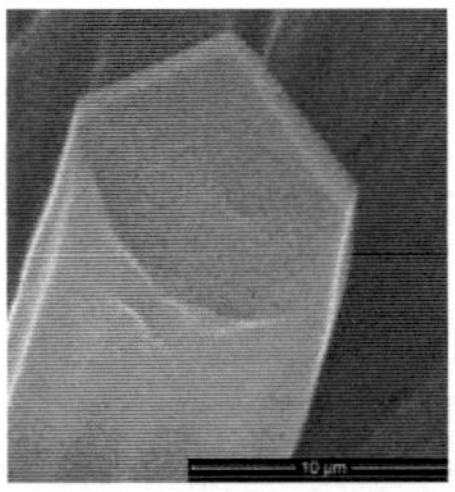

- Average length of fullerene self-assemblies (FSA) at 200 rpm: ~5 µm.

- Good control over size and morphology of fullerene rods at lower rpm (next slide).

THE UNIVERSITY OF ARIZONA.

Size Analysis : Projected Area and Length Distributions of C_{60} rods on Cold Rolled Cu Substrate

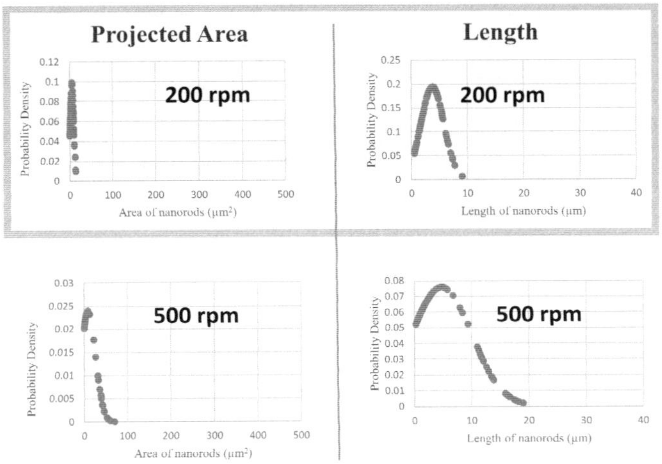

- Best distribution of rods at 200 rpm.
- Average length of fullerene self-assemblies at 200 rpm is ~5 µm.

THE UNIVERSITY OF ARIZONA.

Results and discussions

Substrates	Solution
Cold rolled Cu	▪ Fullerene dispersed in Toluene (2mg/ml)
Annealed Cu	
Electropolished Cu	
Graphene coated Cu	

THE UNIVERSITY OF ARIZONA.

C_{60} Self-Assemblies on Annealed Cu Substrate: larger variations in size and morphology

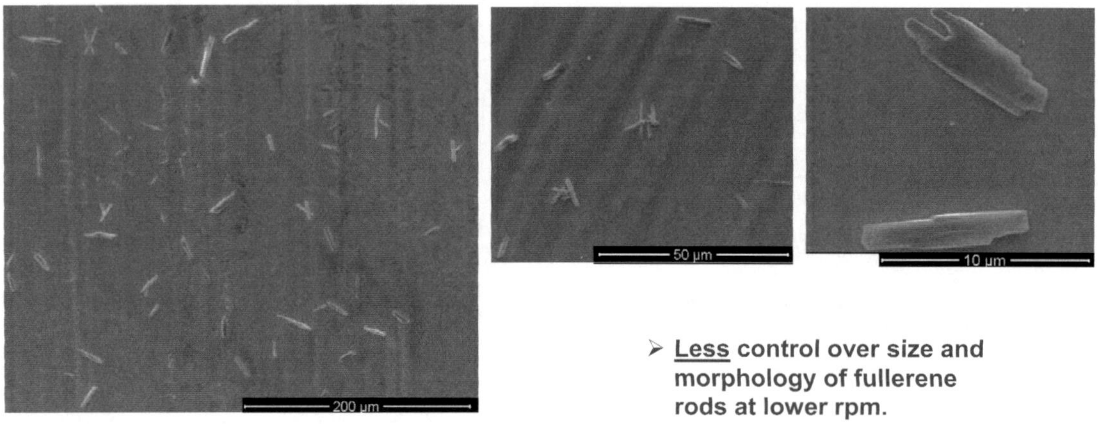

- Average length of FSA is ~5 µm.

➢ Less control over size and morphology of fullerene rods at lower rpm.

THE UNIVERSITY OF ARIZONA.

Size Analysis: Projected Area and Length Distribution of C_{60} self-assemblies on Annealed Cu Substrate

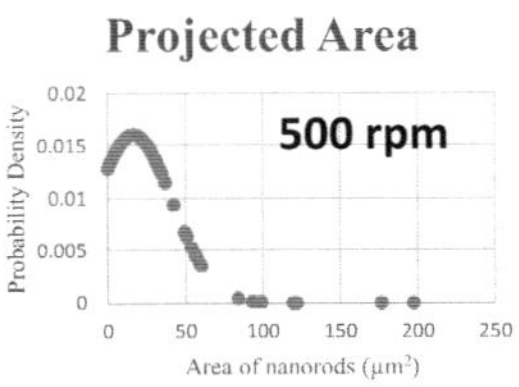

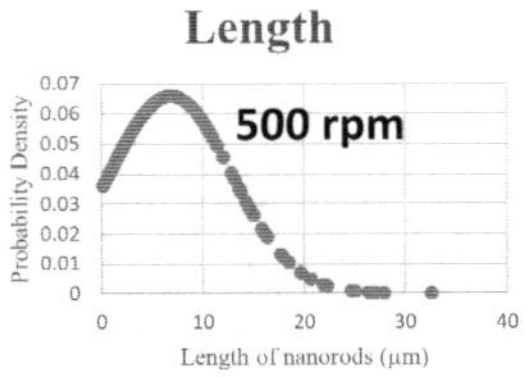

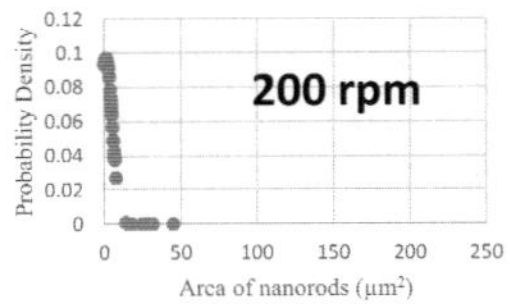

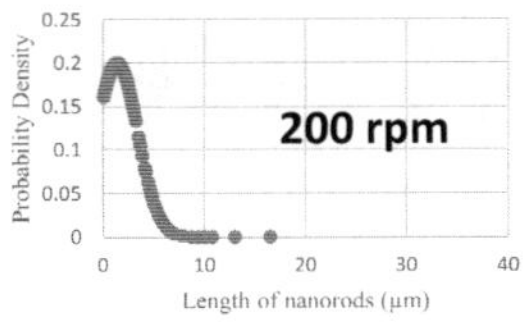

- Poor distribution at low and high spinning speeds.
- Average length of 5 μm and projected area of 18 $μm^2$ at 200 rpm.

THE UNIVERSITY OF ARIZONA.

Results and discussions

Substrates	Solution
Cold rolled Cu	▪ Fullerene dispersed in Toluene (2mg/ml)
Annealed Cu	
Graphene coated Cu	
Electropolished Cu	

C_{60} rods on Graphene Coated Cu Substrate: Two distinct morphologies

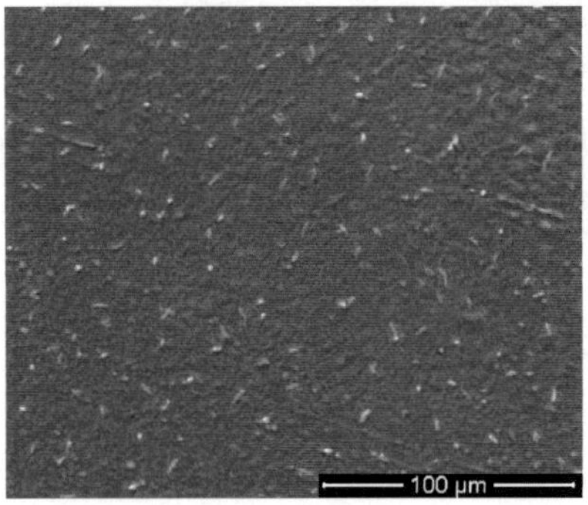

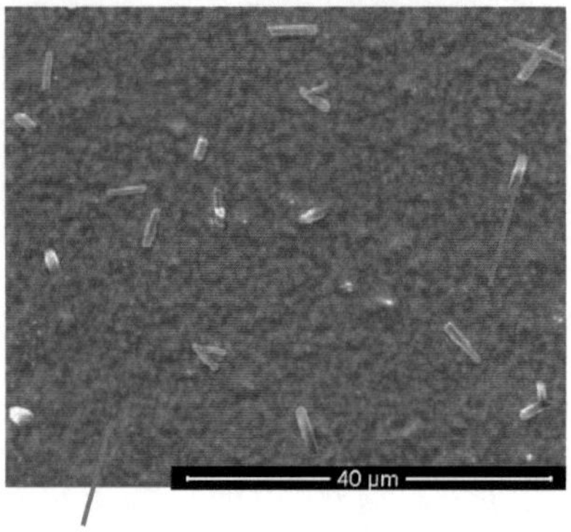

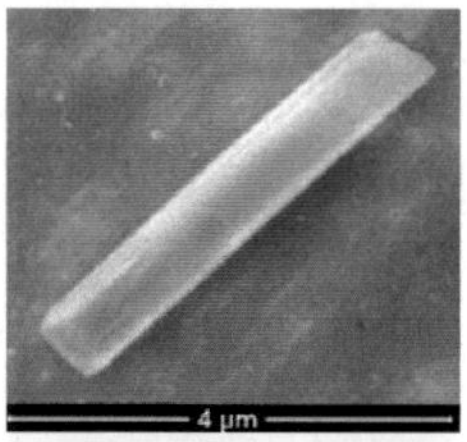

- **Good** control over distribution and morphology of fullerene nanorods at 200 rpm.

- **average length ~ 5 µm.**

THE UNIVERSITY OF ARIZONA.

Size distribution of larger rods: Projected Area and Length Distribution on Graphene Coated Cu Substrate

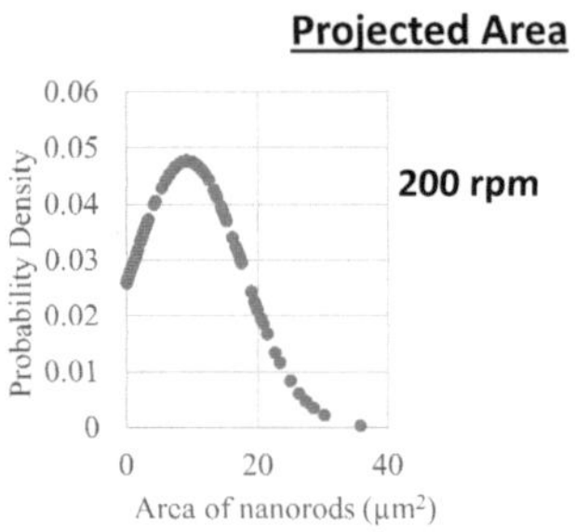

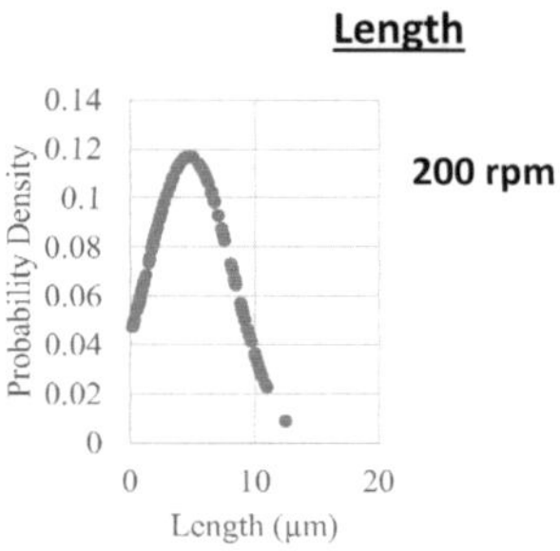

- Projected area of 10 μm^2 with length of 5 µm.

THE UNIVERSITY OF ARIZONA.

Results and discussions

Substrates	Solution
Cold rolled Cu	▪ Fullerene dispersed in Toluene (2mg/ml)
Annealed Cu	
Graphene coated Cu	
Electropolished Cu	

THE UNIVERSITY OF ARIZONA.

Highly controllable synthesis of C_{60} rods on Electropolished Cu Substrate

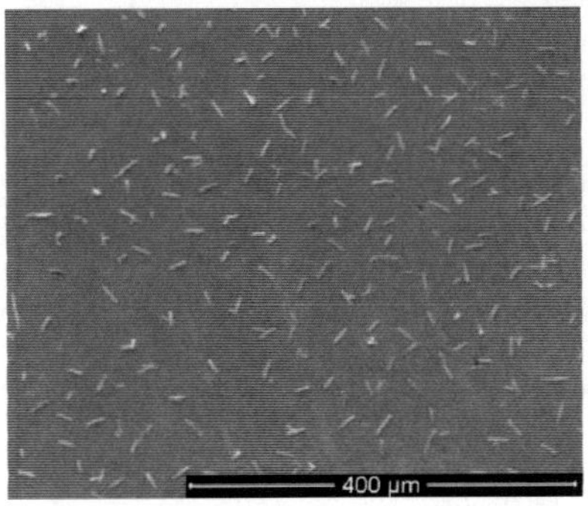

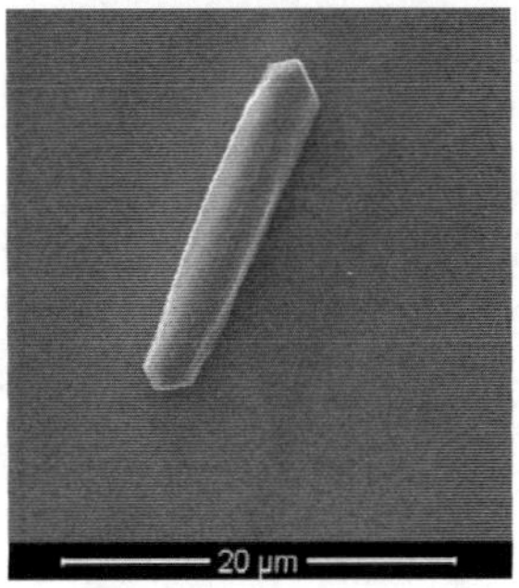

➢ Excellent control over size and morphology of fullerene nanorods at lower rpm.

▪ Length of rod : ~10 µm

C_{60} rods on Electropolished Cu (200 rpm) by AFM Analysis

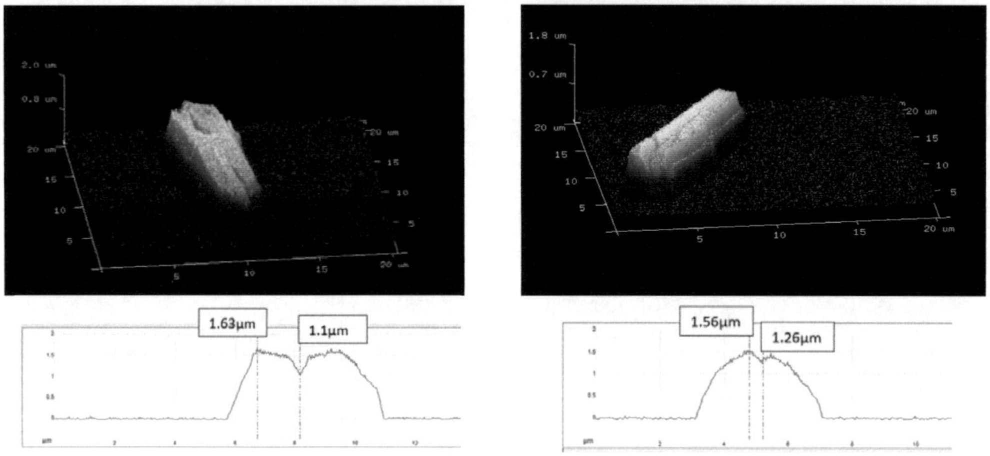

- Length of nanorods is ~14 µm with diameter of ~1.5 µm.

Size analysis: Projected Area and Length Distribution of rods on Electropolished Cu Substrate

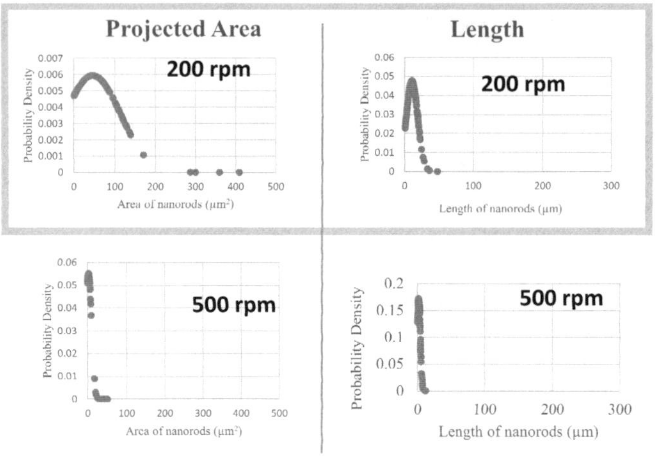

- **Best distribution and morphology of rods at 200 rpm.**
- **Length of fullerene self-assemblies is ~10 µm.**

THE UNIVERSITY OF ARIZONA.

Distinct Nanowire bundles in areas without C_{60} rods on electropolished Cu

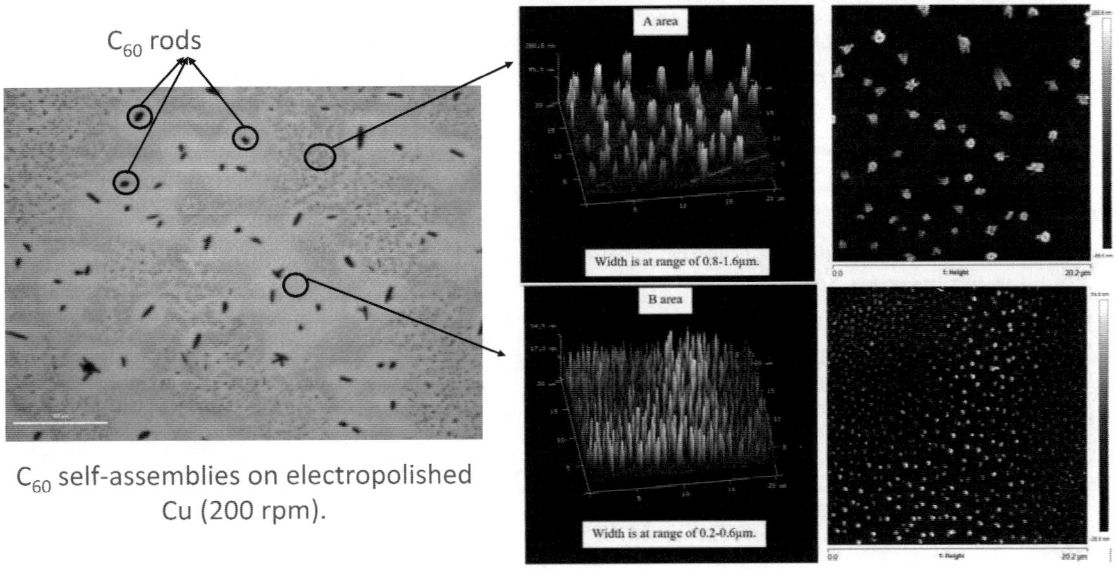

C_{60} self-assemblies on electropolished Cu (200 rpm).

THE UNIVERSITY OF ARIZONA.

In contrast, No well-defined nanowires in Graphene Coated Cu (200 rpm)

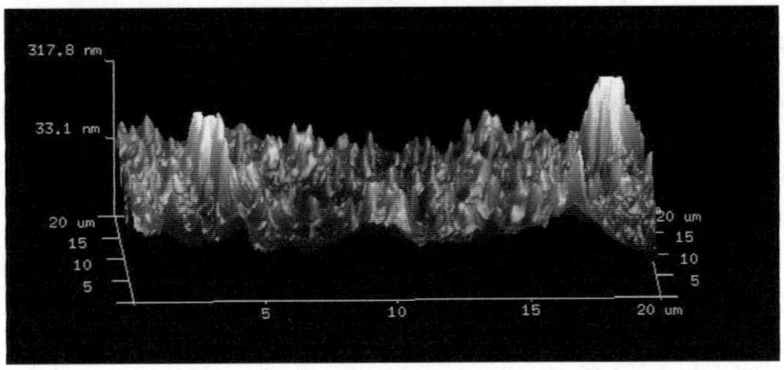

Areas without nanorods are not well defined as one on EP Cu substrate.

Discussion

- **It has been shown from DFT calculations that surface defects on Cu as well as graphene corrugations serve as strong adsorption sites for C_{60} molecules**
  - Well defined adsorption sites on electropolished Cu and graphene on Cu lead to more control on size, shape, morphology of C_{60} rods
  - Bigger rods are formed via nucleation and growth during the waiting time of 1 minute prior to ramp-up spreading
  - The nanowires are formed after the ramp-up as a result of 'coalescence' between the remaining C_{60} molecules
  - The coalescence is initiated due to C_{60}-C_{60} interaction arising as a result of spinning
  - The presence of intrinsic surface defects on annealed and cold-rolled Cu leads to less control of C_{60} morphology

Conclusions and Future Outlook

- A simple, wet-chemistry based spin-coating method developed for obtaining c_{60} rods and nanowires in a controllable fashion
- The size, shape and morphology of the C_{60} structures are intimately linked to the substrate on which they are formed.
- The developed method provides a new avenue to achieve high aspect ratio C_{60} molecular wires based interconnects as well as devices with tunable electrical properties (see next slide)

C60 milli-rods: electrical conductivity

Electrical conductivity of millimeter long C_{60} rods (aspect ratio = 10:1) formed from CS_2 solution within a glass/polycarbonate vial

$$\sigma_{rod} = 0.1(\Omega cm)^{-1}$$

$$\sigma_{C60\,film} = 10^{-6}(\Omega cm)^{-1}$$

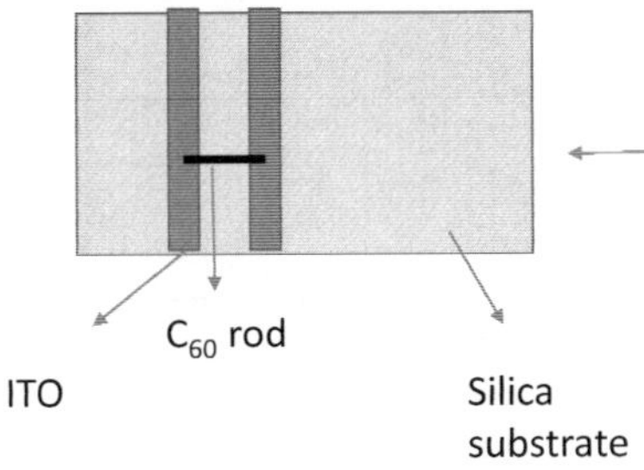

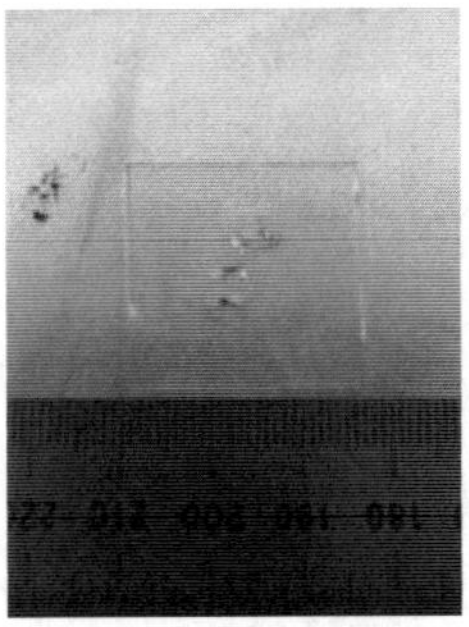

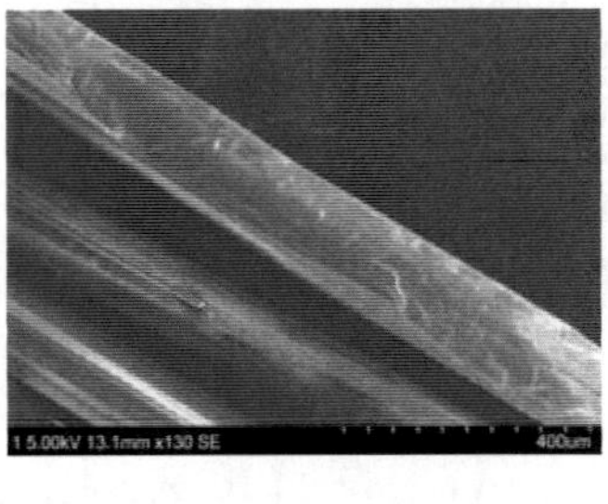

Novel Reactive Chemistry Sources for Surface Passivation of Future Generation Channel Materials

Presented by

Daniel Alvarez 858-259-1220 dalvarez@rasirc.com

Daniel Alvarez; Jeffrey Spiegelman; Russell Holmes; Chris Ramos
RASIRC, INC., San Diego, CA, 92126

Andrew C. Kummel, Mary Edmonds, Steven Wolf
University of California, San Diego, 92093

RASIRC

Overview

- Challenges for passivation of new channel materials
- Why Hydrogen Peroxide & Hydrazine?
- RASIRC BRUTE™ Technology
- Passivation approach to InGaAs(001)
- Passivation approach to SiGe(110)
- Conclusions

RASIRC

Challenges for Passivation on InGaAs and SiGe Channel Materials

- In Situ methods are desired
- Passivate surface dangling bonds
- Maintain an electrically unpinned surface Fermi Level ready for subsequent high-K gate oxide nucleation
- Passivation layer must prevent atomic migration into subsequent layers
- Underlying substrate must not be damaged
- Low thermal budget constraints (<400C)

RASIRC

Why Gas Phase Hydrogen Peroxide?

- Ge dosing experiments
- HOOH/H_2O will nucleate the surface more efficiently than H_2O

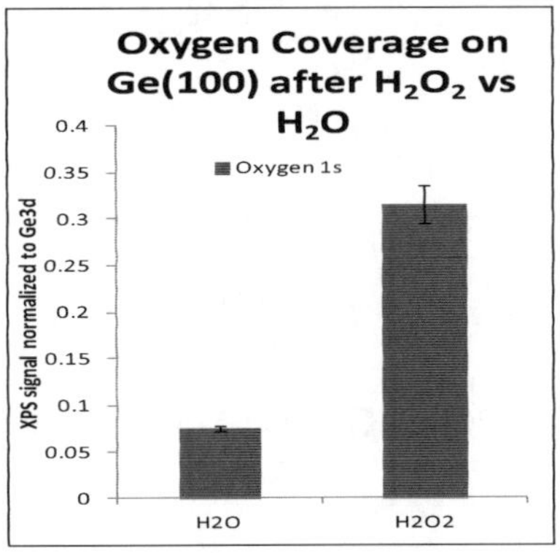

RASIRC

Why Gas Phase Hydrazine?

- Weakness of the N-N bond strength leads to high reactivity on Metal surfaces
- New Channel Materials have limited thermal budgets (typical <400C)
- Early studies (1992) by Slaughter and Gland show H_2NNH_2 to be more reactive than NH_3 on an Si(100) surface

RASIRC

Challenges

- Anhydrous Hydrogen Peroxide liquid is difficult to handle and may rapidly decompose leading to explosion

$$HOOH_{(l)} \rightarrow H_2O_{(g)} + O_{2(g)}$$

- Anhydrous Hydrazine has a low flash point of 37C and is highly toxic
  - Current commercial sources lack sufficient purity

RASIRC

RASIRC Membrane Technology

The liquid (solution surrounds the membrane tubes, passes thru the membrane walls and gets picked up by the carrier gas

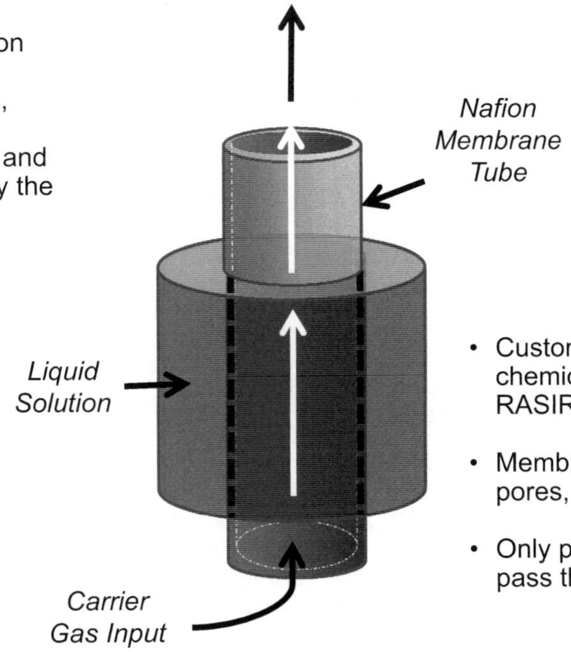

- Custom extruded and chemically modified by RASIRC
- Membrane has 10 to 50 nm pores, ionically charged
- Only polar molecules can pass through

RASIRC

Ionically Charged Channels

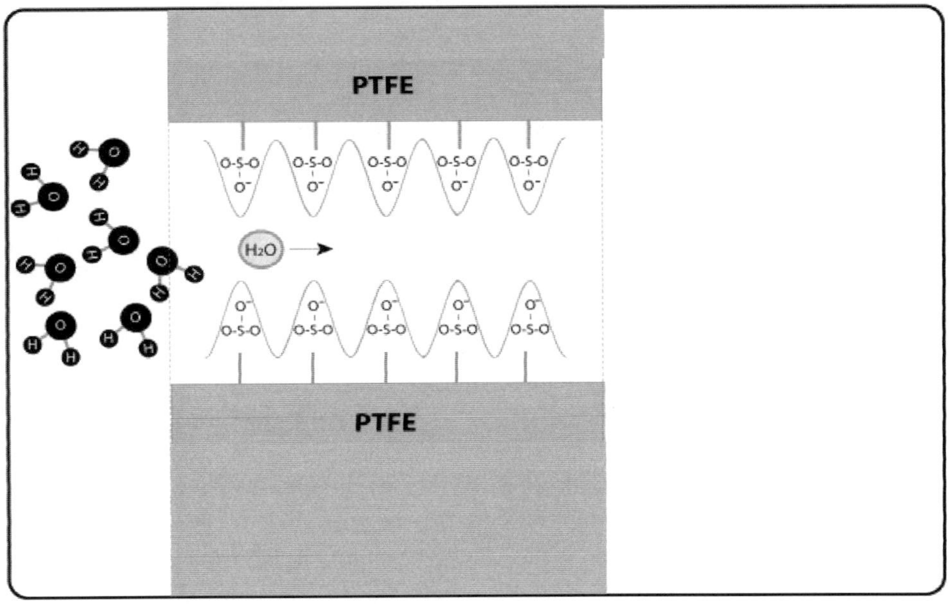

RASIRC

Approach to Anhydrous Delivery

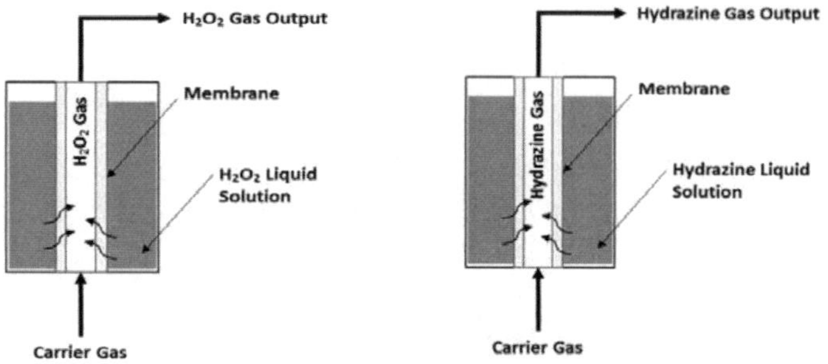

- Purify and Isolate Ultra High Purity Anhydrous HOOH (l) or H2NNH2 (l)
- Reactive chemical is stabilized by mixing with a proprietary solvent
- Tubular Membrane/Carrier Gas
  - Selective for Hydrogen Peroxide or Hydrazine molecules
- Desired Molecule permeates membrane and is delivered to process
- Solvent remains on liquid side of membrane & does not enter gas stream

RASIRC

Anhydrous Peroxide

Video Presentation in
Original e-media

RASIRC

H_2O_2 in Solvent

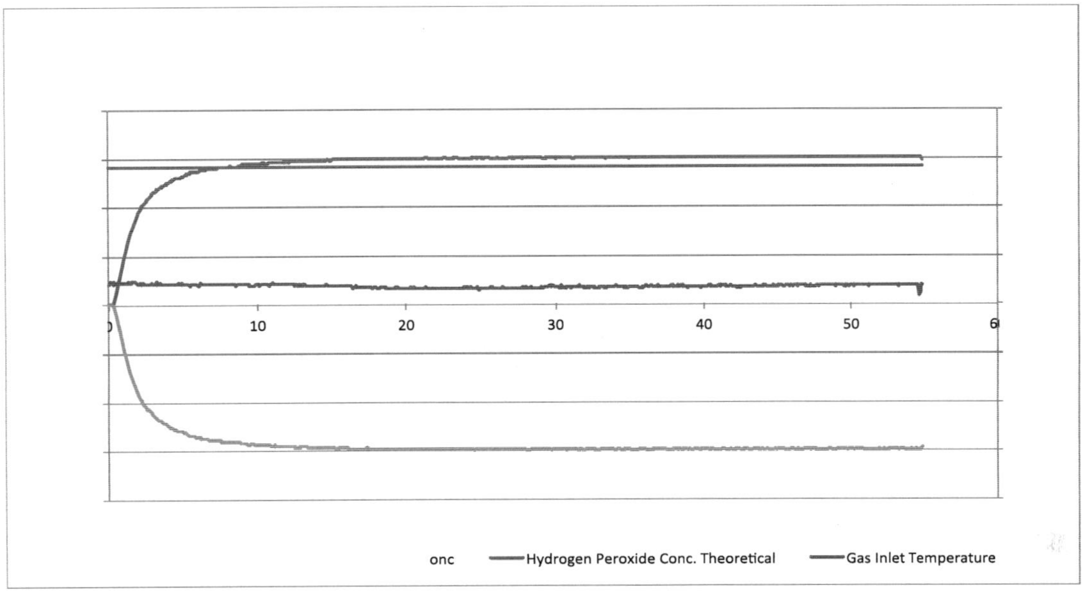

RASIRC

H_2NNH_2 in Solvent

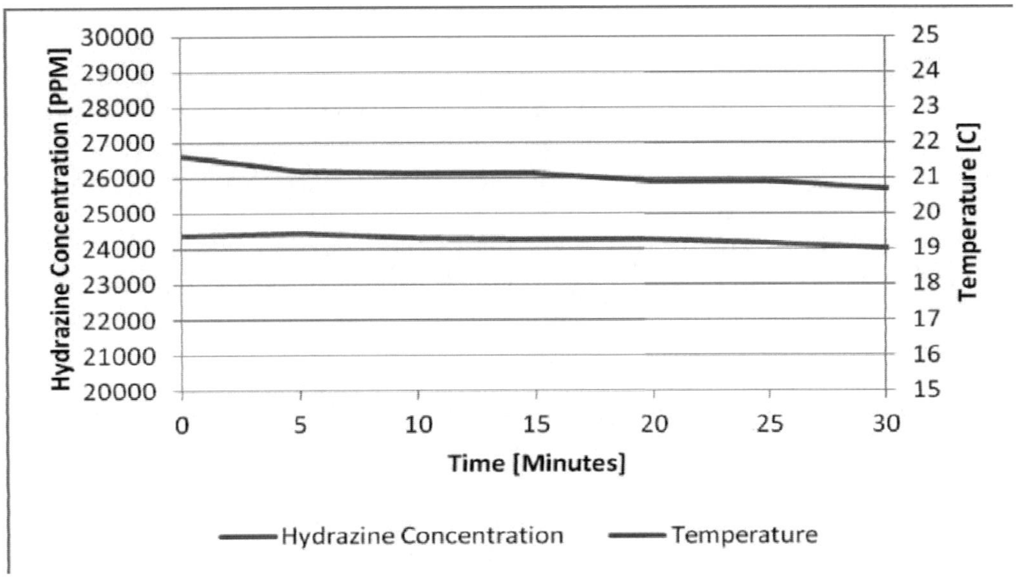

RASIRC

Approach to InGaAs Passivation/Functionalization

- Decapped InGaAs surface
- InGaAs + Si_2Cl_6 → InGaAs/Si + $GaCl_3$
- InGaAs/Si + HOOH → InGaAs/Si(O)OH
- InGaAs/Si(O)OH + TMA → InGaAs/SiO/Al_2O_3

RASIRC

Self Limiting CVD – Si_2Cl_6/InGaAs(001)-(2x4)

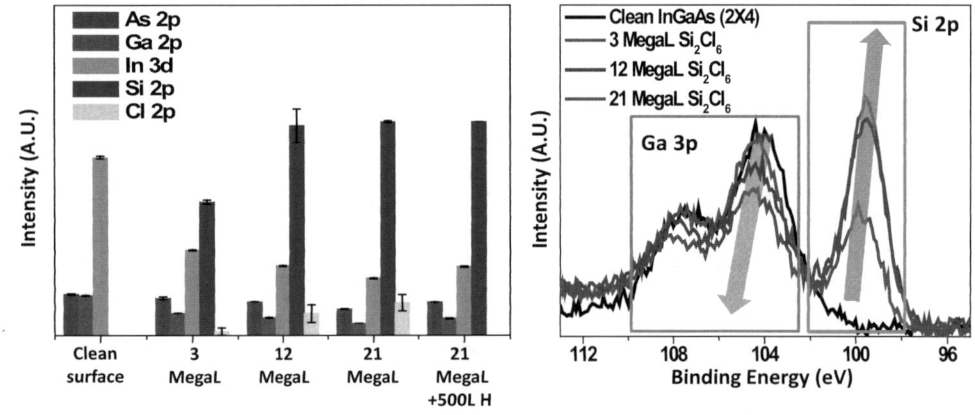

Left: XPS Spectra collected at 30° glancing angle. Corrected XPS peak areas for clean (2x4), and following 3, 12, and 21 MegaLangmuir total Si_2Cl_6 doses at 350°C.(Si_2Cl_6 dosed at 10 second pulses of 0.025 Torr).

Right: Raw XPS peak areas for Ga 3p and Si 2p on clean InGaAs(2x4), and following 3, 12, and 21 MegaLangmuir total Si_2Cl_6 doses at 350°C.

➢ Desorption limited CVD of Si_2Cl_6 at 350°C until no clean InGaAs surface sites available

UCSD RASIRC

Self Limiting CVD – SiO_x/InGaAs(001)-(2x4)

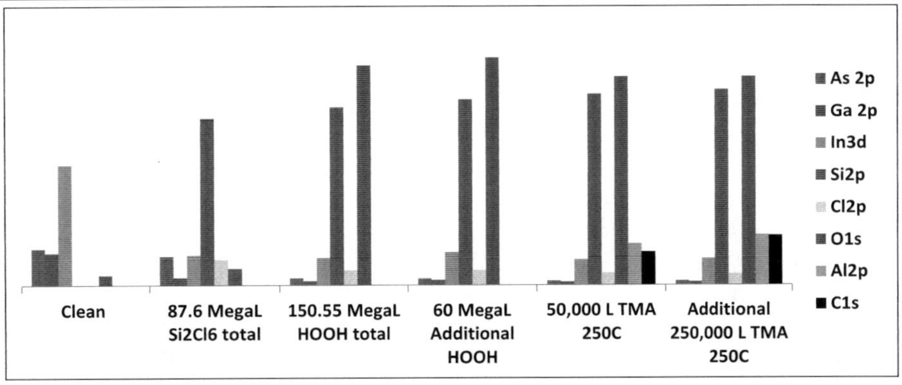

XPS Spectra collected at 30° glancing angle. Corrected XPS peak areas for clean (2x4), and following 87.6 MegaLangmuir Si_2Cl_6 at 350°C, 150.55 MegaLangmuir HOOH at 350°C, 60 MegaLangmuir additional HOOH at 350°C, 50,000 L TMA at 250°C, and additional 250,000 L TMA at 250°C. Si_2Cl_6 pulses at 2.5x10^-2 Torr . HOOH pulses at 1x10^-2Torr . TMA pulses at 5x10^-3 Torr .

- Anhydrous HOOH does not diffuse through SiOx layer and does not attack substrate
- TMA reacts and saturates on the saturated HOOH surface at 250C
  - SiO_x control layer on InGaAs nucleates high-K gate oxide growth

UCSD RASIRC

Self Limiting CVD – SiO_x/InGaAs(001)-(2x4)

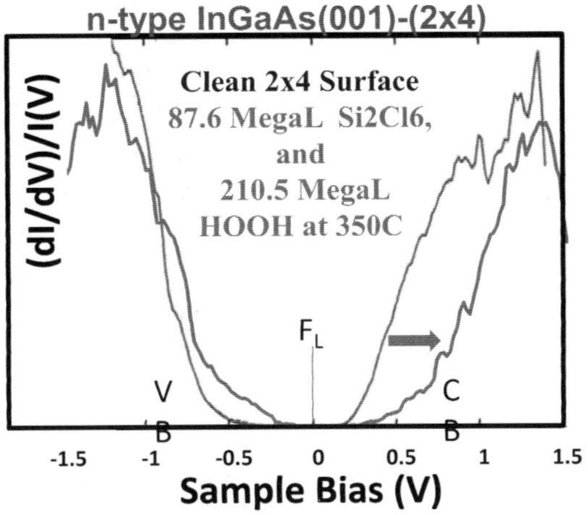

(Left) **STS on n-type InGaAs: clean n-type InGaAs (2x4), :** 87.6 MegaL Si_2Cl_6 and 210.5 MegaL anhydrous HOOH at 350C (ave of 7 curves)

Right: **Filled State STM image following: 87.6 MegaL Si_2Cl_6 and 210.5 MegaL anhydrous HOOH at 350C on n-type InGaAs(2x4)**

- Surface Fermi level shifts towards valence band from –OH and –O induced surface dipole

UCSD RASIRC

TMA Nucleation on SiO_x/InGaAs(001)-(2x4)

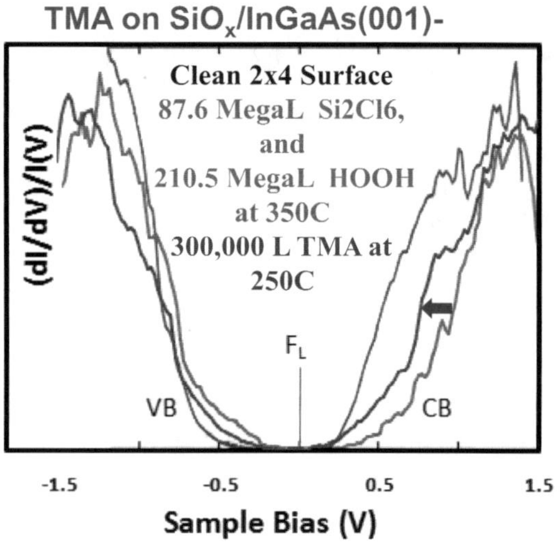

Left: **STS on n-type InGaAs: clean n-type InGaAs (2x4),** 87.6 MegaL Si2Cl6 + 210.5 MegaL HOOH at 350C (ave of 7 curves), **300,000 L TMA at 250C (ave of 11 Curves)**

- **Surface Fermi Level shifts back near conduction band following TMA dose at 250C**

UCSD RASIRC

Off-the-Shelf 30% HOOH at 350C

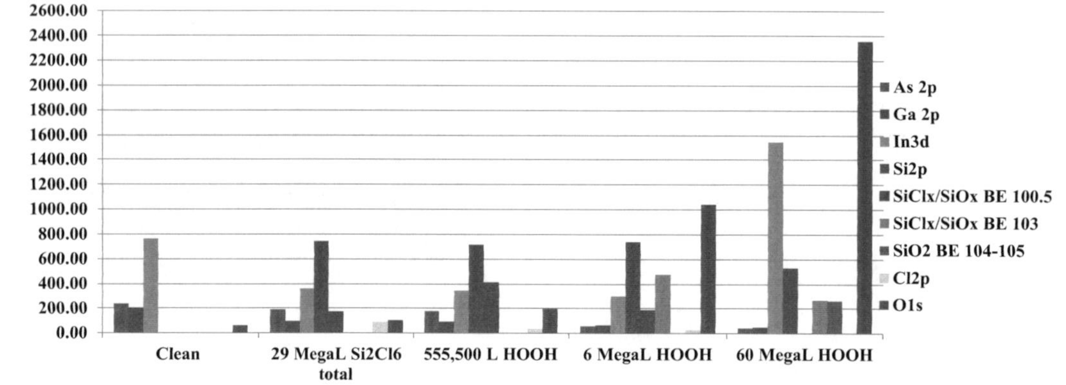

- **29 MegaLangmuir Si_2Cl_6** = Dosed 8 MegaLangmuir total of Si_2Cl_6 at 350°C at P=1x10^-2 Torr and then adjusted dosing conditions and dosed total 21 MegaLangmuir Si_2Cl_6 at P=2.5x10^-2 Torr at 350°C. **About 1.2 Monolayers of Si coverage**
- **555,500 Langmuir HOOH** = 500 L HOOH, additional 5,000 L HOOH, additional 50,000 L HOOH, and then 500,000 L HOOH all at 350°C (1x10^-3 Torr for 500 sec).
- **Additional 6 MegaL HOOH** = 3x10^-2 Torr for 200 seconds at 350°C. **60 MegaL HOOH**= 3x10^-2 Torr for 33 min at 350°C.
- See no shift in BE of In,Ga,As peaks following 6 MegaL HOOH dose.
- See noticeable shift of Si 2p peak to higher BE components: ~101.5 eV and ~103 eV
- Following 60 MegaL HOOH dose at 350C indium diffuses to surface – large InOx peak seen
- Going to establish standard Si2Cl6 dose to maintain same SiClx coverage every time

UCSD RASIRC

Low Temperature Passivation on SiGe(110) via Plasma Free Process

- Atomic H clean p-type $Si_{0.5}Ge_{0.5}(110)$
- Passivate $Si_{0.5}Ge_{0.}$ (110) dangling bonds with anhydrous hydrazine
- Maintain an electrically unpinned surface Fermi Level ready for subsequent functionalization
- Functionalize the surface and fabricate a MOSCAP with strong performance

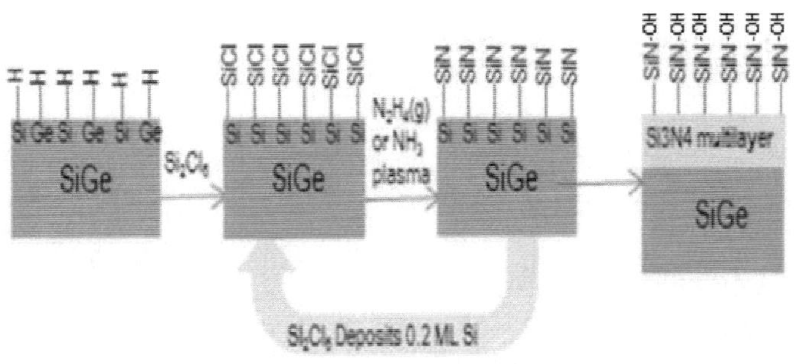

- **Subsequent doses of anhydrous hydrazine and hexachlorodisilane can further increase the amount of SiN_x on the surface**
- **Final treatment with HOOH can prepare the surface for High *k* deposition**

UCSD RASIRC

Passivation of SiGe(110) at 275C

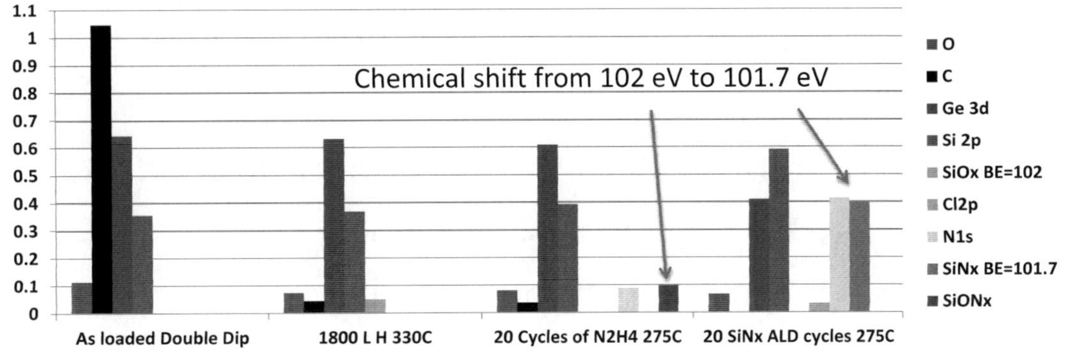

- **XPS corrected peak areas. $SiGe_{0.5}(110)$ p-type sample underwent ex-situ and atomic H cleaning procedures.**
- **20 cycles of N_2H_4at 275°C: Each cycle = 20 MegaL N_2H_4 (400MegaL total)**
- **20 SiN_x ALD cycles at 275°C: Each cycle = 13.5 MegaL Si_2Cl_6 followed by 20 MegaL N_2H_4**
- **See large SiN_x peak with no increase in oxygen signal with no evidence of contamination in chamber**
- **Higher binding energy silicon peak shift from 102 to 101.7 eV—consistent with Si_3N_4 film growth**
- **Estimated 3-4 monolayers of silicon nitride overlayer**
- **Growth rate of ~0.4 A / ALD cycle**

UCSD RASIRC

STS of Atomic H vs SiN_x Passivated Surface

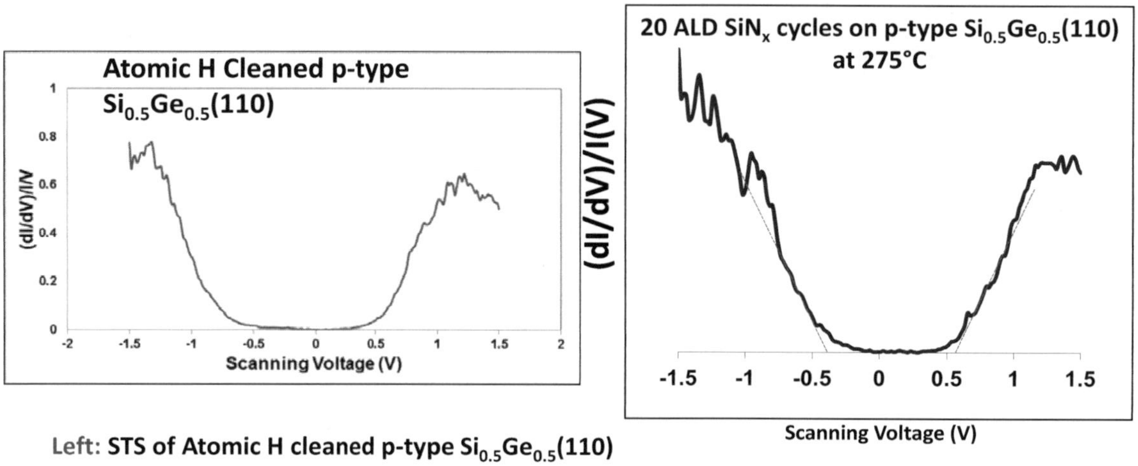

Left: **STS of Atomic H cleaned p-type $Si_{0.5}Ge_{0.5}(110)$**

Right: **STS following 20 SiN_x cycles at 275°C on $Si_{0.5}Ge_{0.5}(110)$ p-type sample following 1800L atomic H dose at 330°C and 400 MegaL hydrazine prepulse at 275°C**

- Atomic H cleaned surface Fermi level is at the midgap
- SiN_x on SiGe(110) surface still looks slightly more p-type with bandgapsize of ~0.8 – 0.9 eV

UCSD RASIRC

Conclusion

- Demonstrated stable delivery anhydrous hydrogen peroxide
- Demonstrated Si(O)OH passivation of InGaAs(001)
- Underlying InGaAs(001) is not damaged, Si oxidation with HOOH appears to be self-limiting
- Demonstrated low temperature nitride passivation of SiGe(110)
- MOSCAP studies are underway

RASIRC

Collaborators

- Prof. Andy Kummel – UC San Diego
  - Mary Edmonds
  - Steve Wolf
- Dan Tempel – Matheson Tri-Gas
- Hank Simidzu – Matheson Tri-Gas

RASIRC

Thank You

RASIRC

A Comparison of Sulfur-Based Chemistries to Passivate the (100) Surfaces of SiGe 25% and 75%

Zhonghao Zhang, Stacy L. Heslop, and Anthony J. Muscat

Department of Chemical and Environmental Engineering
University of Arizona, Tucson, Arizona, 85721

Contact Info:
muscat@email.arizona.edu
520-621-6162

THE UNIVERSITY OF ARIZONA

1

Prospect and Challenge on SiGe

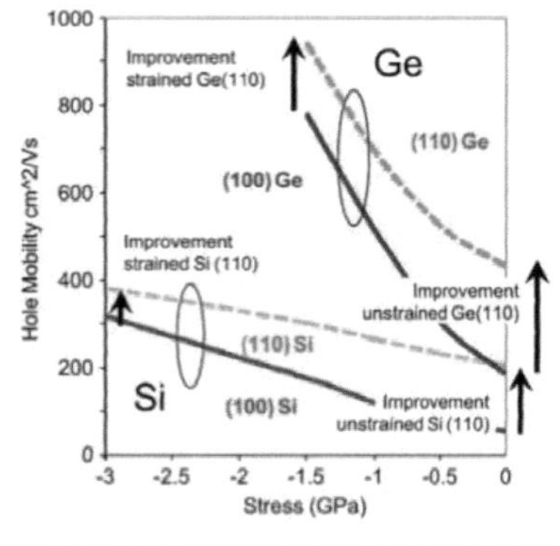

K.J. Kuhn, *IEEE Trans. Electron Devices*, **59**(7), 1813 (2012)

Prospect:
- Bandgap and carrier mobility of SiGe alloys can be tuned by varying Ge content.
- SiGe can be more easily integrated into Si processes.

Challenge:
Unstable Ge oxides formed on SiGe surfaces are detrimental to device performance.

Solution:

2

Passivate Ge and SiGe with Sulfur

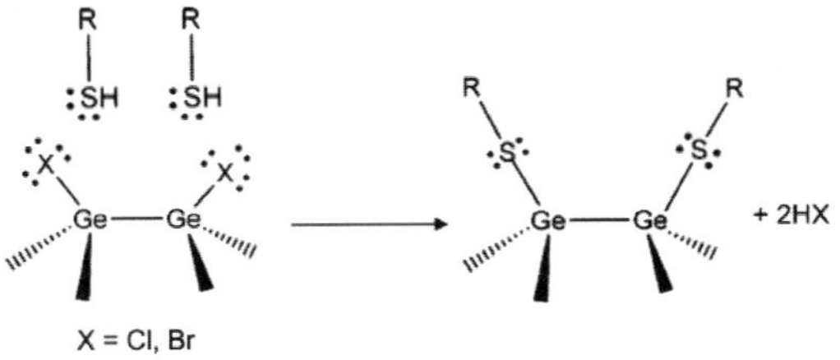

P. Ardalan *et al.*, *Langmuir*, **26**(11), 8419(2010)

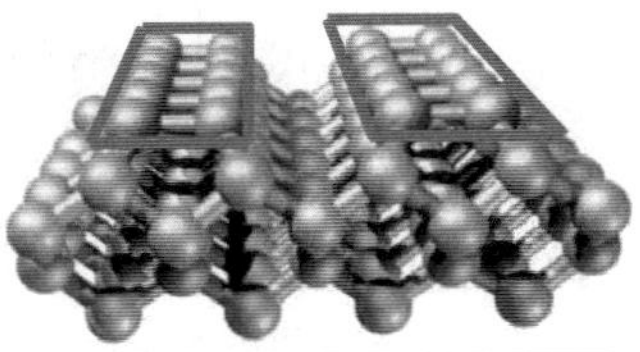

Ge (100) Surface

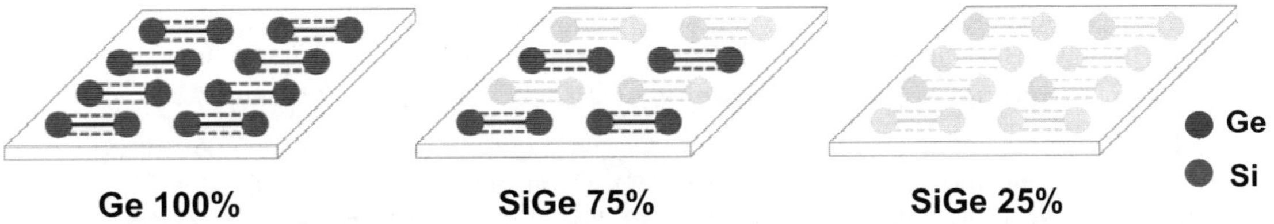

Goal: Explore sulfur chemistries to passivate SiGe 25% and 75% surface.

- Deposit sulfur on SiGe 25% and 75% (100) surface .
- Avoid surface re-oxidation during chemical treatment.

3

SiGe Cleaning and Passivation Strategy

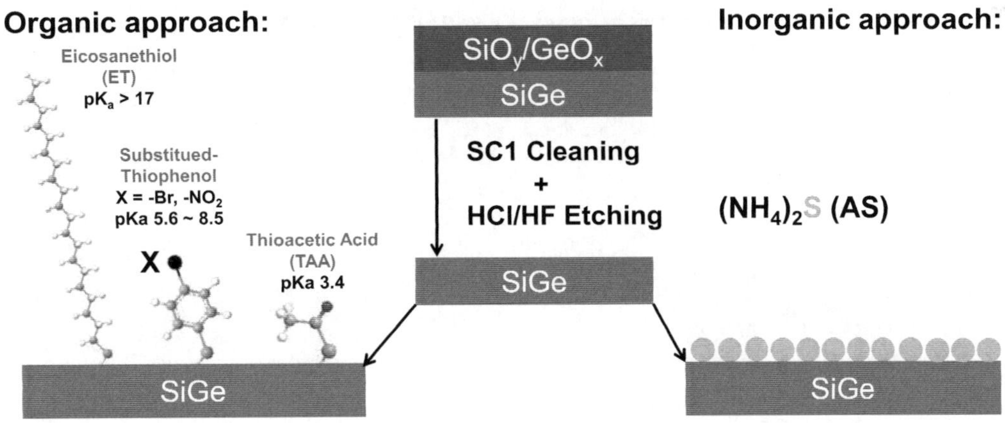

- **Long-Chain Alkylthiols:**
  Can potentially form a self-assembled monolayer on surface.
- **Low pK_a Thiols:**
  Generate higher concentration of active sulfur species in solution.

Good compatibility with aqueous process.

4

Experimental Procedures

SC1 Cleaning (25 °C): Immerse sample in SC1 solution (H_2O_2:NH_4OH:Water = 1:1:500) with stirring for 2 min.

Rinse (25 °C): Immerse sample in ultra pure water for 1 min without stirring. Dry slowly with N_2 afterward.

HF/HCl Clean (25 °C): Immerse sample in HF/HCl solution (HF:HCl:Water = 1:3:300) with stirring for 5 min. No rinse or dry afterward. Immediately immerse sample into passivation solution.

Passivation Treatment (25 °C): Immerse sample in passivation solution with stirring for desired time (24 hr in ET solution, 20 min in AS solutions and low pK_a thiol solutions).

Rinse (25 °C): Rinse the sample in water (for ammonium sulfide passivation) or ethanol (for thiol passivation) for 30 s.

5

Passivation on SiGe 25% Surface (Inorganic Approach)

SiGe 25% SiGe 75%

6

Effect of SC-1 and HF/HCl Cleaning

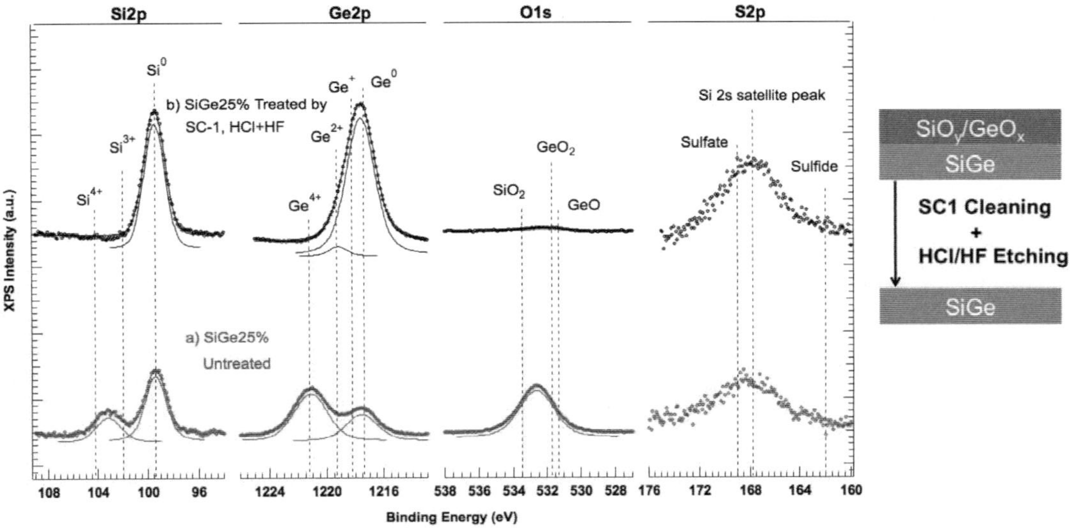

- Surface oxide was successfully removed by SC-1 cleaning followed by HF/HCl cleaning.

7

$(NH_4)_2S$ (AS) Treatment (20 min) on SiGe 25%

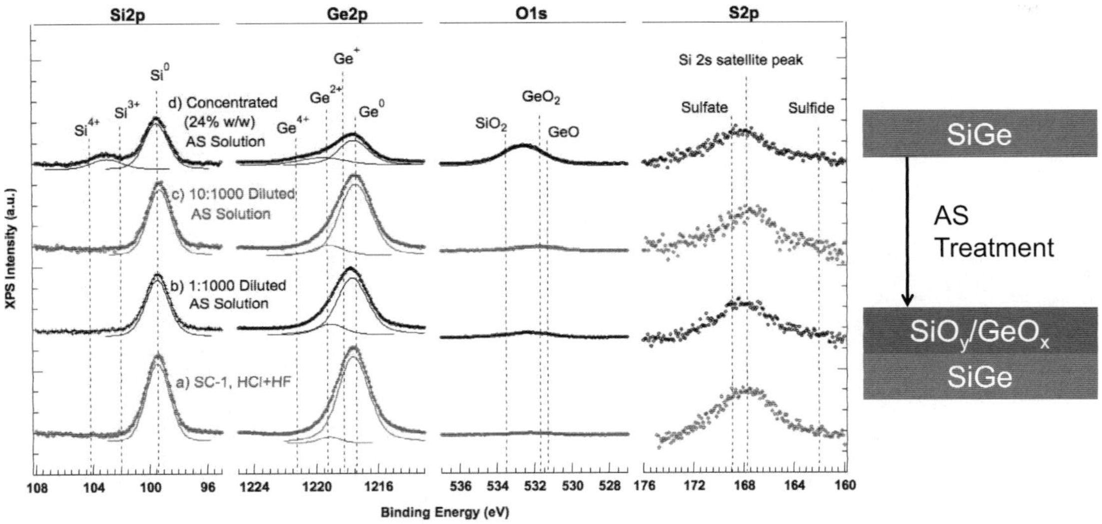

- No covalent Ge-S bond formation was observed.
- Instead of passivation, the surface was re-oxidized during treatment.

8

$(NH_4)_2S$ Treatment with Elemental S Added (20 min) on SiGe 25%

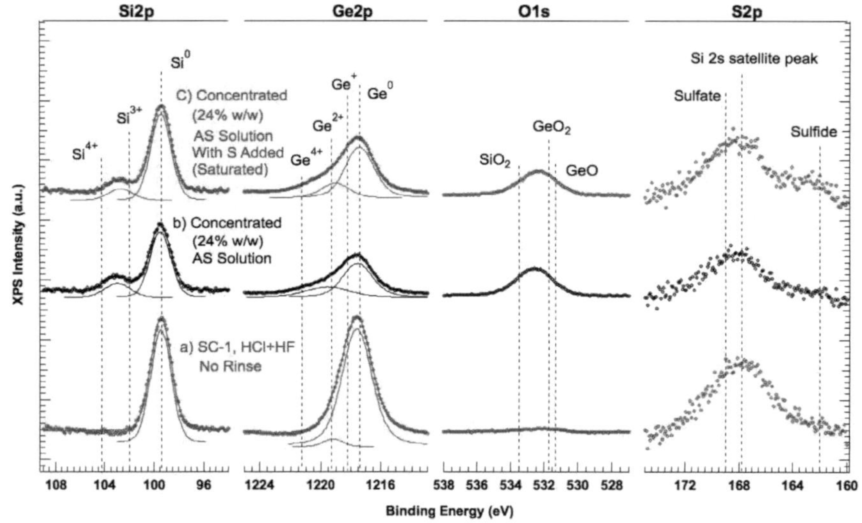

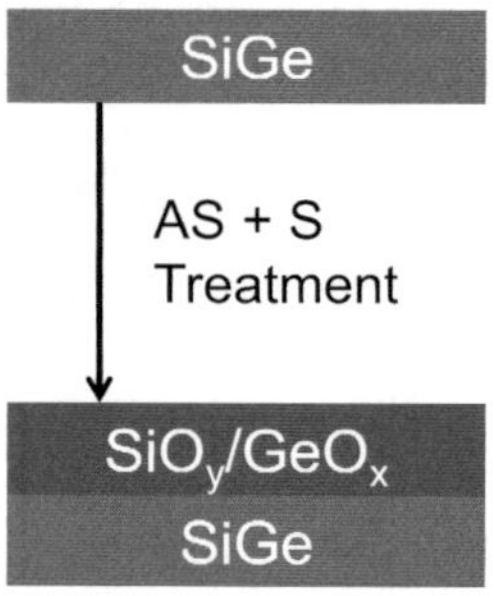

- Very weak sulfide peak was observed in S2p region.
- The surface was re-oxidized during treatment

9

$(NH_4)_2S$ Treatment with Acid Added (20 min) on SiGe 25%

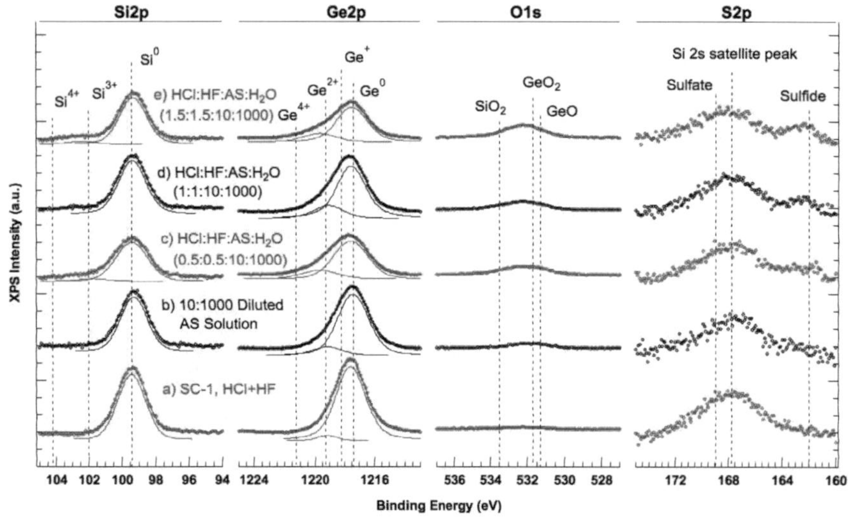

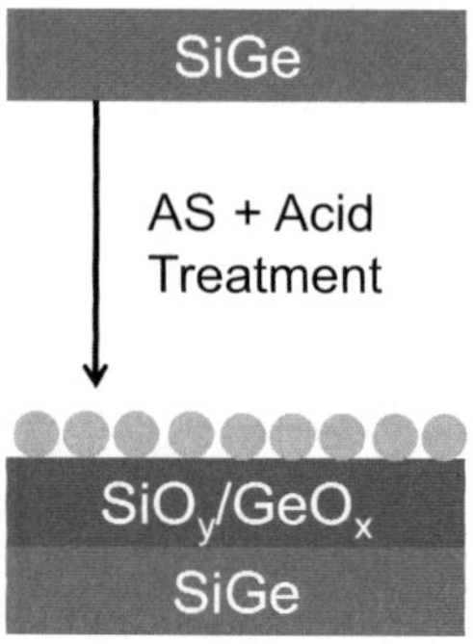

- Sulfide bond formation was observed.
- The surface was re-oxidized during treatment.

10

Summary of Inorganic Approach on SiGe 25%

Experiment	Final Effect	
	Sulfur Deposition	Surface Re-oxidation
$(NH_4)_2S$ Only	No	Yes
$(NH_4)_2S$ + Elemental S	Minimal	Yes
$(NH_4)_2S + H^+$ (Acid)	Yes	Yes

Sulfur Deposition Hypothesis:

$$(NH_4)_2S + 2H^+ \longrightarrow H_2S + 2(NH_4)^+$$

Surface Re-oxidation Hypothesis:

- **Oxidation of $(NH_4)_2S$ Solution During Storage**

$$2(NH_4)_2S + 2O_2 + H_2O \longrightarrow (NH_4)_2S_2O_3 + 2(NH_4)OH$$

- **Hypothesis: NH_3 Promotes Si Oxidation**

$$(NH_4)_2S \longrightarrow 2NH_3 + H_2S$$

11

Future Work of Inorganic Approach on SiGe 25%

- **Use aqueous H_2S (prepared by bubbling H_2S into water).**
- **Purify ammonium sulfide to remove $(NH_4)_2S_2O_3$:**

$$S_2O_3^{2-} + 2H^+ \longrightarrow \text{S(Solid Precipitate)} + SO_2 + H_2O$$

$$SO_2 + 2S^{2-} + 4H^+ \longrightarrow \text{3S(Solid Precipitate)} + 2H_2O$$

- **Use alternative sulfide molecules.**

12

Passivation on SiGe 25% Surface (Organic Approach)

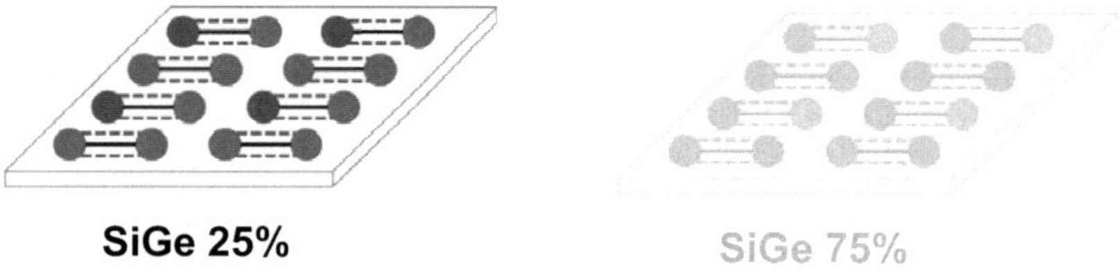

13

Long Chained Alkylthiol Treatment (24 hr) on SiGe 25%

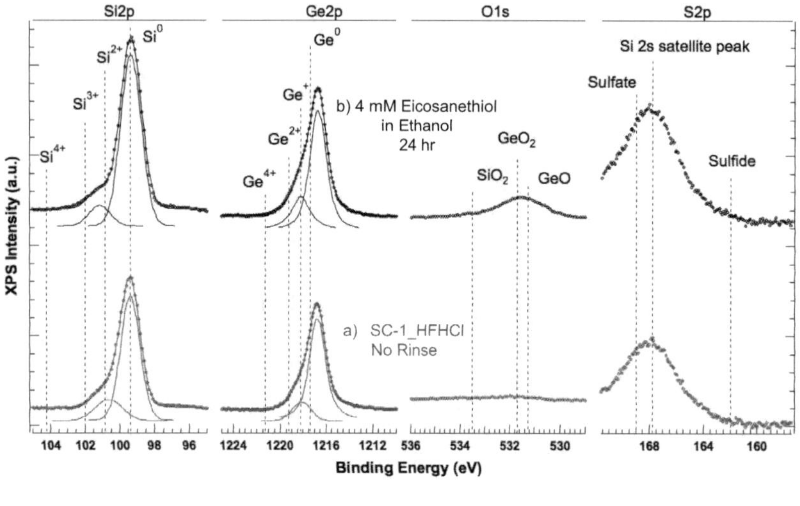

Eicosanethiol (ET)
$pK_a > 17$

- ET did not deposit on SiGe 25% surface.
- Surface was re-oxidized during treatment.

14

S-H Bond Dissociation and pK_a of Thiol Molecules

R-SH ⟶ R-S^- + H^+

(Thiol Molecule) — Active Species for S Deposition

$$K_a = \frac{[RS^-][H^+]}{[RSH]} \qquad pK_a = -log_{10}K_a$$

- **Thiol molecules with lower pK_a values have higher tendency for S-H bond dissociation.**
- **Higher tendency of S-H bond dissociation can potentially facilitate Ge-S bond formation.**

15

Low pK_a Thiol Treatment (20 min) on SiGe 25%

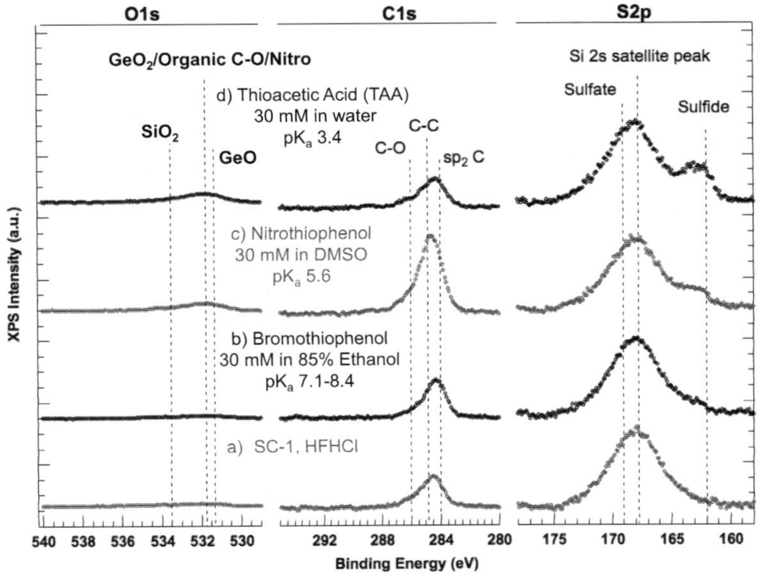

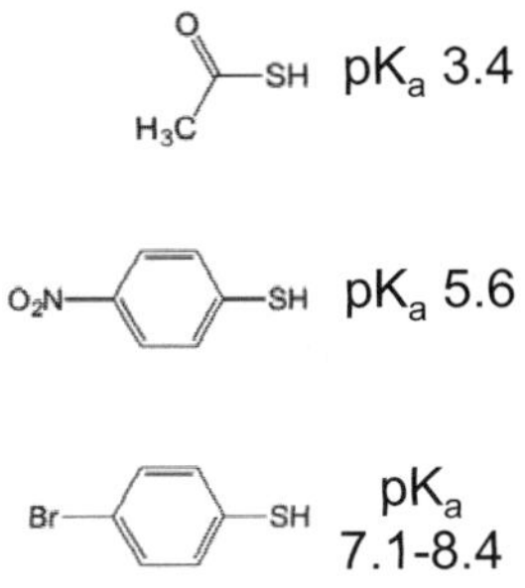

- Thiol molecules with lower pK_a values deposited more S on SiGe 25%.
- Oxygen was observed after treatment.

16

Where did the Oxygen Come From

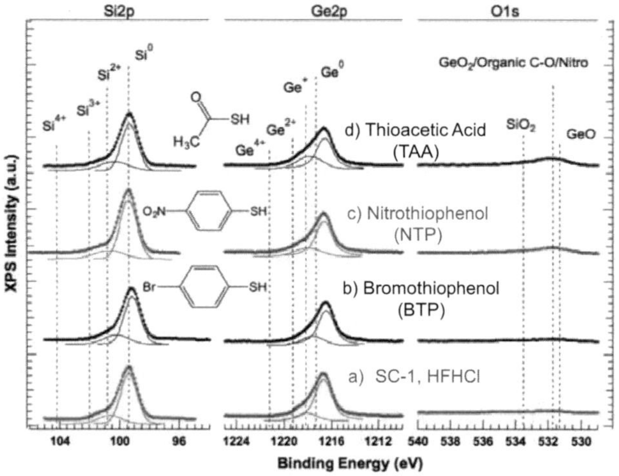

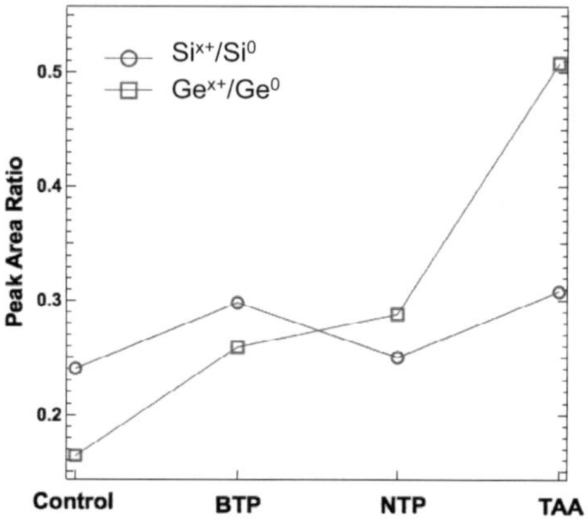

- Oxygen mainly came from oxygen containing groups in thiol ligands.

17

Surface Re-Oxidation on TAA Treated SiGe 25%

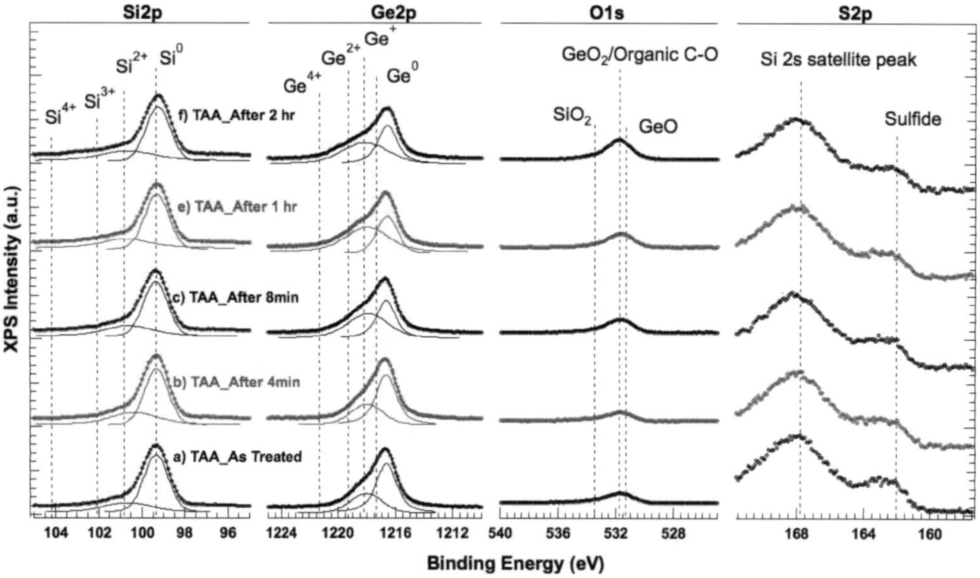

- Surface was re-oxidized over time after treatment.
- The short chain of TAA cannot protect the surface.

18

Conclusion of Organic Approach on SiGe 25%

- Long alkyl chain eicosanethiol (ET) did not deposit on SiGe 25% surface even after prolonged (24 hr) treatment, possibly due to its high pK_a value.
- Thiol molecules with lower pK_a values more efficiently deposited onto SiGe 25% surface.
- Thioacetic acid (TAA) did not protect SiGe25% surface against oxidation after treatment, possibly due to its short carbon chain.

19

Future Work of Organic Approach on SiGe 25%

- Study surface re-oxidation on nitrothiophenol (NTP) treated SiGe 25% surface.
- Explore low pK_a thiol molecules with long alkyl chain, which can both effectively deposit on SiGe 25% surface and potentially form thicker and/or denser organic layer.

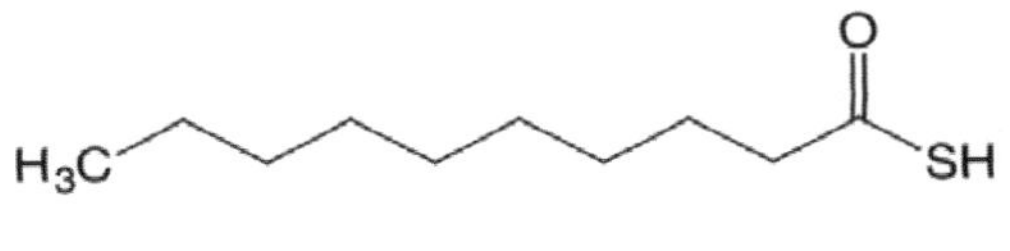

20

Passivation on SiGe 75% Surface (Organic Approach)

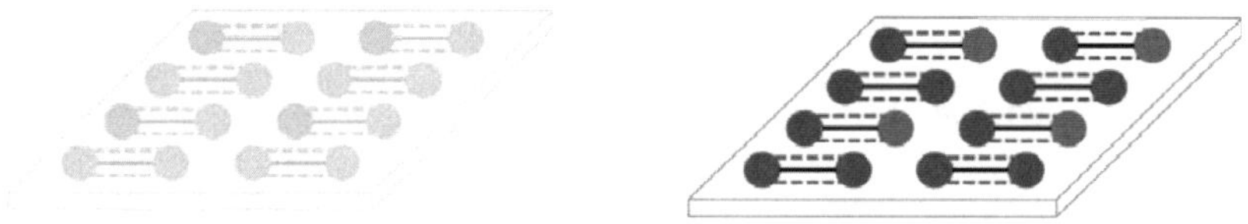

21

Organic Thiol Treatment on SiGe 75%

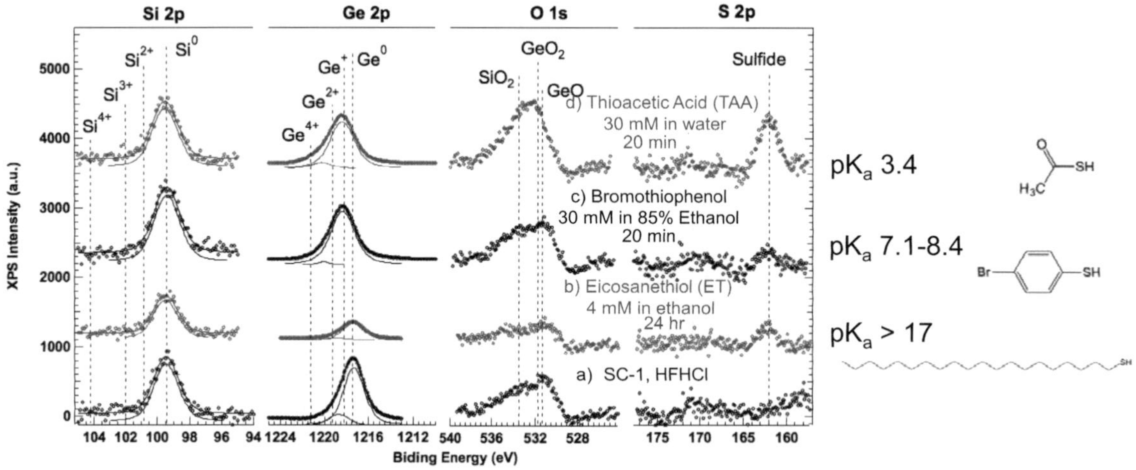

- ET successfully deposited on SiGe 75% surface.
- Lower pK_a thiols deposited more sulfur on SiGe 75% surface.

22

Organic Thiol Treatment on SiGe 75%

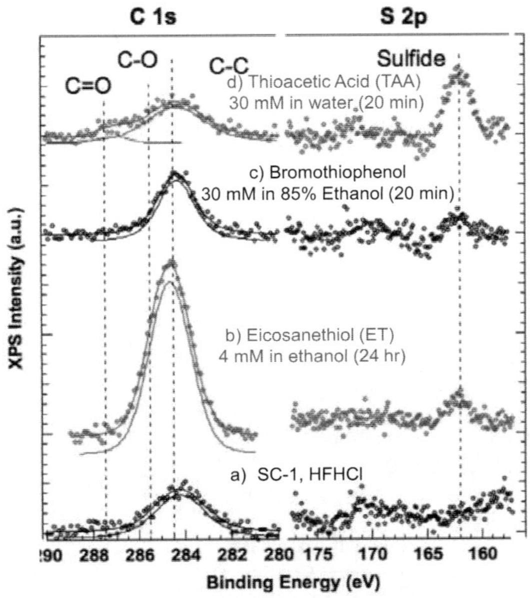

TAA
pK_a 3.4

BTP
pK_a 7.1-8.4

ET
pK_a > 17

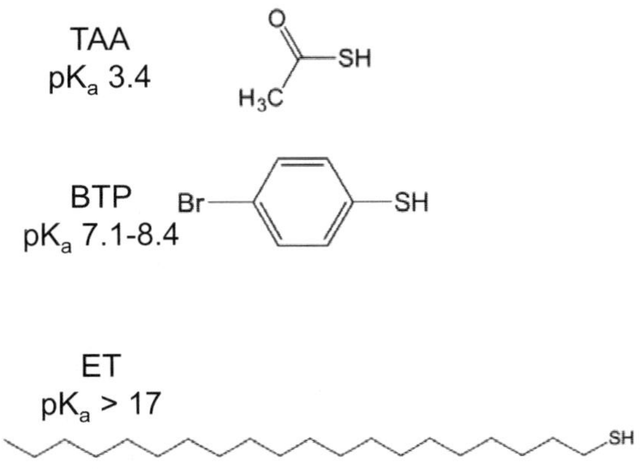

- ET deposited the thickest and/or most dense organic layer.

23

Conclusion and Future Work of Organic Approach on SiGe 75%

Conclusion:

- Thiol molecules with lower pK_a values deposited more sulfur on SiGe 75% surface.
- Although eicosanethiol did not deposit the most sulfur on SiGe 75% surface due to its high pK_a, it deposited the highest amount of organic layer on SiGe 75% surface due to its long alkyl chain. This can potentially provide the best protection on SiGe 75% surface against oxidation.

Future Work:

- Study SiGe 75% surface re-oxidation after ET treatment.

24

Acknowledgement

Muscat Research Group

- Yissel Contreras
- Lauren Peckler
- Pablo Mancheno
- Shawn Miller
- Adam Hinckley
- Gabriela Diaz
- Jimmy Hackett
- Lance Hubbard
- Shuo Yang
- Ruoyun Xiao

Lam Research Corporation

- Dr. Nerissa Draeger
- Dr. Reza Arghavani

THE UNIVERSITY OF ARIZONA

Lam RESEARCH

25

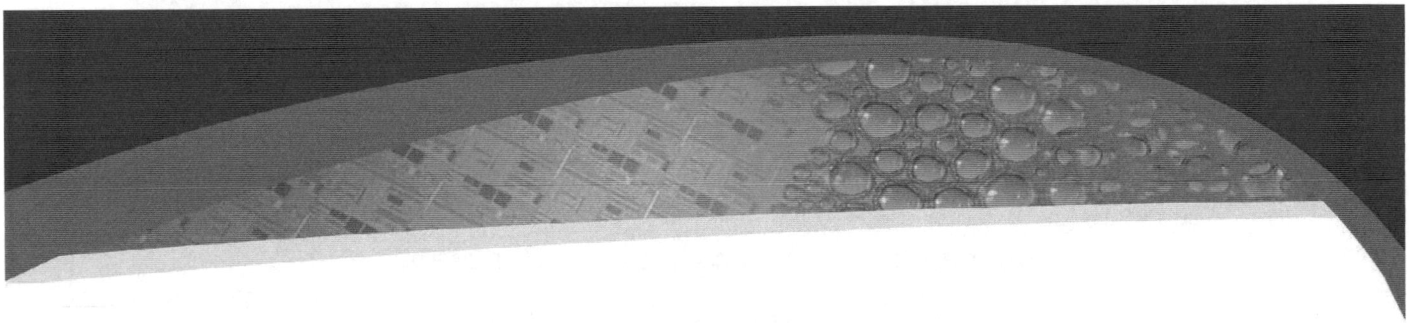

2016 Surface Preparation and Cleaning Conference

Please contact the author for a copy of this presentation:

shariq@us.ibm.com

High Temperature Water as a Clean and Etch of SiO_2 Films

Joshua Barclay, Oseoghaghare Okobiah,
Tina Sengphanlaya, Rick Reidy
University of North Texas

UNT UNIVERSITY OF NORTH TEXAS

Rationale

- High temperature water (HTW) has demonstrated some interesting capabilities etching SiN (SPCC 2015)
- Given its enhanced reactivity, we were interested in the selectivity of HTW for SiN versus SiO_2

UNT UNIVERSITY OF NORTH TEXAS

Outline

- Description of Experimental Apparatus and Conditions
- Characterization of SiO_2 wafers to examine the behavior following exposure to HTW
  - Ellipsometry—determine thickness changes
  - FTIR—changes in the chemistry and thickness
  - Profilometry—define boundaries and approximate changes in thickness

UNT UNIVERSITY OF NORTH TEXAS

Experimental Apparatus

Heating mantle and reactor

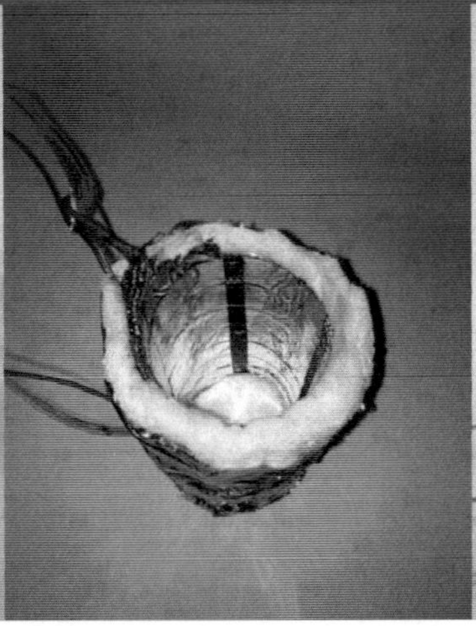

Reactor can be easily removed from mantle

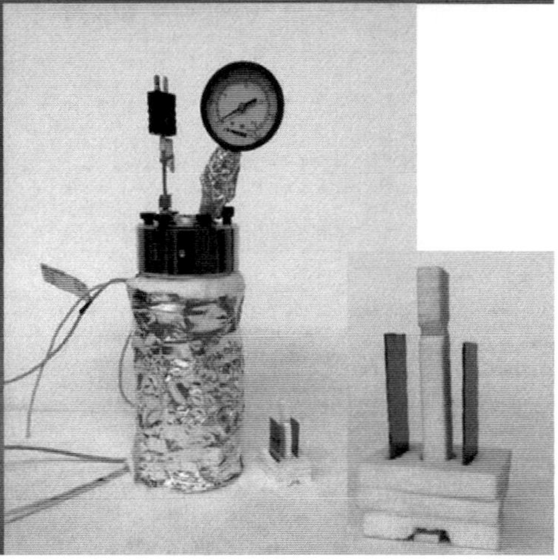

Reactor and sample holder

UNT UNIVERSITY OF NORTH TEXAS

Experimental Conditions

- DI water heated to 98°C
- SiO_2/Si samples placed in the reactor and reactor closed (~45sec)
- Temperature set to 160°C for 5, 10, and 20 minutes (heat up takes ~17min)
- Reactor removed from heat and quenched in water for 2 minutes
- Reactor opened (~45sec)
- Sample stand is removed
- Samples rinsed with DI water and allowed to dry.

UNT UNIVERSITY OF NORTH TEXAS

What are the Experimental Conditions?

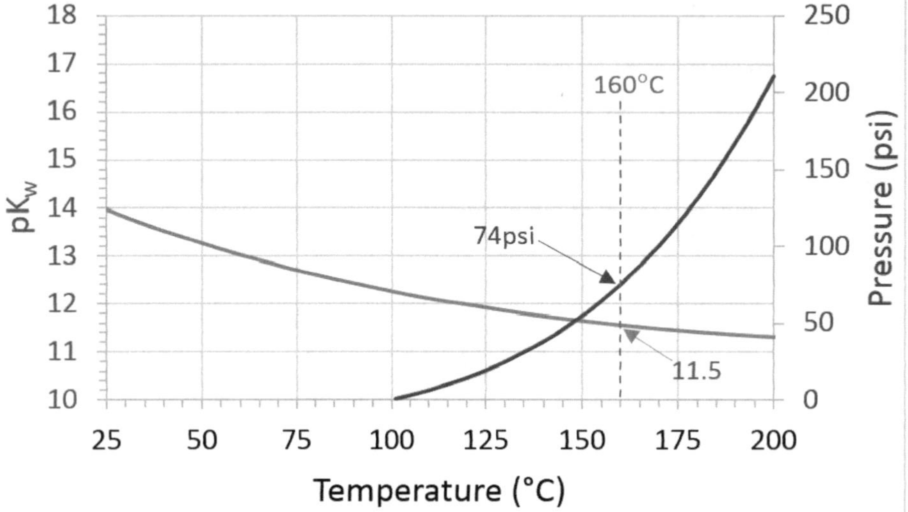

- $pK_w = -\log[K_w] = -\log[H^+][OH^-]$: at 160°C, water has more ionic species than at room temp

UNT UNIVERSITY OF NORTH TEXAS

FTIR of oxide exposed to 160°C steam

- FTIR (trans) of the top of the sample (above the water level)
- Shows SiO_2 thickness increasing with longer exposures

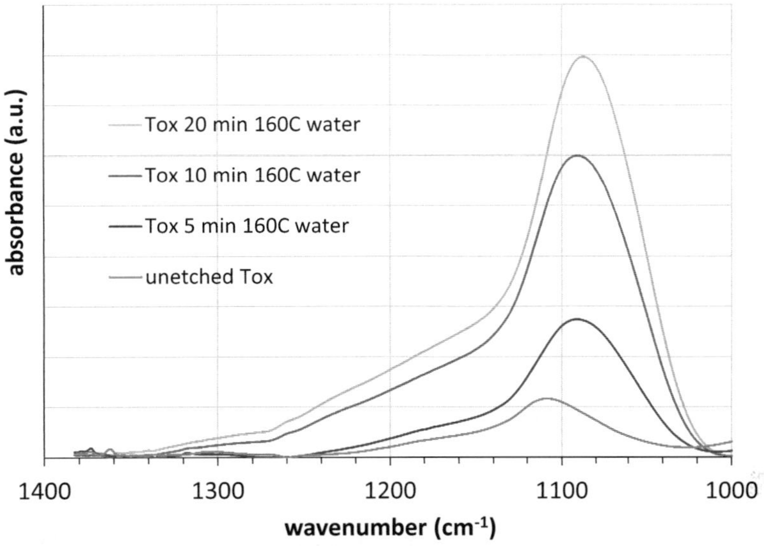

UNT UNIVERSITY OF NORTH TEXAS

Ellipsometry: oxide exposed to 160°C steam

- Tox oxidized and ↑ thickness
- Δthickness after 5 minutes → ~13nm
- Samples were too roughened after 10 and 20 minutes to model

UNT UNIVERSITY OF NORTH TEXAS

Profilometry Profile: from protected to exposed to 160°C Water

- Region exposed to HTW showed etching
- Etch depth increased with time

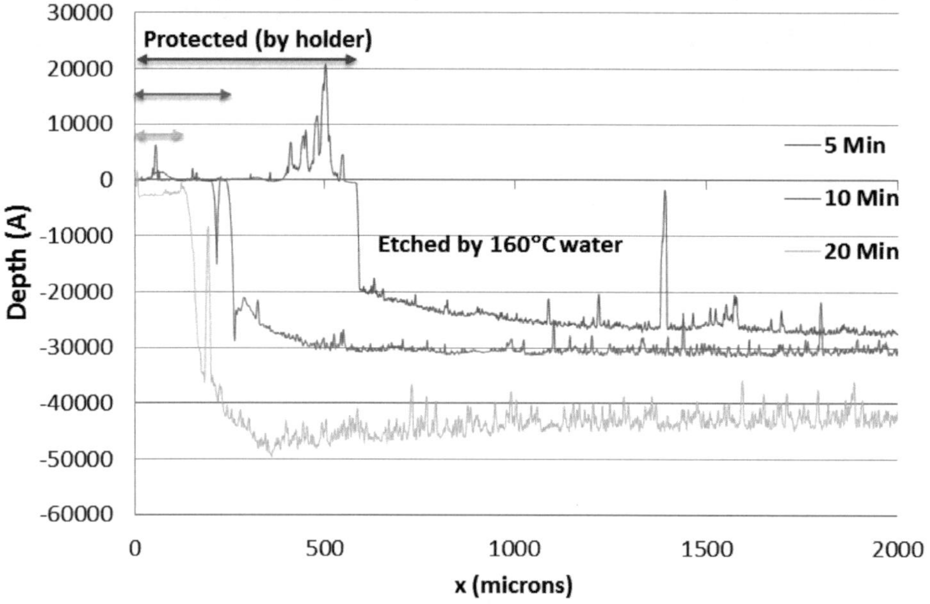

UNT UNIVERSITY OF NORTH TEXAS

Profilometry Profile: from 160°C Steam to 160°C Water region

- More gradual profile change compared to film regions exposed to HTW

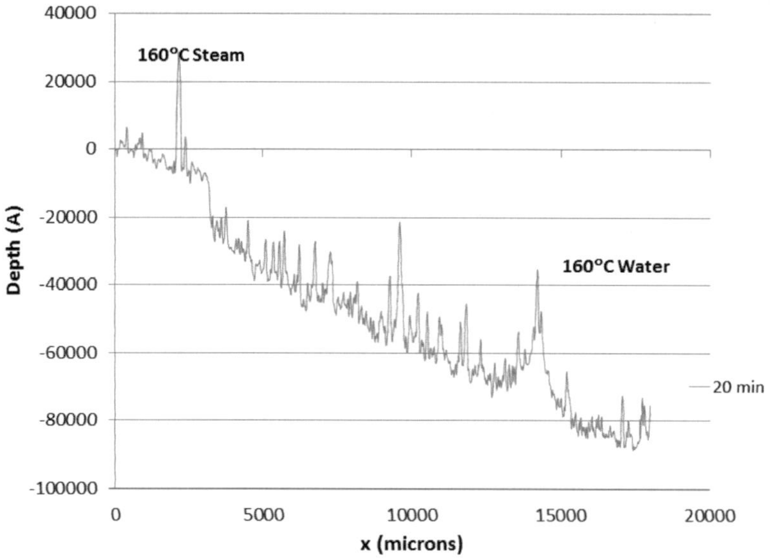

UNT UNIVERSITY OF NORTH TEXAS

Previous work on SiO_2 in HTW

- Solubility increases ~6x at 160°C
- This solubility is sufficient to account for the loss of >1000 nm of oxide

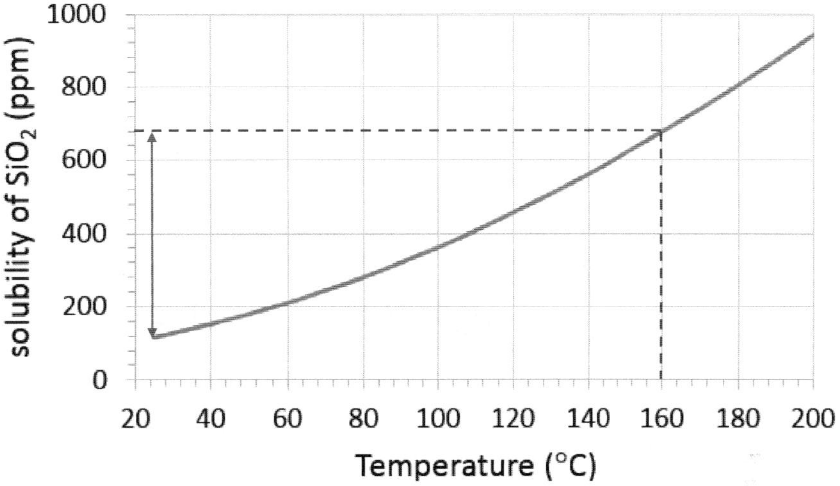

RO Fournier and JJ Rowe, Amer. Miner, 62 1052-1056 1977

UNT UNIVERSITY OF NORTH TEXAS

Ionization of Silica in HTW

- Ionization of silica at 160°C: ~10x greater than room temperature

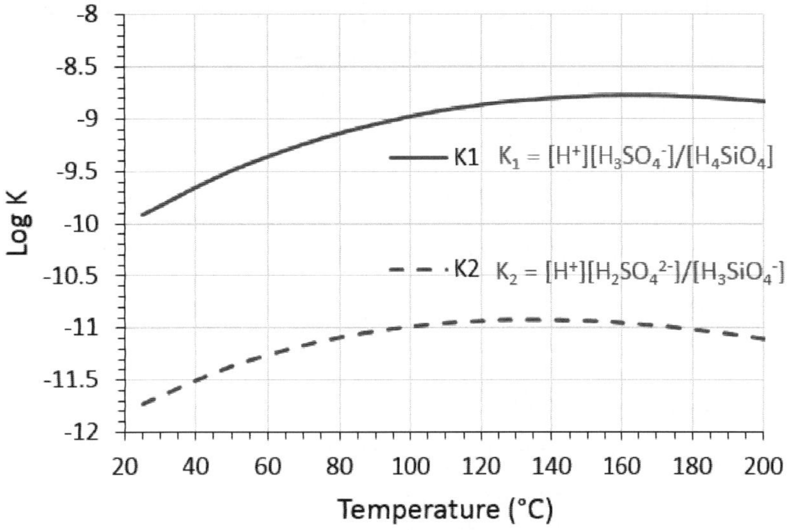

Arnorsson, Sigurdsson, Svavarsson, 1982

UNT UNIVERSITY OF NORTH TEXAS

What is Happening?

- Oxide Growth on top of sample—steam oxidation of silicon
- Regions exposed to HTW—water is more reactive (higher: pKw, solubility, ionization; lower: viscosity and surface tension)→ etching
- Because the heating time is relatively short, most of the reaction is occurring at 160°C

UNT UNIVERSITY OF NORTH TEXAS

Summary

- The oxide sample appears to be oxidized and etched in the same reactor at the same time.
  - Steam→ oxidation
  - HTW→ etching
- Steam is like other oxidizing steams
- HTW is different from T < 100°C water
- Interesting but reaction may not be easy to control like SiN
- Future work:
  - Confirm etching mechanism
  - Determine etching rate as a function of temperature
  - Determine oxidation rate as a function of temperature

UNT UNIVERSITY OF NORTH TEXAS

Acknowledgements

- Sematech
- UNT Center of Advanced Research and Technology (CART)
- UNT Advanced Materials and Manufacturing Processes Institute

UNT UNIVERSITY OF NORTH TEXAS

Hilltop Economics

A Consensus Outlook of the Global Economy

And What It Implies for Semiconductors

Surface Preparation and Cleaning Conference (SPCC)
April 2016

Duncan Meldrum
Chief Economist, Hilltop Economics, LLC

duncan.meldrum@hilltopeconomics.com
610-709-7951

**A global outlook based on Consensus Economics, Inc. CONSENSUS FORECASTS®*

Hilltop Economics

Outline

- Why the semiconductor industry should care about the broader economy (real GDP)
- Global outlook and how we got here
- Semi MSI outlook (ESF March 2016)
- Risks in the economy

Hilltop Economics

Semiconductor Industry Measure: SEMI.ORG's MSI

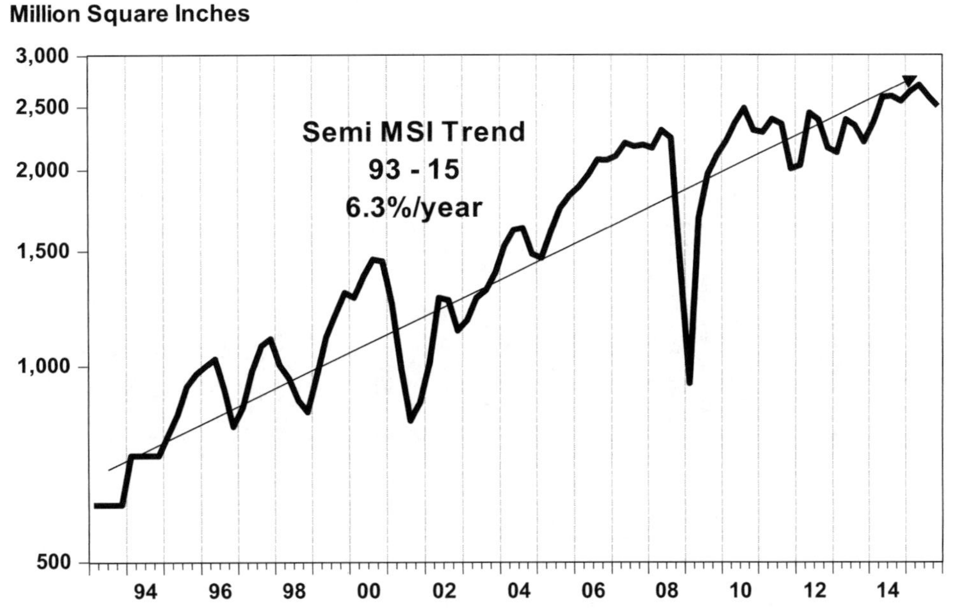

Duncan Meldrum
SPCC April 19-20 2016

3

Hilltop Economics

MSI = Real GDP Since the Recession

Chips are so pervasive in the economy, demand is now driven by the economy

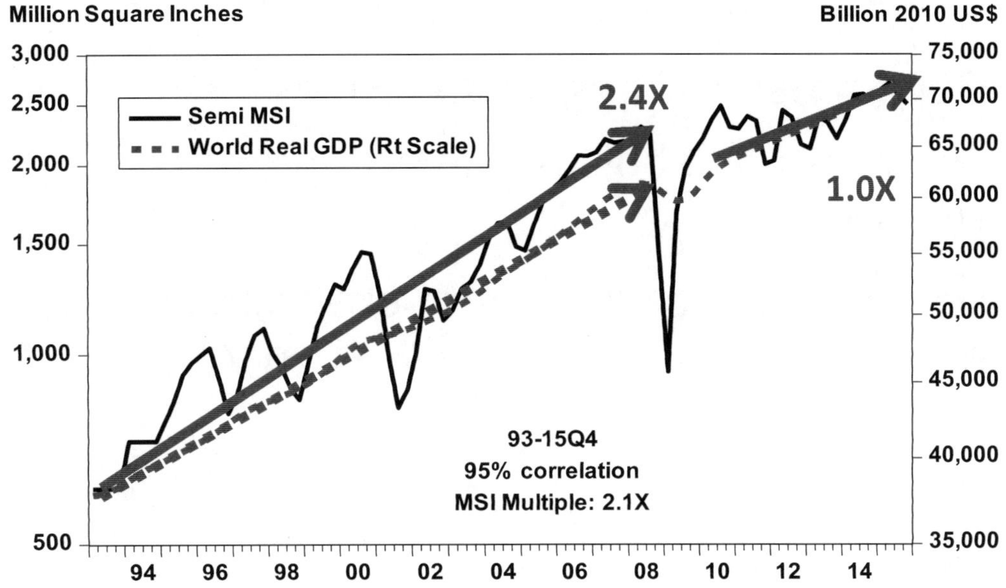

Total real (inflation-removed) spending on finished goods and services by businesses, governments and consumers in 85 of the largest economies: highly correlated with Semi MSI

Duncan Meldrum
SPCC April 19-20 2016

4

Hilltop Economics

World Real GDP Growth Below Par 2012 - 2017

World* Par ~3.2%/Year

2014: +2.6% 2015: +2.4% 2016: +2.4% 2017: +2.8%

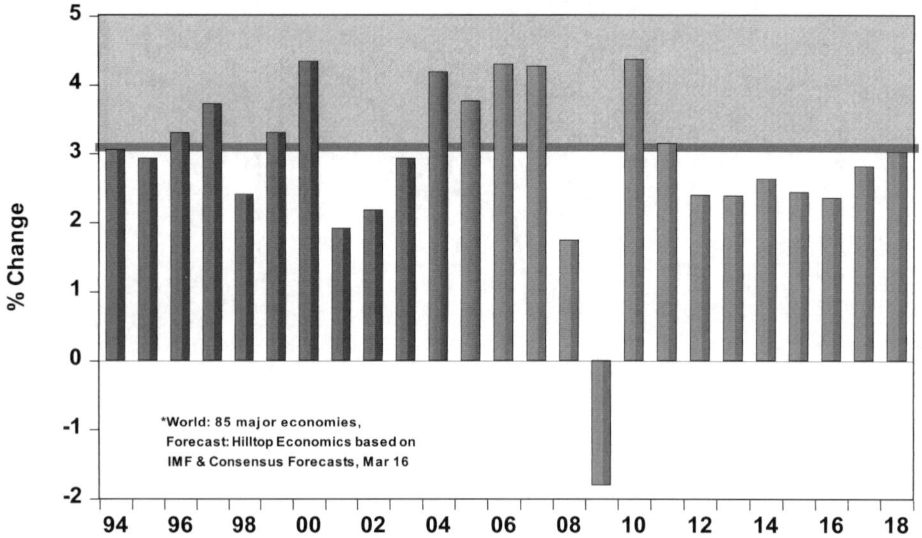

Duncan Meldrum
SPCC April 19-20 2016

5

Hilltop Economics

Why Slow and Will It Continue?

- Governments changed the "Rules"
- Consumer and business aversion to risk skyrocketed
- Demand did not recover to pre-crisis growth rates once the crisis passed (by 2010)

Duncan Meldrum
SPCC April 19-20 2016

6

Hilltop Economics

Why Slow and Will It Continue?-I

- Governments changed the "Rules"
  - Economic policies & regulations to "fix" the financial system
  - Monetary policy: low interest rates, oceans of money
  - Economic policies to cope with demand collapse
  - Creating high economic policy uncertainty as policies & regulations continue to adjust
- Expect policies to SLOWLY stabilize but…
- Economic policy uncertainty will remain high until the preponderance of polices stabilize
- Watch: Economic Policy Uncertainty Indexes

Duncan Meldrum
SPCC April 19-20 2016

7

Hilltop Economics

Unprecedented Period of Uncertainty After 2008

Economic Policy Uncertainty

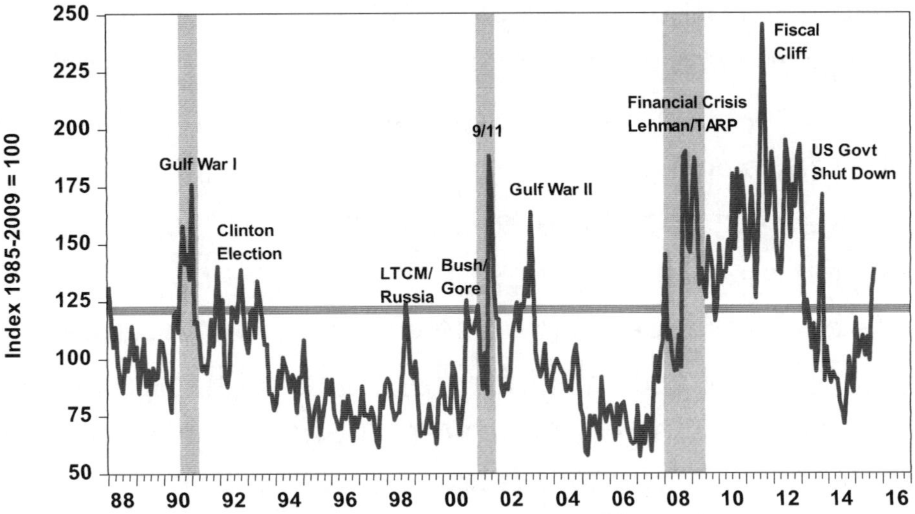

Source: Scott Baker, Nicholas Bloom and Steven J. Davis at www.PolicyUncertainty.com
Shaded areas indicate official recessions

Duncan Meldrum
SPCC April 19-20 2016

8

Hilltop Economics

Uncertainty In Key Economies Remains A Dampening Factor on Growth: Expect Very Slow Improvement

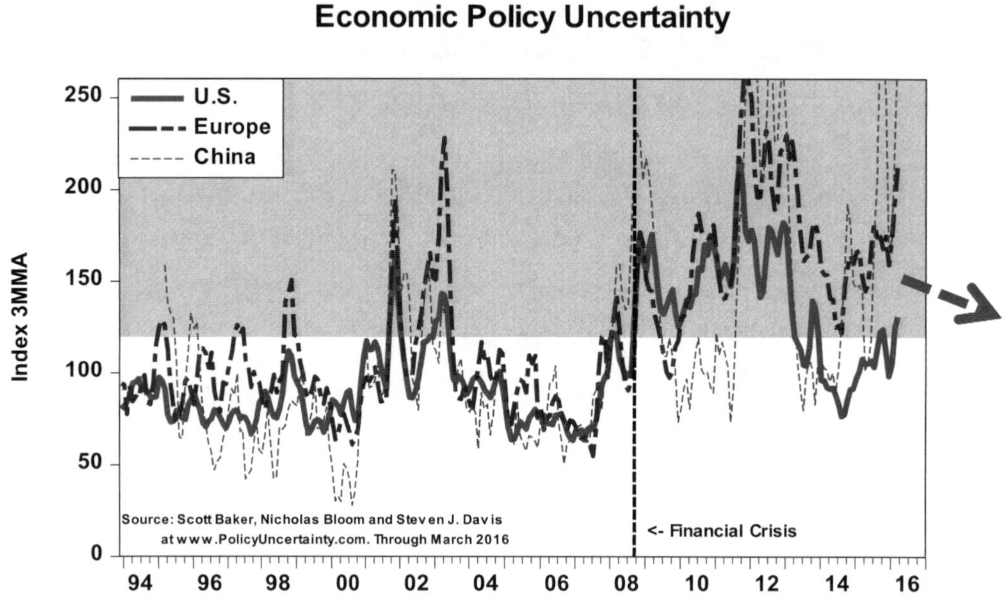

Duncan Meldrum
SPCC April 19-20 2016

9

Hilltop Economics

Why Slow and Will It Continue?- II

- Consumer and business aversion to risk skyrocketed
  - Stopped borrowing to spend
  - Paid down debt even as governments took on debt
  - Raised cash
- Risk aversion should recede slowly:
  - As jobs recover
  - As businesses begin to perceive growth opportunities
- Forecast Risk: potential for permanent change in consumer, business and government behavior with all the negative feedbacks (e.g. the Depression Generation, part II)

Duncan Meldrum
SPCC April 19-20 2016

10

Hilltop Economics

U.S. Household Debt Drops as Government Rises

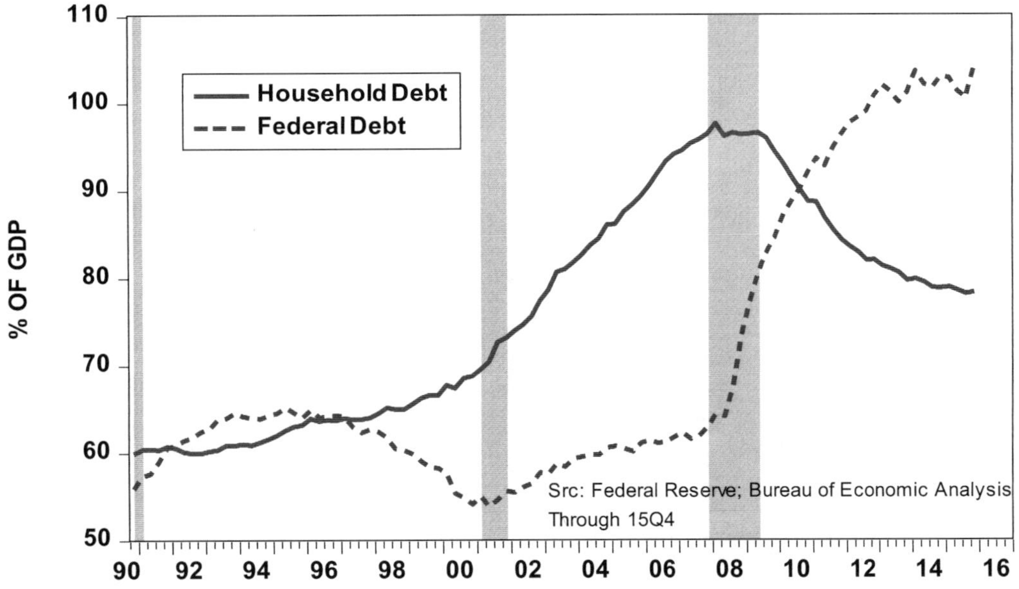

Duncan Meldrum
SPCC April 19-20 2016

11

Hilltop Economics

Why Slow and Will It Continue? - III

- Demand did not recover to pre-crisis growth rates
  - Consumers remain cautious
  - Investment remains very weak
    - Slow real GDP + low inflation = weak topline growth
    - High uncertainty
    - Excess capacity (until output gap closes all the way)
  - Cheap capital (low interest rates) not enough to stimulate investment or consumption.
    - Stimulates M&A activity
- Demand is forecast to recover by 2017-2018

Duncan Meldrum
SPCC April 19-20 2016

12

Hilltop Economics

What the Global Economy Can Produce: Much Slower Growth Since the Financial Crisis

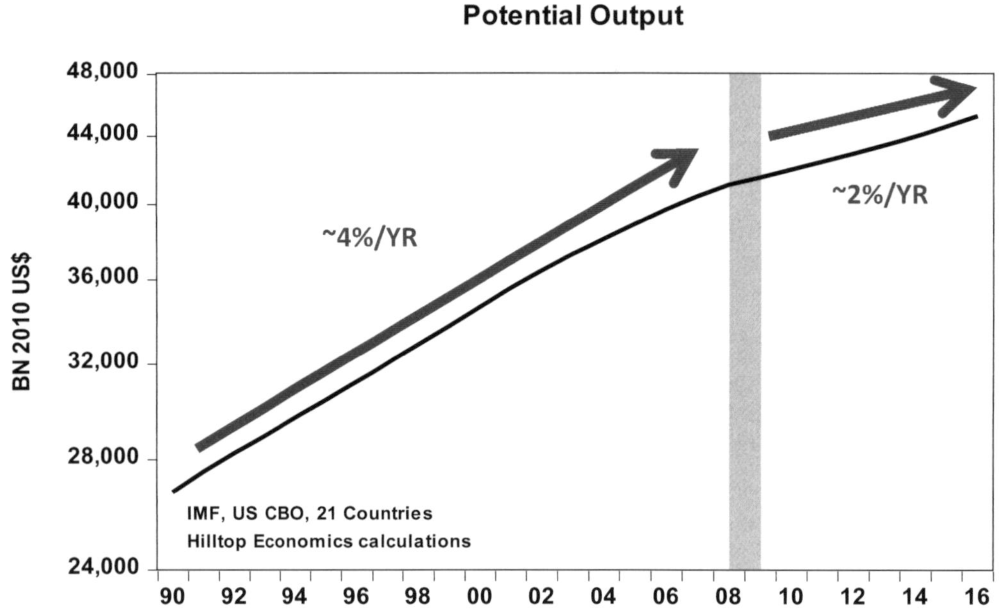

Duncan Meldrum
SPCC April 19-20 2016

13

Hilltop Economics

Slower Potential Growth Has Closed the Output Gap Giving Support to the Forecast for Investment in 2017

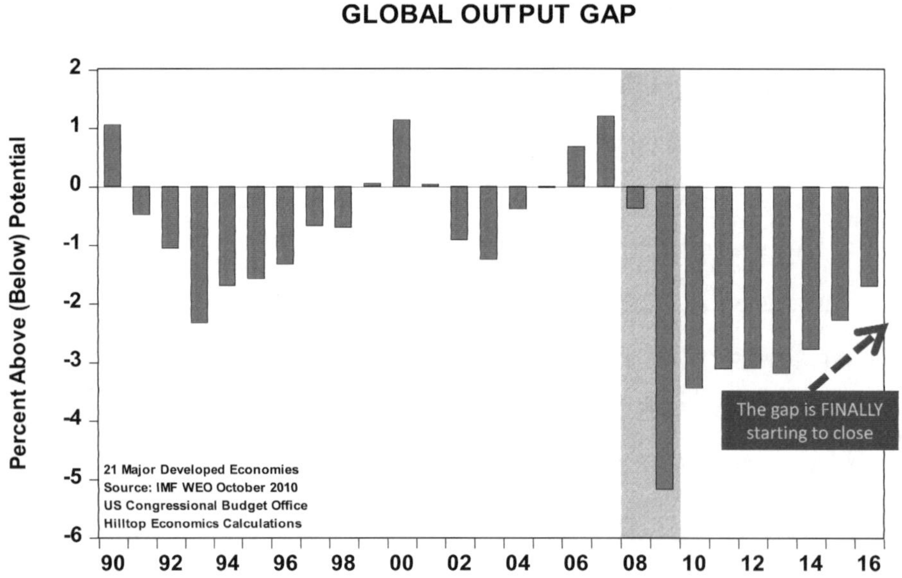

Duncan Meldrum
SPCC April 19-20 2016

14

Hilltop Economics

Global Investment Continues To Grow Too Slowly

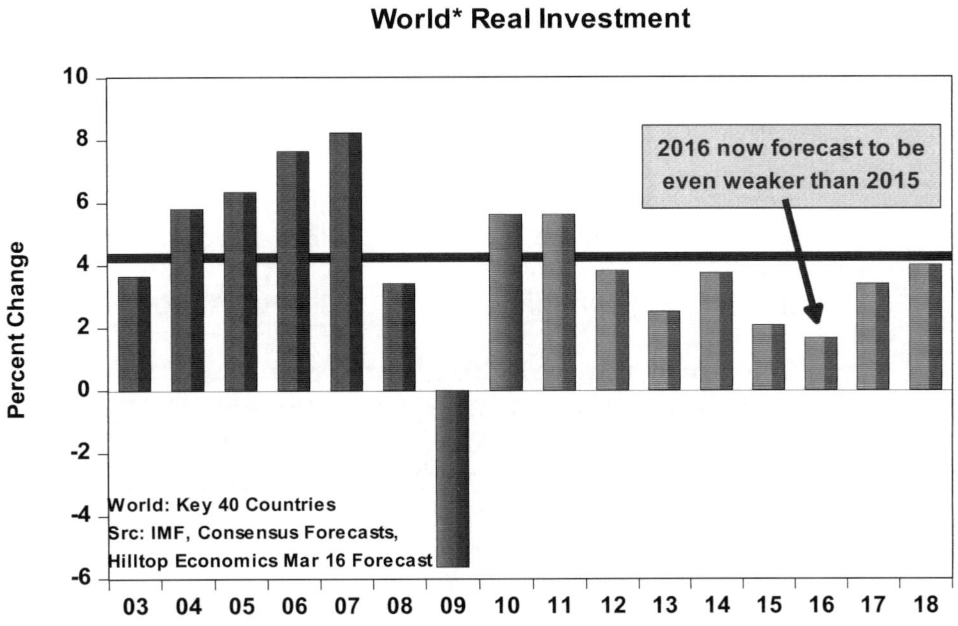

Duncan Meldrum
SPCC April 19-20 2016

15

Hilltop Economics

Real Investment Critical for Semiconductors

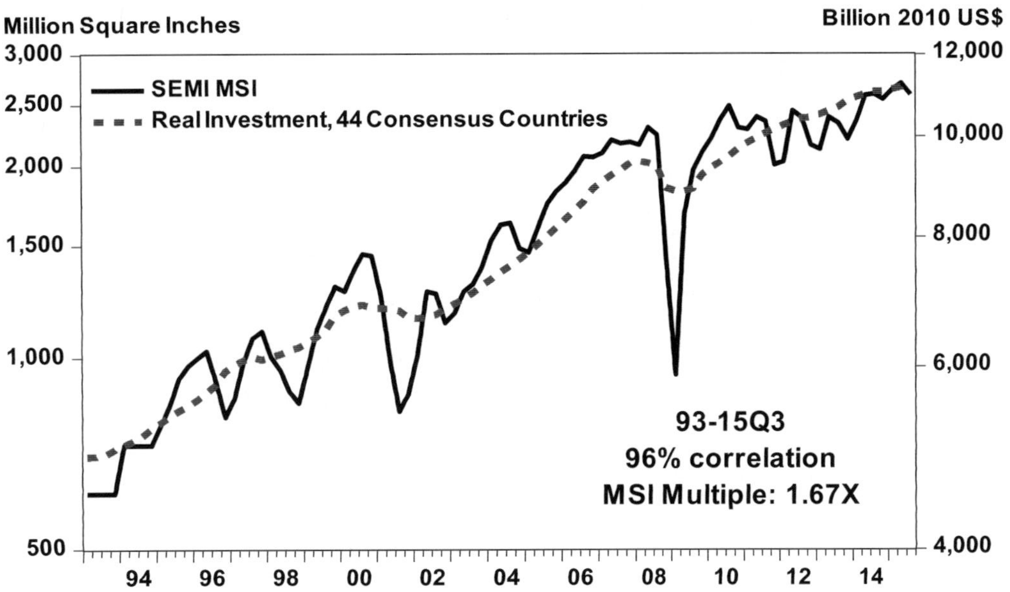

Final spending on fixed investment – plant, equipment & structures - is even more highly related to MSI over the past 23 years than total real GDP

Duncan Meldrum
SPCC April 19-20 2016

16

Hilltop Economics

Semiconductor MSI Outlook Summary

March Update to ESF 2016 Q1	2014	2015	2016	2017	2018
MSI	10,097	**10,434**	**10,465**	**11,264**	**11,744**
%Change	11.4%	**3.3%**	**0.3%**	**7.6%**	**4.3%**

- ***MSI: Flat in 16 on poor investment outlook, snaps back in 2017, falls to around trend in 2018***
- **Global final demand grows marginally in 2016, better in 2017 & 2018**
  - **Real GDP should increase in 2016, reach "trend" by 2017**
  - **Business investment stalls in 2016, begins to recover in 2017**
  - **Consumer spending rises back toward trend of 3%**
  - **PC unit sales down sharply in 2015; down modestly in 2016; increase slightly in 2017**
  - Mobile phone sales growth slows
  - **No major new markets**

Duncan Meldrum
SPCC April 19-20 2016

17

Hilltop Economics

Does the "final demand" forecast modeling technique work?

Linx-Consulting Econometric Semiconductor Forecast (ESF)

The First ESF Quarterly Forecast:
Overforecast Actual MSI by 0.4% (13 Qtr Trend)

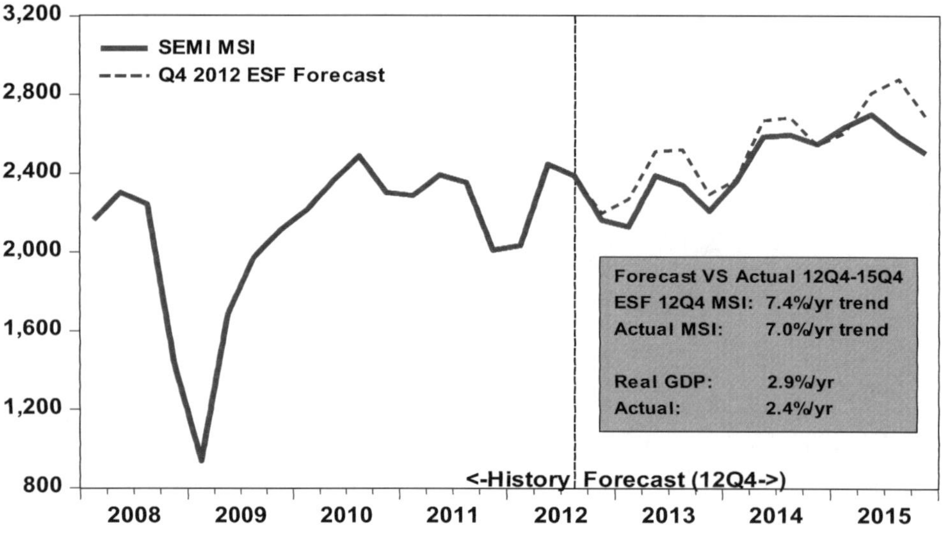

Duncan Meldrum
SPCC April 19-20 2016

18

Hilltop Economics

Does the "final demand" forecast modeling technique work?

Forecast Performance for ESF Three Years, Real Time

Final demand approach provided superior forecasts. Longer term error attributed to over-forecasts of economic growth, not model error. Pervasiveness of semiconductors key to success of the "final demand" approach to forecasting semiconductors

Cumulative Error on LEVELS	1 Quarter Ahead	2 Quarters Ahead	3 Quarters Ahead	4 Quarters Ahead	8 Quarters Ahead
Ave Percent Error	**-0.5%**	**0.0%**	**0.0%**	**-0.1%**	**0.8%**
Mean Absolute Percent Error	**2.5%**	**2.8%**	**3.0%**	**2.9%**	**2.6%**
# Forecasts	13	12	11	10	6

Duncan Meldrum
SPCC April 19-20 2016

19

Hilltop Economics

Things to Watch: Risks

- Yield Curve (expected to stay near current levels)
- Federal Reserve interest rate increases impact on financial markets
  - (forecasters don't expect more than two in 2016)
- Magnitude of oil price increase (expected to gradually rise)
- Value of the Dollar (expect to strengthen modestly)
- Inflation
  - (expected to slowly rise to ~2% in US, lower in other parts of the developed world)
- Jobs keep growing and unemployment rates hold
  - (forecasters expect unemployment rate in US to stay in the range of 4.5% and 4.9% in 2016)
- Forecasts getting lowered in a systematic fashion through the year

Duncan Meldrum
SPCC April 19-20 2016

20

Hilltop Economics

The Best Single Recession Predictor: Treasury Yield Curve Inverts (Short Rates > Long Rates)

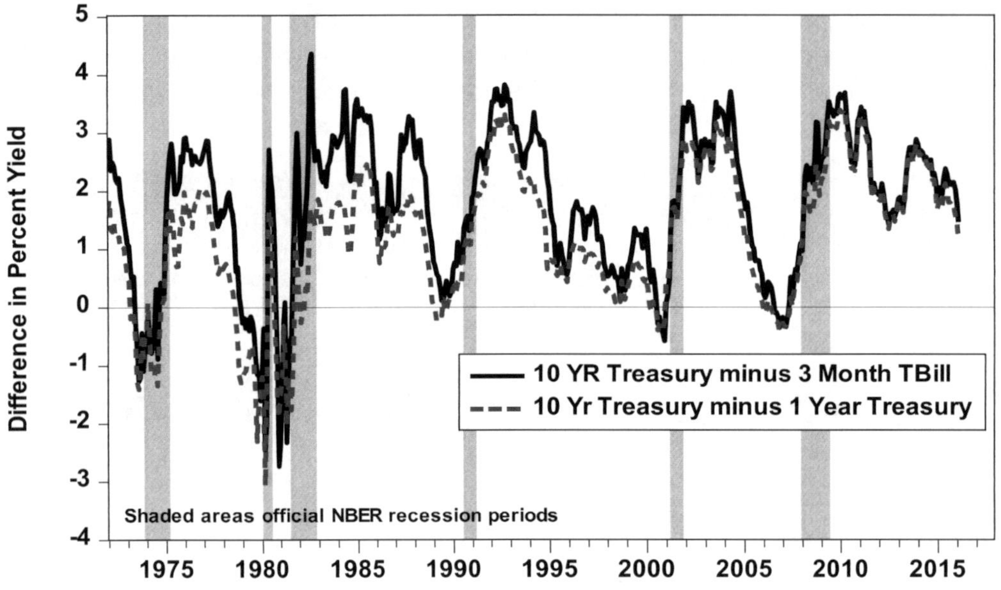

Duncan Meldrum
SPCC April 19-20 2016

21

Hilltop Economics

Stock Market: One Place All the Money Went

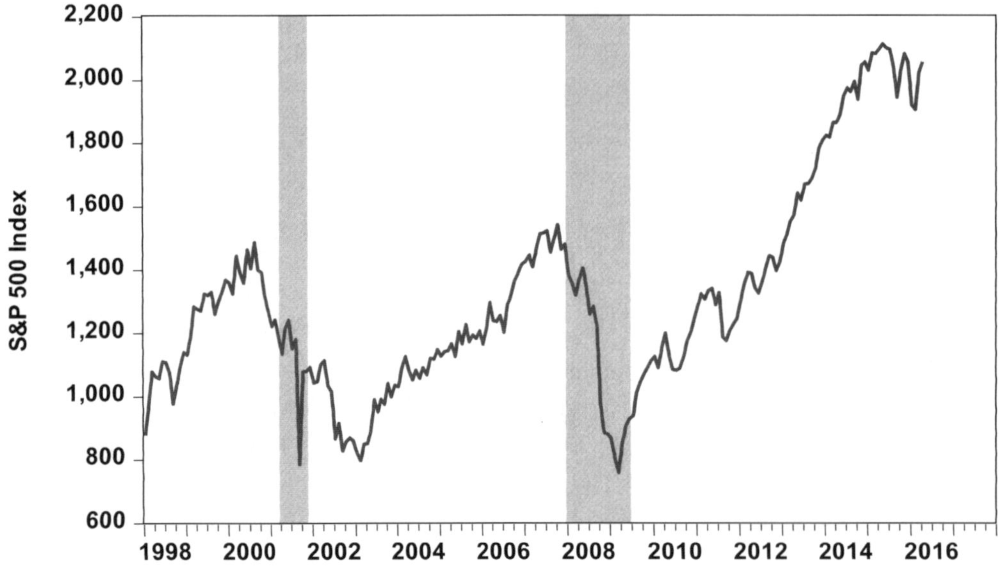

Duncan Meldrum
SPCC April 19-20 2016

22

Hilltop Economics

Forecasts always seem to expect "next year" to be back to "normal"

Steady reduction in '15 and '16 forecasts

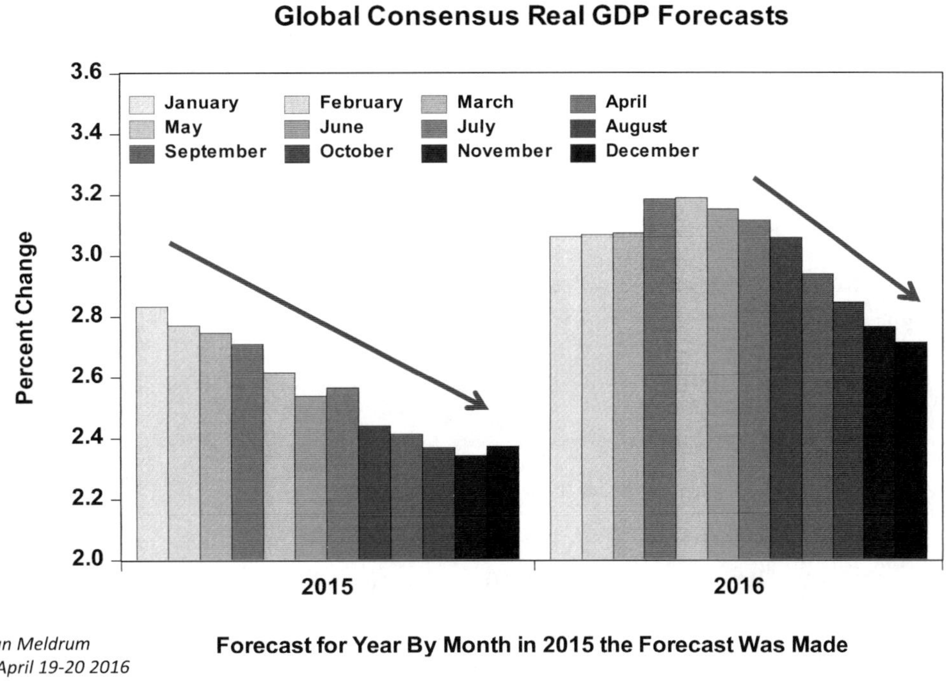

Duncan Meldrum
SPCC April 19-20 2016

23

Hilltop Economics

Forecasters continue the trend: "next year" gets closer to "normal" but 2017 already below "normal" 3+% trend

Steady reduction in '16 and '17 forecasts

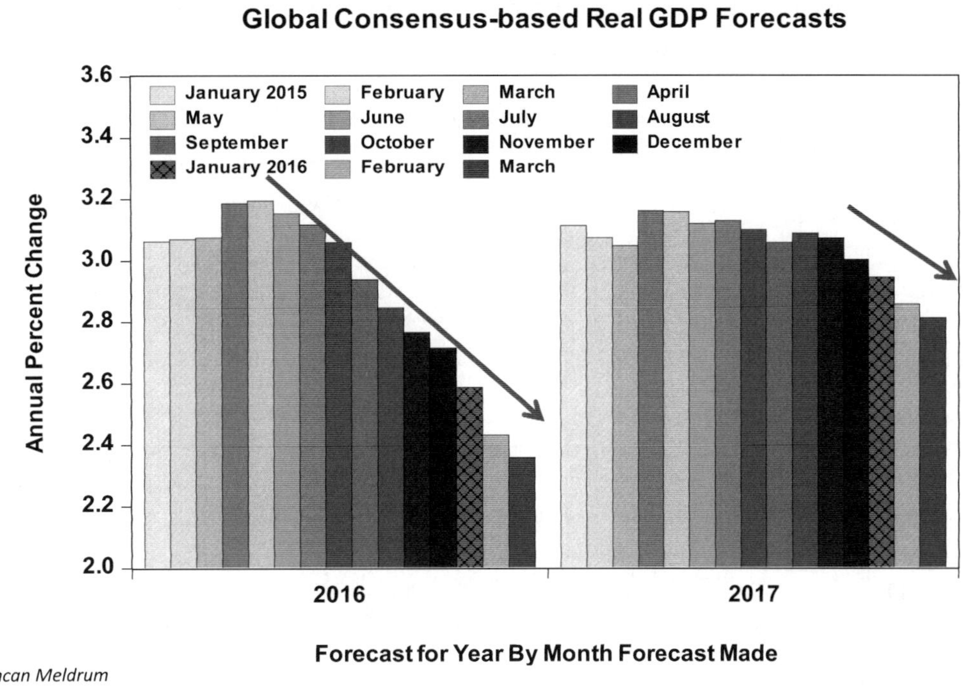

Duncan Meldrum
SPCC April 19-20 2016

24

Hilltop Economics

Oil Prices Rise a Little

Modest demand growth plus plenty of supply limit increases

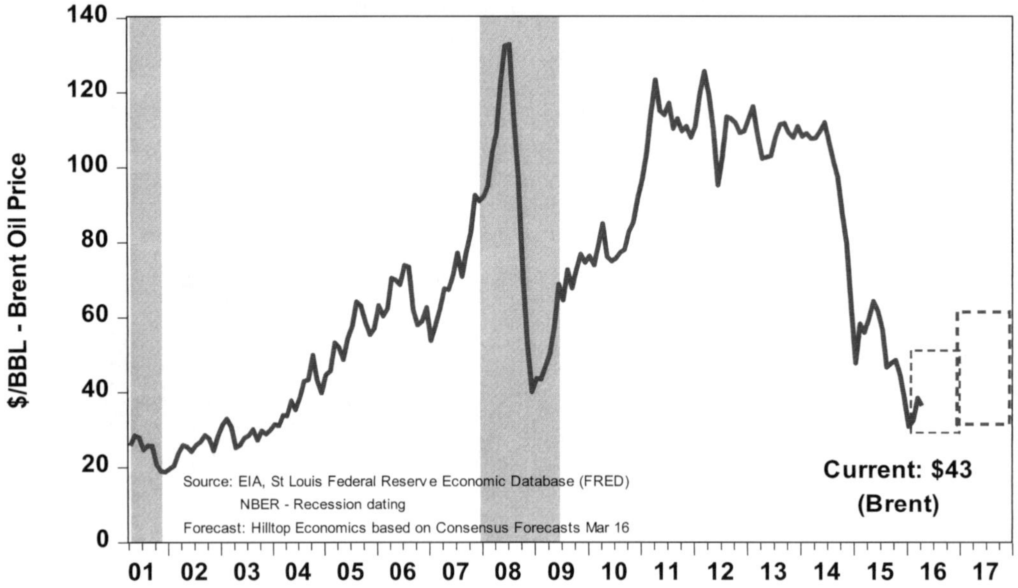

Duncan Meldrum
SPCC April 19-20 2016

25

Hilltop Economics

Consumers Globally Benefit From Low Energy Prices

(U.S. consumers alone gained almost 2% of disposable income since the end of 2013: almost $250 BN)

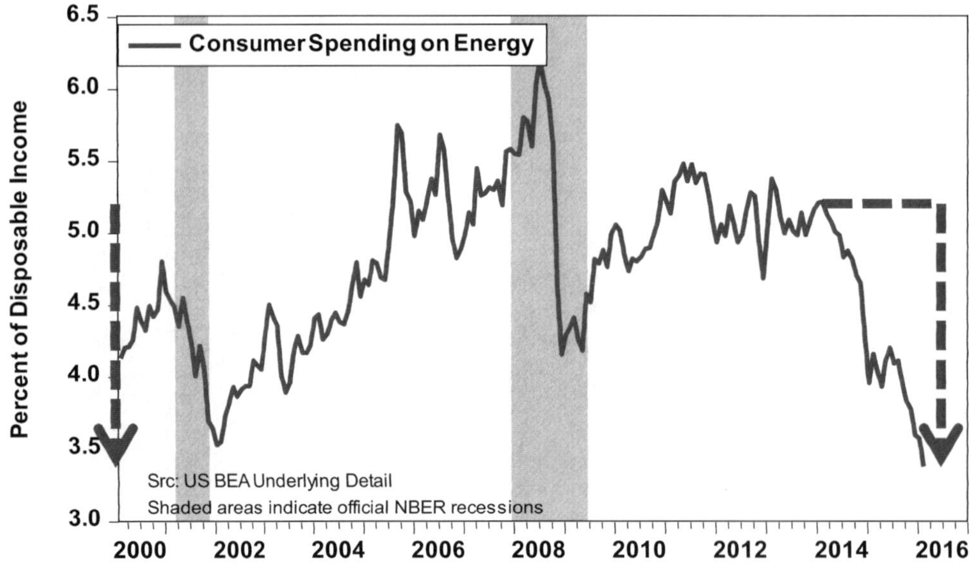

Duncan Meldrum
SPCC April 19-20 2016

26

Hilltop Economics

Recent Dollar Retreat Expected to Reverse

U.S. Interest rates are rising, key trading partners (Europe, China, Japan) are keeping rates steady or cutting: capital will flow to U.S., pushing dollar stronger

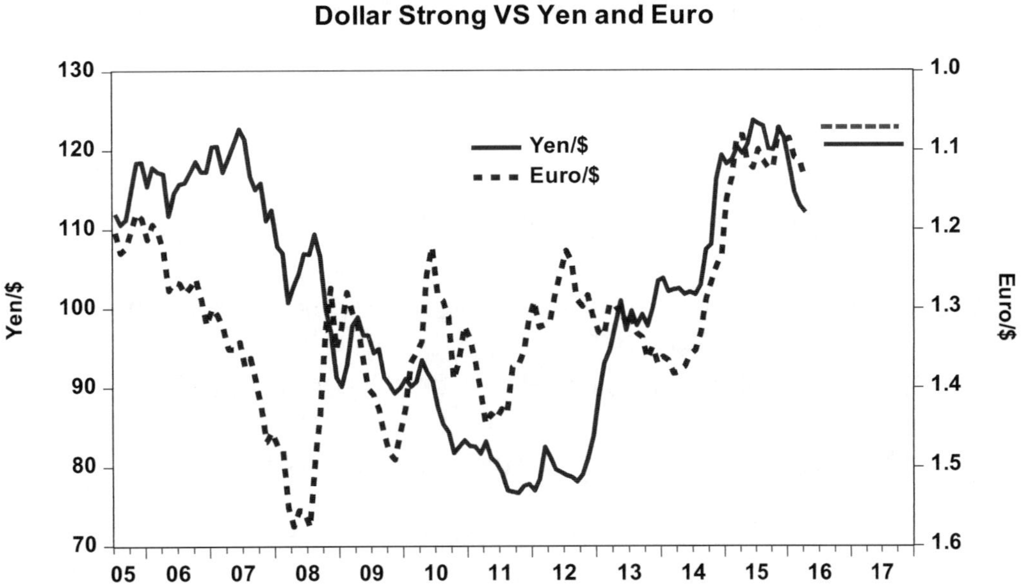

Duncan Meldrum
SPCC April 19-20 2016

27

Hilltop Economics

Duncan Meldrum
Chief Economist
Hilltop Economics, LLC
duncan.meldrum@hilltopeconomics.com

610.709.7951

Duncan Meldrum
SPCC April 19-20 2016

28

Powerchip

Novel STI Step-height Uniformity Control by Wet Etch Process in 4xnm CMOS Device

H. W. Ho[a], Y. M. Liao*[a], M. C. Lu[b], T. H. Ying[b]

[a]Wet Etch Technology Group, Process Technology Division, Powerchip Technology Corp.
[b]Process Technology Division, Powerchip Technology Corp.

No. 12, Li-Hsin RD. 1, Hsinchu, Taiwan, R.O.C

*: thenova@powerchip.com
+886-3-5795000 ext 8690

Powerchip

Outline

- **Introduction**
  - HARP & HDP materials characteristic comparison
  - The importance of step height uniformity control
  - STI HARP wet etching uniformity performance by liquid HF
- **Step Height Uniformity Improvement**
  - Annealing effect
  - Liquid HF (LHF) & gas HF (GHF) etching mechanism
  - Process optimization result
- **Conclusion**

Powerchip STI Materials Comparison (HDP & HARP)

❖ High aspect ratio process (HARP) has been applied in shallow trench isolation (STI) for 45nm CMOS and beyond due to better gap fill ability.

◆ STI (Shallow Trench Isolation) Material Road Map

Generation		130 nm	65 nm	55 nm	4x nm	2x nm
HDP	A/R<4					
HARP	A/R~6					

◆ HDP (High density plasma):

✓ Precursor: SiH_4 (silane)

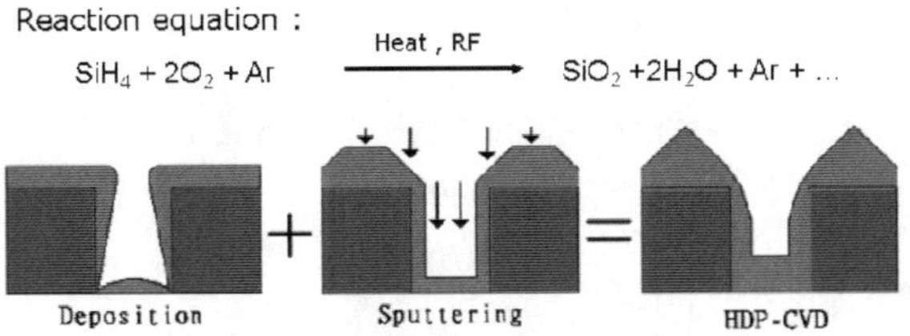

✓ SiO_2 cross-linking **before** annealing
✓ Poor gap fill ability (A/R<4)

◆ HARP (High aspect ratio process):

✓ Precursor: O_3 + TEOS ($Si(OC_2H_5)_4$)

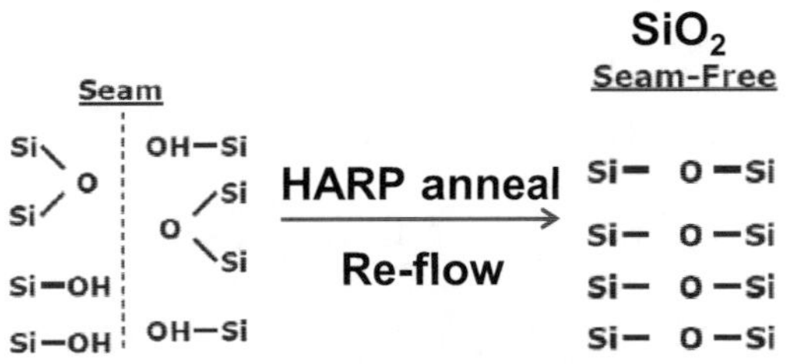

✓ SiO_2 cross-linking **after** annealing
✓ Better gap fill ability (A/R~6)

Powerchip Worse Etch Uniformity of HARP by LHF

❖ HARP wet etching rate (E/R) is harder to be controlled than HDP in different STI width by conventional liquid HF (LHF).

◆ HDP :

✓ Good step height uniformity

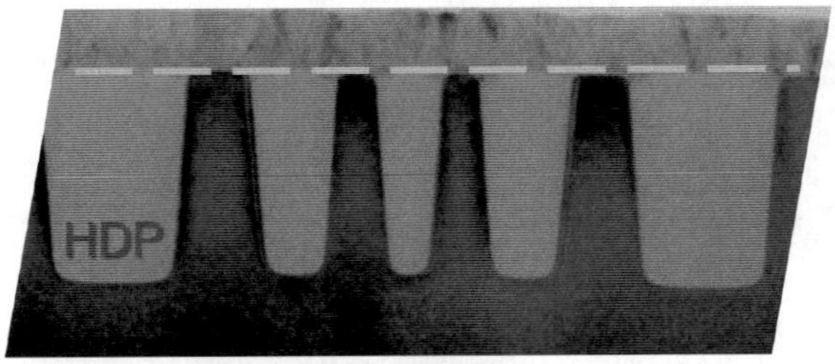

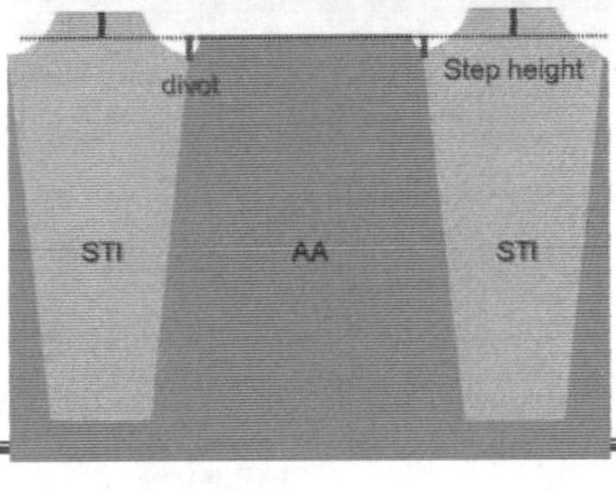

◆ HARP:

✓ Worse step height uniformity

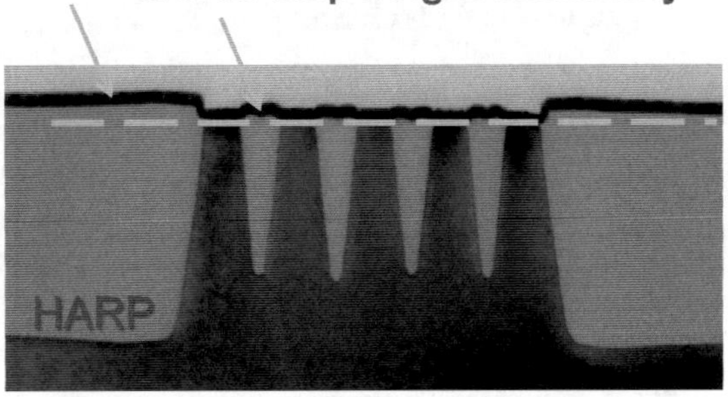

Powerchip The Definition & Importance of Step height & Divot

❖ Step height control is necessary to avoid poly residue issue and Y% loss.

Before poly gate etch ⇒ **After poly gate etch**

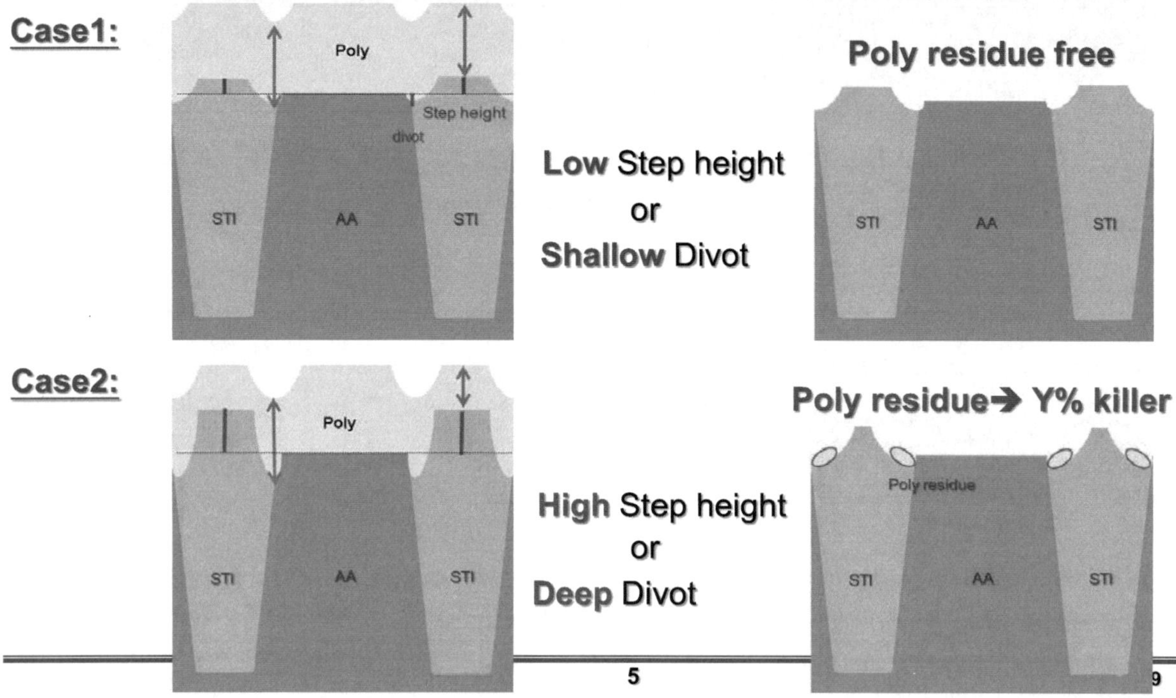

Powerchip Worse Step height Uniformity by LHF

❖ Step height (S/H) uniformity in different STI width is worse under LHF etching during well implant and gate oxide formation process.
❖ S/H in narrow STI width is lower than wide one and the bias is18.5nm.

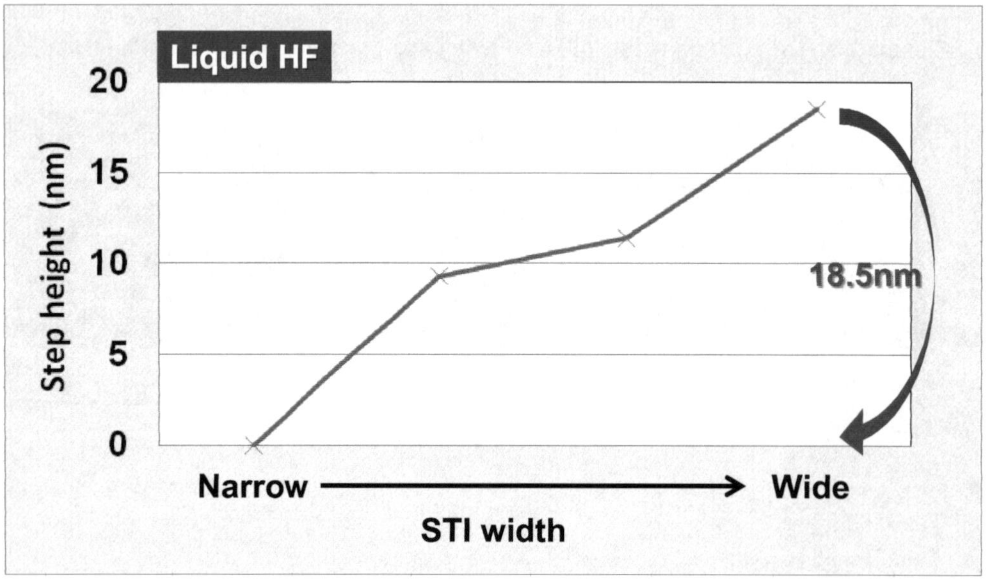

Powerchip — Anneal Effect on Wet Etch Uniformity

❖ Annealed HARP quality is different from narrow to wide STI width.
❖ Etch amount (E/A) uniformity is worse in annealed HARP than without annealed one.

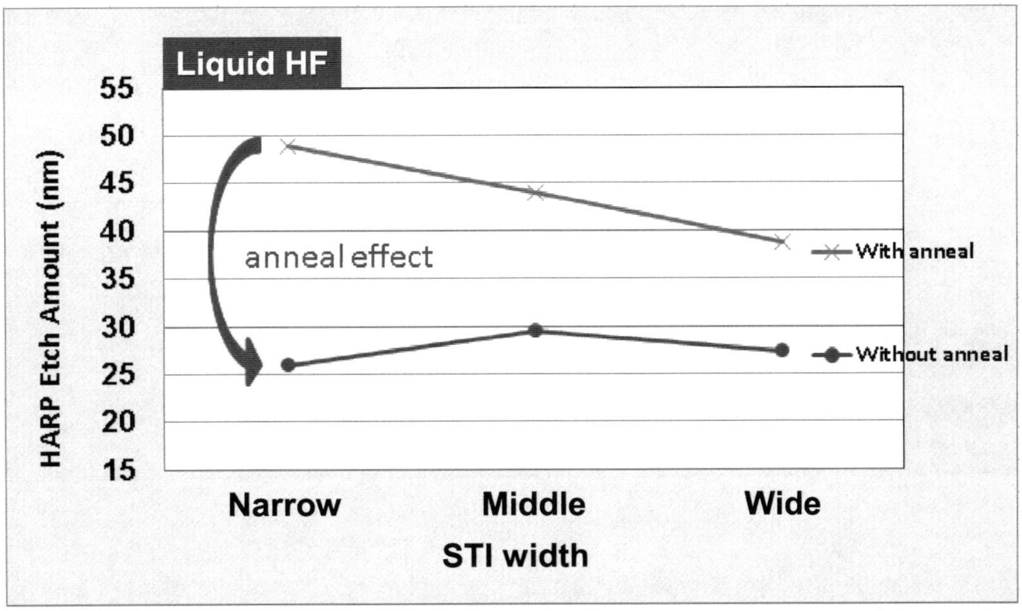

Powerchip — Annealed HARP Quality in Different STI Width

❖ Narrow STI width area: weak cross-linking & high impurity.
❖ Wide STI width area: strong cross-linking & low impurity.

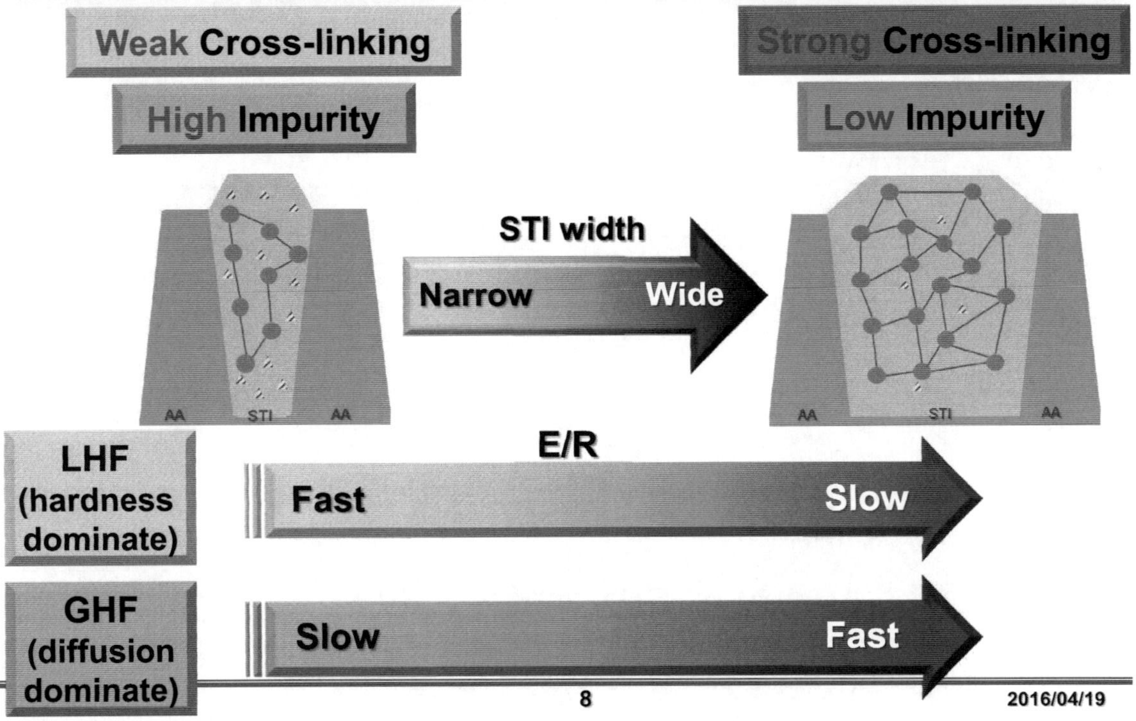

Powerchip — LHF & GHF Etching Mechanism

- ❖ LHF: HARP cross-linking (hardness) dominate.
- ❖ GHF: gas diffusion is limited by HARP impurity.

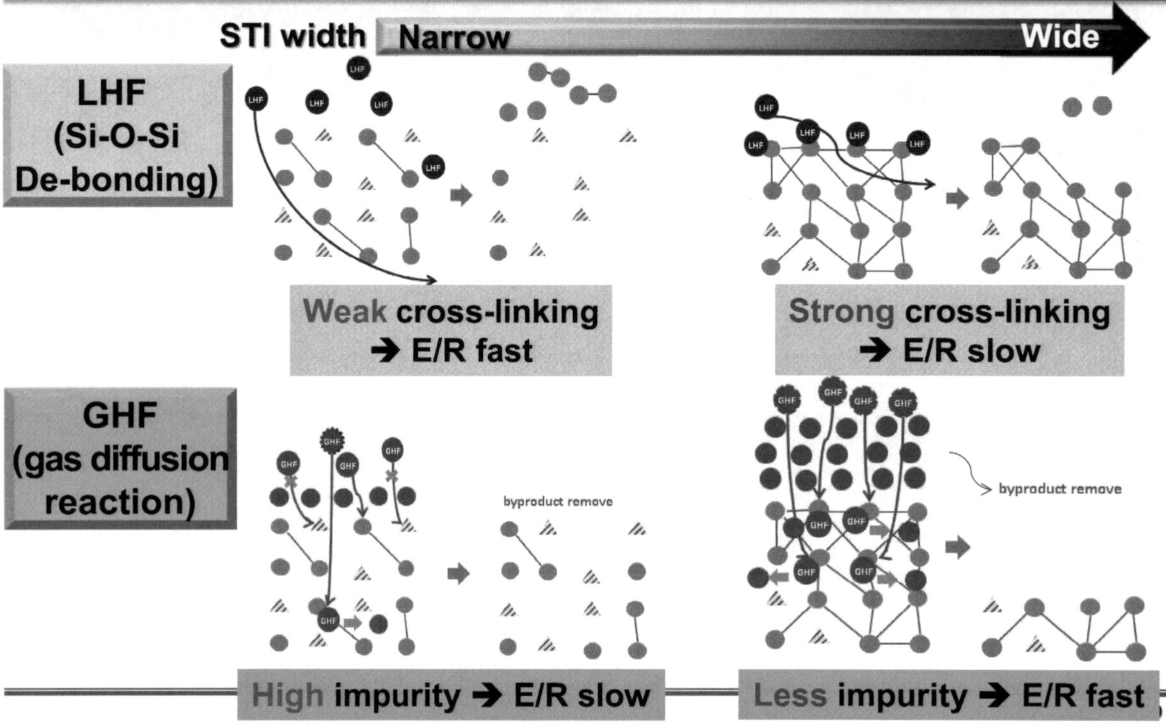

Powerchip — GHF & LHF E/A Performance Comparison

- ❖ LHF and GHF E/R in different STI width is opposite.

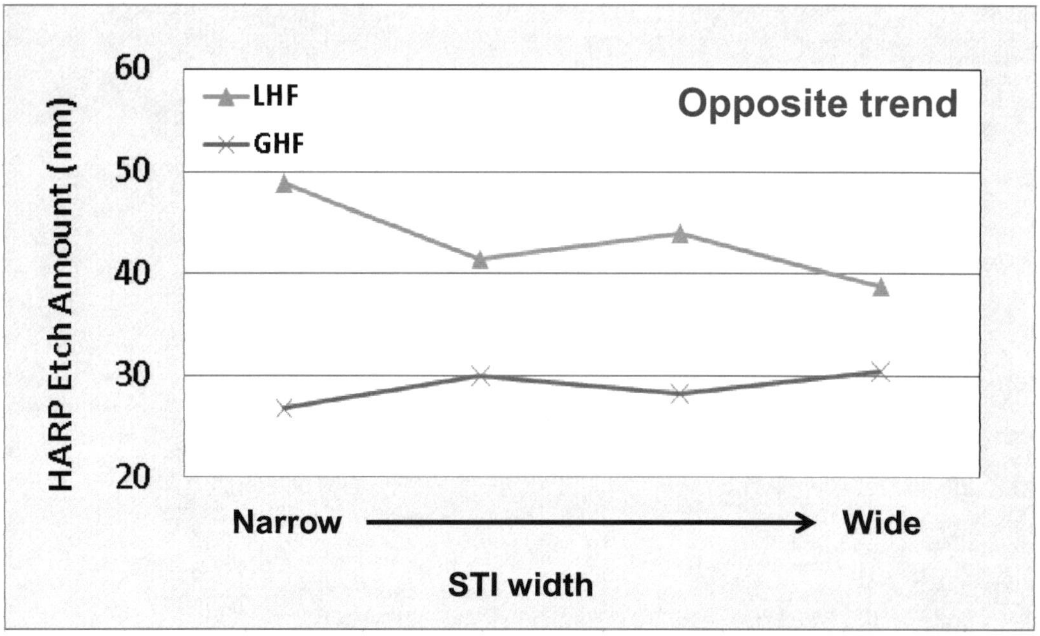

Powerchip — Process Optimization Result

❖ After optimizing the process flow by combining LHF and GHF, S/H bias through all STI width can be reduced from 18.2nm to 4.5nm

❖ The approach is very helpful for process window enlargement in following gate etch step.

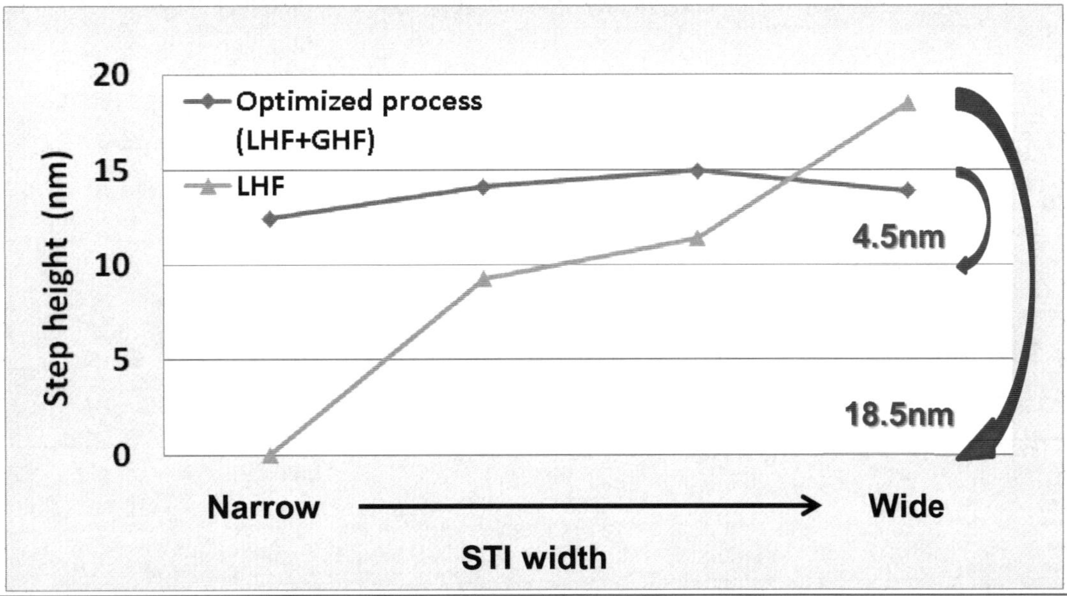

Powerchip — TEM Profile

TEM Profile			
STI width	Narrow	Middle	Wide
LHF only	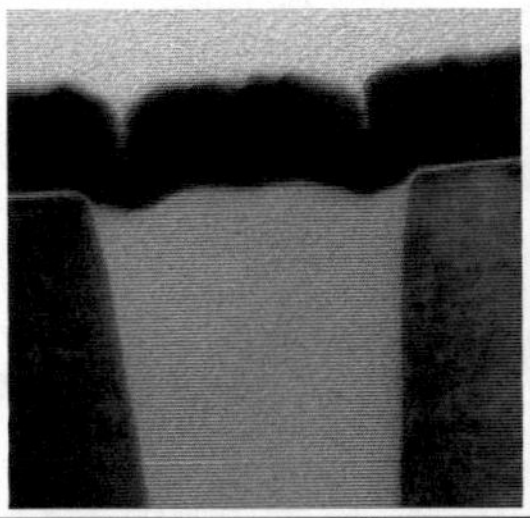		
Step height (nm)	0	11.4	18.5
Optimized process (LHF + GHF)	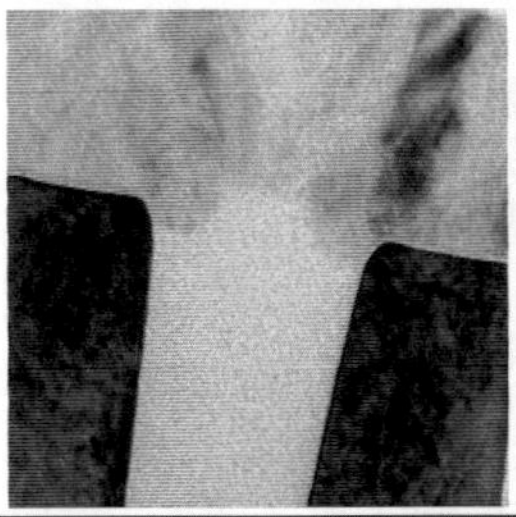	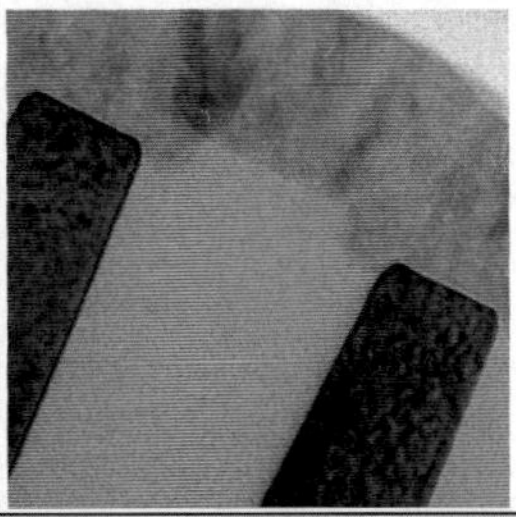	
Step height (nm)	12.4	14.1	14.9

Powerchip

Conclusion

- The cross-linking and impurity content of annealed HARP is different between STI width, which is key impact factors of STI wet etching uniformity.
- E/R of LHF is related to oxide film hardness, on the other hand, E/R of GHF is limited by HARP impurity. Therefore, different HARP quality between STI width leads opposite E/A trend in LHF and GHF.
- Low S/H bias (<5nm) can be achieved by combining LHF and GHF, and poly gate etch process window can be enlarged by this fine-tuned STI profile.

13 2016/04/19

Powerchip

Thank you for your attention

14 2016/04/19

TEL
TOKYO ELECTRON

Improved Cryogenic Gas Cleaning for Nanoparticle Removal

SPCC 2016
April 19-20, 2016

Chimaobi Mbanaso & Jeffery W. Butterbaugh
Process Technology Department
TEL FSI, Inc.
3455 Lyman Boulevard, Chaska, Minnesota
Chimaobi.Mbanaso@us.tel.com. (952) 361 7254

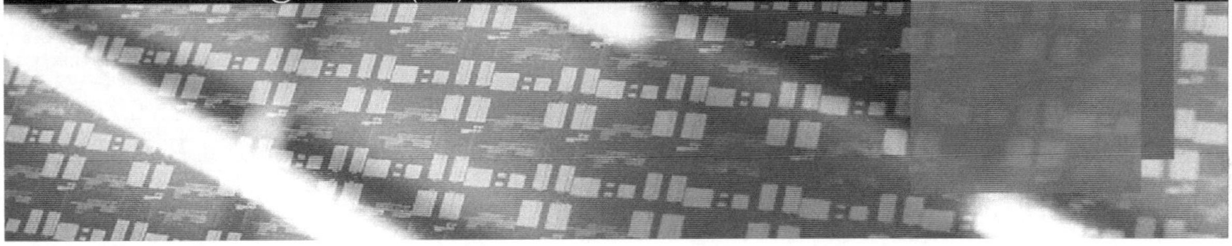

Outline

- Nanoparticle Removal Roadmap and Challenges
- Dry Particle Removal Using Cryogenic Aerosol Technology
- Recent Advances in Cryogenic Aerosol Technology for Nanoparticle Removal
- Conclusions

Nanoparticle Removal Roadmap and Challenges

- ITRS 2015 roadmap critical size = 11 nm in 2016, 7.5 nm in 2020
- Consideration of agglomeration and/or fragmentation
- Multi-patterning defectivity → quad-patterning = "quad defects"

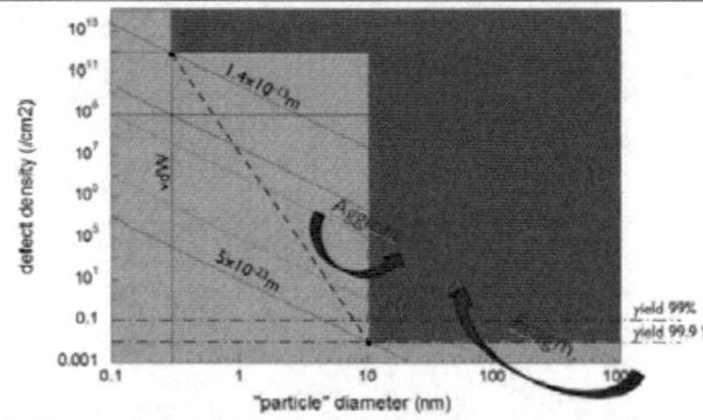

Figure 2: Diagram showing the surface concentration versus diameter of contamination particles. 2 yield lines are added for the case of an active area of 0.1 cm^2. The particle specification is represented by the point (10nm, 0.01 /cm^2). The specification for O and C on the surface is represented by the point (0.3nm, 10^{12}/cm^2). The color red indicates out of specification, while green means within specification. The slanted (power 1/3) lines represent contamination isochores.

P.W. Mertens, Contamination Specifications, an Overall Perspective, In ECS Trans. 69(8), 87-93(2015).

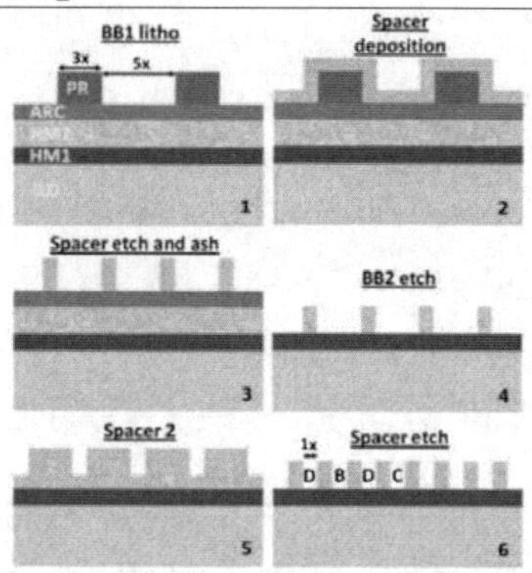

Figure 1: Schematic of a pitch quartering flow using two spacer depositions. The sequential processing steps are labeled 1-6.

M. van Veenhuizen et al., Demonstration of an electrically functional 34nm metal pitch interconnect in ultralow-k ILD using spacer based pitch quartering, in Proc. IITC, 2012, pp. 1–3.

TEL FSI, Inc. / SPCC 2016 / April 19-20, 2016

TEL 3

Motivation for Chemical-Free Dry Particle Removal

- No Corrosion / No Material Loss
  - No Photon Induced Copper Redistribution
  - No Copper Corrosion
  - No Galvanic Corrosion
  - No Fin Loss
- Effective on Hydrophobic Surfaces
- No Adverse Effects on Low-k Films
- No Wetting Issues for Narrow Features

TEL FSI, Inc. / SPCC 2016 / April 19-20, 2016

TEL 4

Cryogenic Ar/N_2 Aerosol Cleaning

- Concept developed in 1990

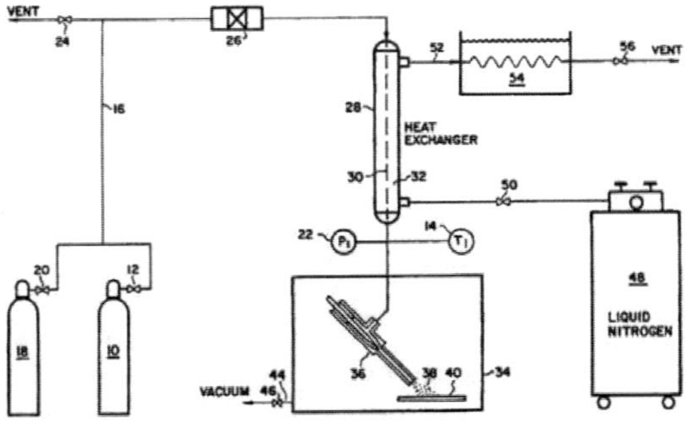

W.T. McDermott et al, US Patent 5,062,898 (1991)

- Applied in IC Manufacturing
  150 nm - 2000 nm size particles removed

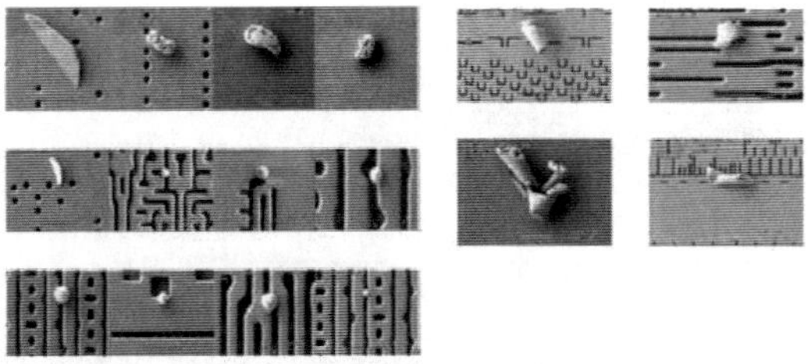

J. Lauerhaas et al., Yield Improvement Using Cryogenic Aerosol for BEOL Defect Removal, in ECS Trans 11(2), 33-39(2007)

Current Cryogenic Ar/N_2 Aerosol Cleaning Principle

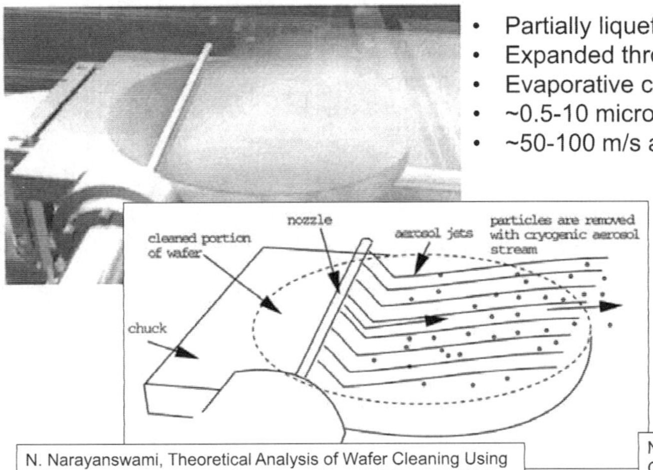

- Partially liquefied gas stream
- Expanded through linear spray nozzle
- Evaporative cooling causes solidification
- ~0.5-10 micron aerosol size
- ~50-100 m/s aerosol velocity

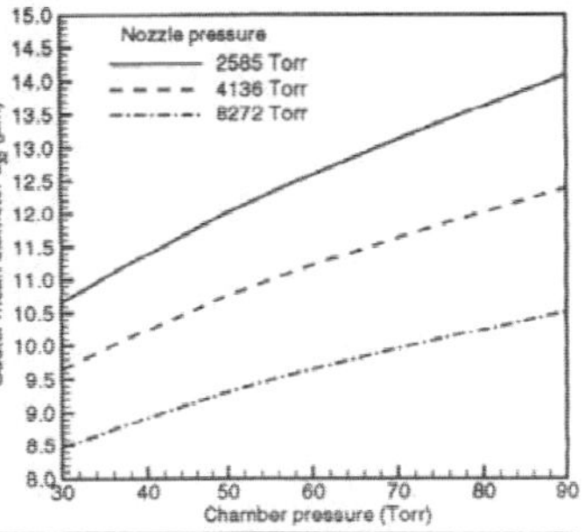

N. Narayanswami, Theoretical Analysis of Wafer Cleaning Using a Cryogenic Aerosol, in J. Electochem. Soc. 146(2), 767(1999)

N. Narayanswami et al., Development and Optimization of a Cryogenic Aerosol-Based Wafer Cleaning System, in Particles on Surfaces 5&6: Detection, Adhesion and Removal, 251(1999)

Damage-Free Particle Removal Window – Cryogenic Ar/N_2 Aerosol

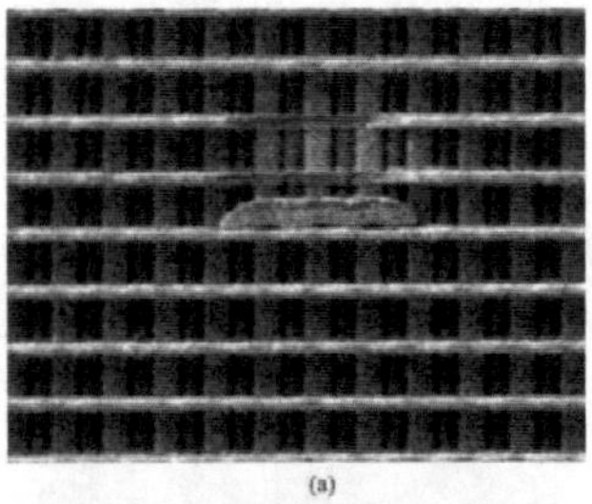

90nm
AR = 1.4:1

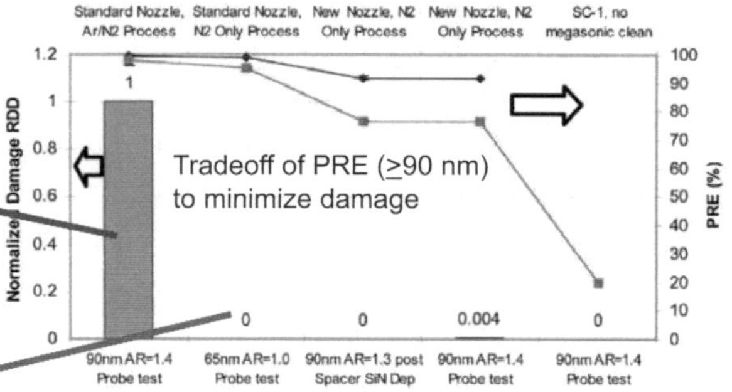

65nm
AR = 1:1

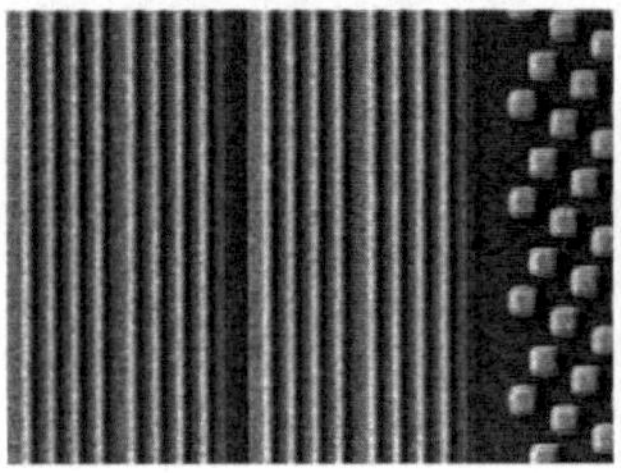

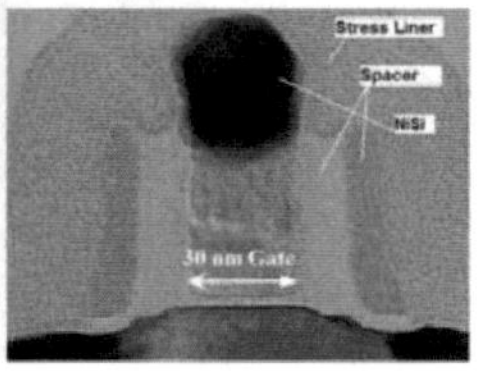

(RDD = Random Defect Density)
(PRE = Particle Removal Efficiency)

H. Lin et al., Damage-Free Cryogenic Aerosol Clean Processes, in IEEE Trans Semi Manuf 20(2), 101(2007)

New Cryogenic Aerosol Process

- Goals:
  - Increase removal efficiency for <40 nm particles, extend to 10 nm and beyond
  - Damage-free for <40 nm features, extend to 10 nm and beyond
  - Remove particles between patterns
- Develop Nozzle and Process Conditions to Achieve:
  - Smaller aerosol size
  - Narrower aerosol size distribution
  - Higher aerosol velocity

New Process Increases Nanoparticle Removal without Damage

- Nanoparticle Removal Efficiency (Dry Deposited Si_3N_4 particles , 36-hour aged)

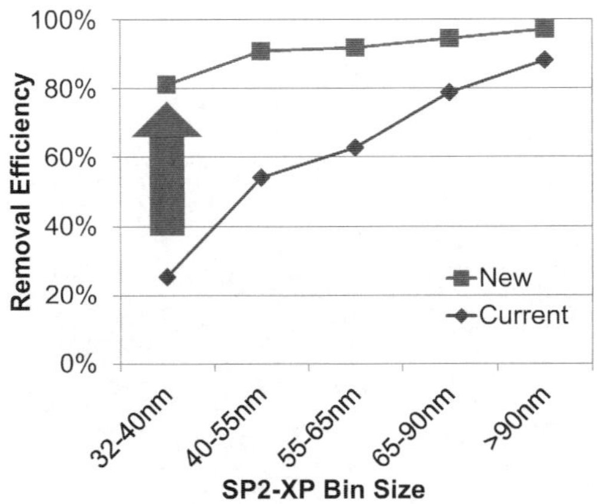

- Pattern Damage Performance

Process	Pattern Features	Post-Process
New	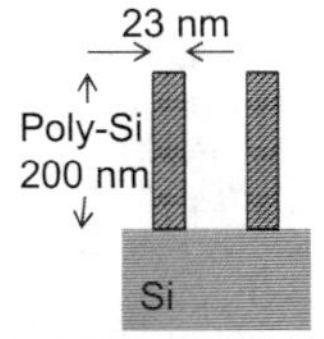	
Current	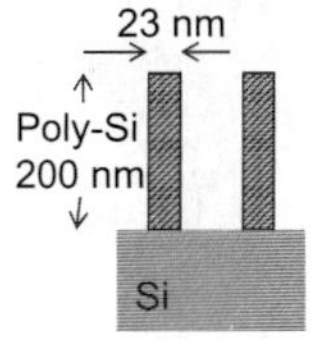	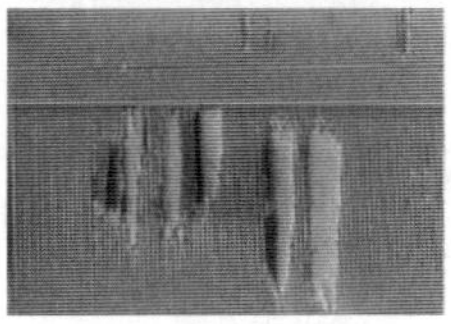

TEL

Nanoparticle Removal Efficiency Based on KLA Surfscan® Measurement

- > 90% removal efficiency as measured by KLA SP2-XP for 40 nm Silica
- > 90% removal efficiency as measured by KLA SP5 for 30 nm Silica

Silica Particle Size	Contaminated Sample (Wet Deposited Silica Particles)	6-Hour Silica Challenge + New Process
40 nm Silica	Particle Size Distribution 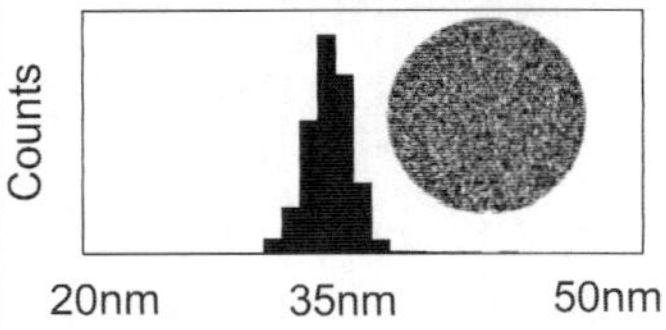	Particle Size Distribution 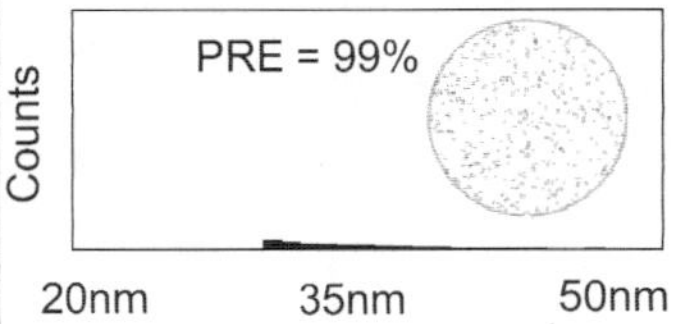
30 nm Silica	Particle Size Distribution 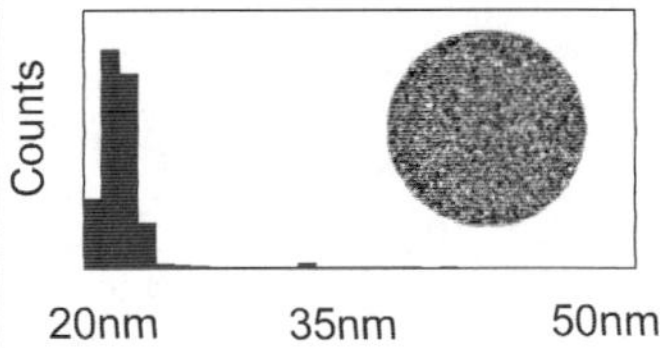	Particle Size Distribution 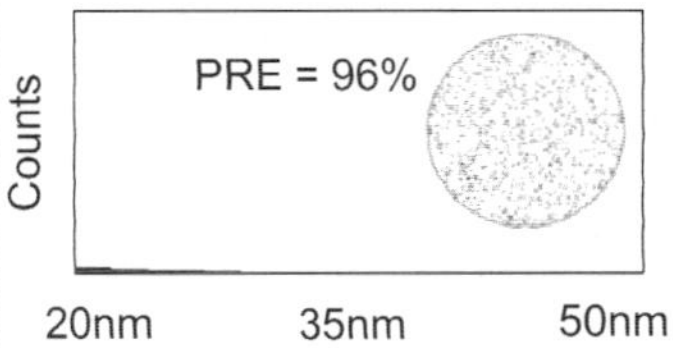

TEL

Nanoparticle Removal Efficiency Based on SEM Inspection

- > 90% removal efficiency as measured by multiple SEM FOV counting
- > 80% removal efficiency as measured by multiple SEM FOV counting

Silica Particle Size	Contaminated Sample (Wet Deposited Silica Particles)	6-Hour Silica Challenge + New Process
30 nm Silica 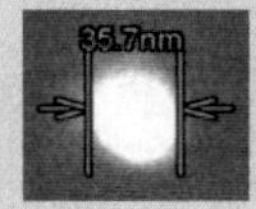		
18 nm Silica 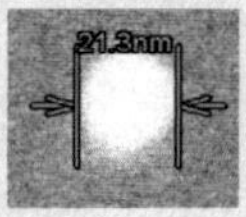	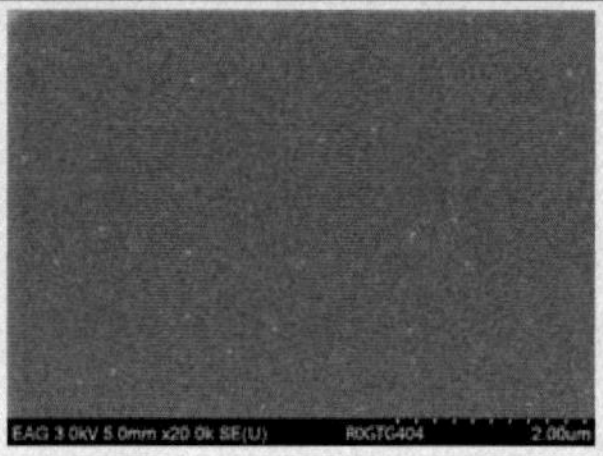	

Light Scattering "Haze" Used for Sub-Resolution Particle Detection

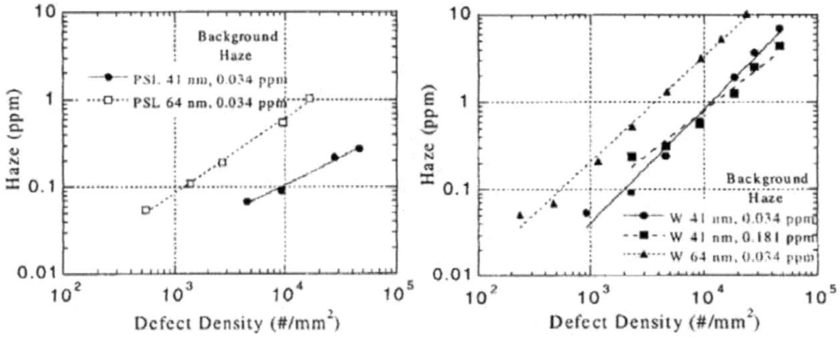

Figure 2. Haze Calibration Curves for PSL Spheres (left) and W Particles (right)

S.H. Yoo et al., Particle Removal Efficiency Evaluation at 40nm Using Haze Particle Standard, in Solid State Phen. Vol 76-77, 259-262(2001)

Nanoparticle Removal Efficiency Based on Haze Response

- 30 nm Silica, wet deposited, 6-hour aged

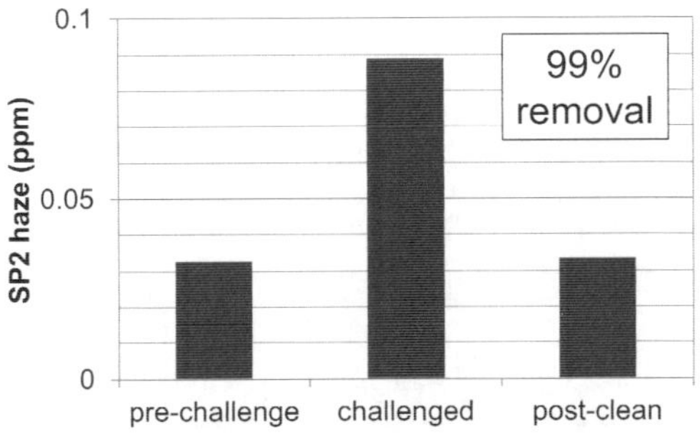

- 18 nm Silica, wet deposited, 6-hour aged

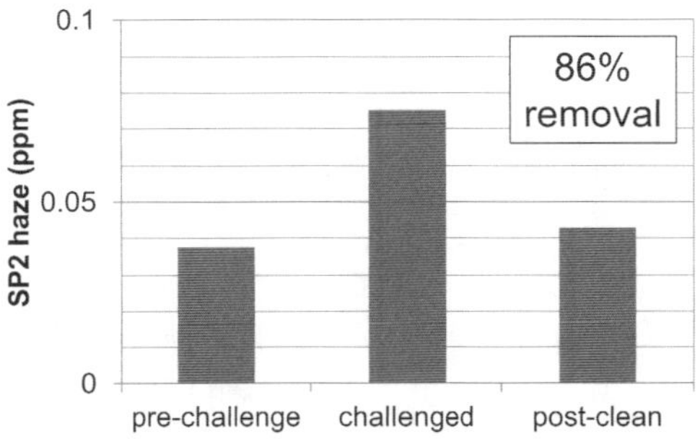

Particle Cleaning from Patterned Si Trenches

- 30 nm Silica particle clusters were dry deposited on patterned Si wafer

Line Width (nm)	Line Space (nm)	Line Height (nm)
45 nm	45-90 nm	100 nm

Particle clusters on patterned silicon surface

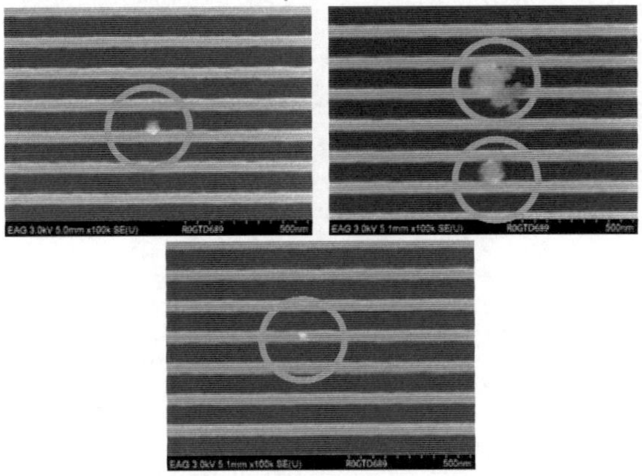

Particle clusters removed after cleaning

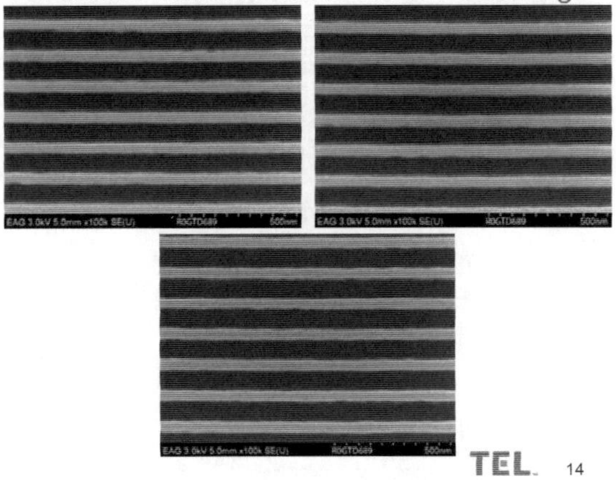

New Cryogenic Process Maintains Compatibility with Low-k Films

- Low-k wafers were processed with cryogenic aerosol processes
- Low-k wafers unaffected after cryogenic aerosol processing

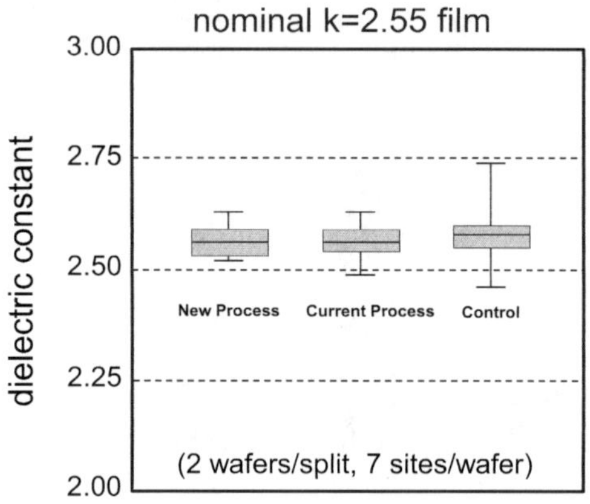

TEL FSI, Inc. / SPCC 2016 / April 19-20, 2016 TEL 15

Summary

- IC ground-rules continue to shrink, creating particle removal challenges
- The damage-free cleaning window is becoming narrower
- Cryogenic aerosol cleaning is a chemical-free, non-corrosive particle removal technology that is proven in production
- New cryogenic nano-aerosol cleaning technology is effective for <40 nm particles and is damage-free for advanced IC manufacturing.

TEL FSI, Inc. / SPCC 2016 / April 19-20, 2016 TEL 16

Acknowledgements

- Greg Thomes, Brent Schwab, TEL FSI, Inc.
- Kazuya Dobashi, Tokyo Electron Limited, Process Development Center
- Koji Kagawa, Tokyo Electron Kyushu Limited, SPE Process Technology Dept.
- Wallace Printz, Derek Basset, Antonio Rotondaro, Tokyo Electron America, Inc. Advanced Technology Group

TEL™
TOKYO ELECTRON

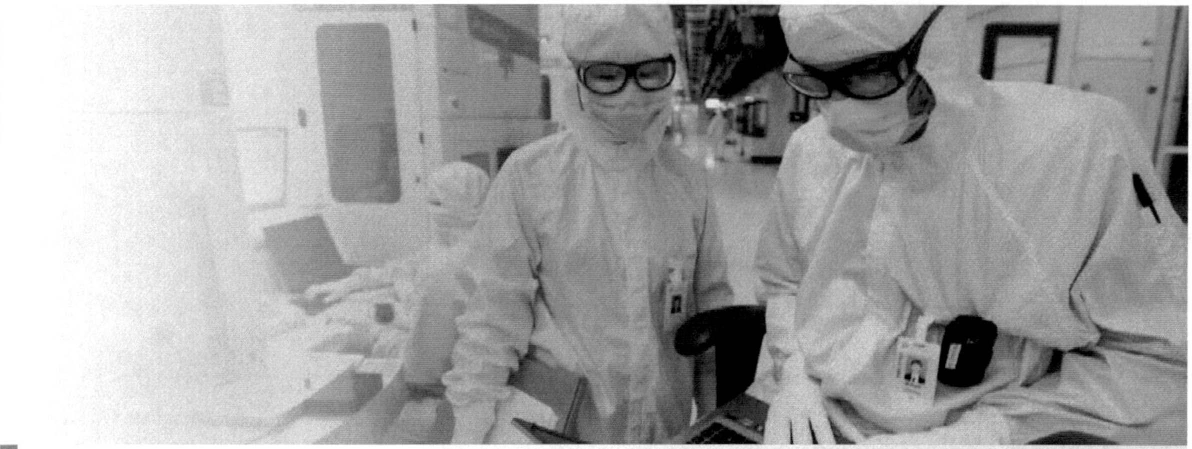

Process Gas Ar Flow Usage Reduction in Aerosol Cleaning

Asha Sharma, Ella Mihevc, Christopher Waitkus, Tsultrim Tharchin and Silas Scott

April 20, 2016

SPCC 2016
Surface Preparation and Cleaning Conference

GLOBALFOUNDRIES

Outline

Background

Particle Removal in Cryogenic Aerosol Cleaning
- Effect of process gas on particles

Methods

Throughput Improvements on Ar Aerosol Cleans
- Tuning recipe parameters
- Ar flow usage reduction, process time & throughput

Results

Throughput Improvements on Aerosol clean @ GLOBALFOUNDRIES
- UPH gain and saving (14 nm)

GLOBALFOUNDRIES

April 14, 2016 2

Outline

Background

Particle Removal in Cryogenic Aerosol Cleaning
- Effect of process gas on particles

Methods

Throughput Improvements on Ar Aerosol Cleans
- Tuning recipe parameters
- Ar flow usage reduction, process time & throughput

Results

Throughput Improvements on Aerosol clean @ GLOBALFOUNDRIES
- UPH gain and saving (14 nm)

GLOBALFOUNDRIES

Particle Removal: Cryogenic Aerosol Clean

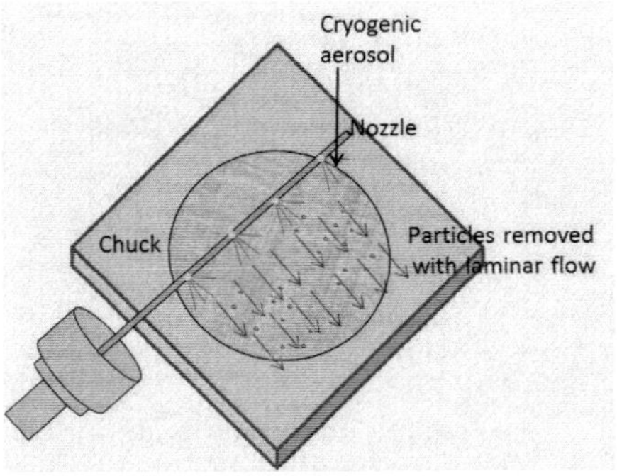

Schematic Cryogenic Aerosol Clean

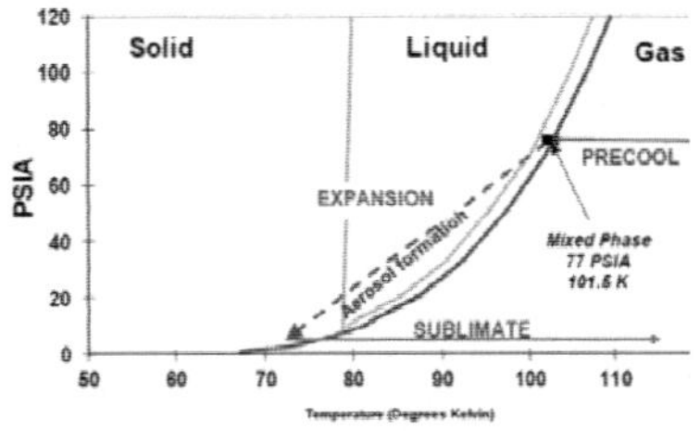

[1,2]Pressure-temperature diagram Ar:N_2 system

Process parameters

- **Process Gas**
- process gas pressure
- chamber pressure
- **Chuck speed/indexing**
- chuck temperature
- Dewar back pressure

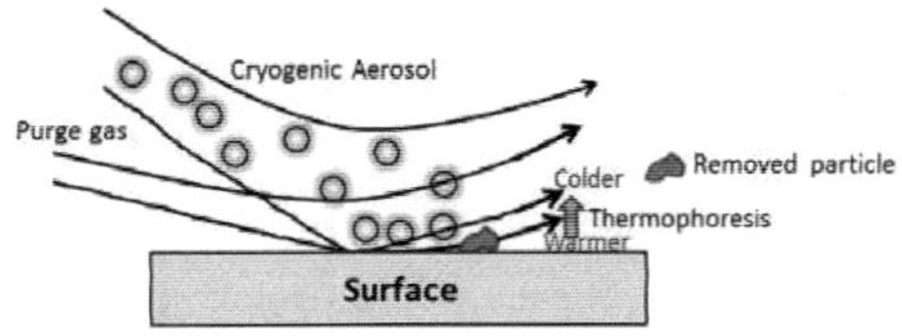

Schematic (redrawn from [1])

[1]Particle Adhesion and Removal, p 460, Wiley & Sons
[2]ANTARES® System, TEL

GLOBALFOUNDRIES

Effect of Process Gas

Particle removal efficiency (@ 32 nm)

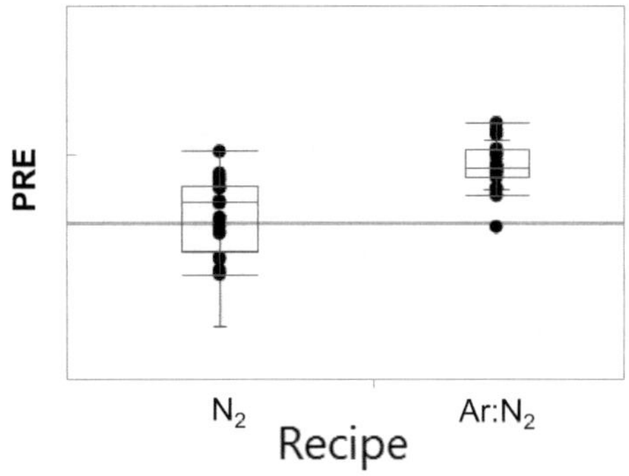

PRE (Particle removal efficiency)
(calculated from pre-post/pre)

Aerosol Cleaning Force[3]:

Momentum, $mv = F.\Delta t$

Collision force, $F = V\rho.\frac{v}{\Delta t}$

where, V is volume and ρ is density of cryogenic aerosol

Atomic mass Ar = 39.948 amu
Molecular weight N_2 = 28.0134 amu

➢ $Ar:N_2$ mixture will have greater momentum transfer than N_2 only

Summary:

- **Particle removal efficiency higher for $Ar:N_2$ mixture**
- **$Ar:N_2$ aerosol - greater momentum transfer than N_2**

3: IEEE TRANSACTIONS ON SEMICONDUCTOR MANUFACTURING, VOL. 20, NO. 2, MAY 2007

April 14, 2016

GLOBALFOUNDRIES

5

Effect of Process Gas on Performance

28 nm: Open and shorted via chains

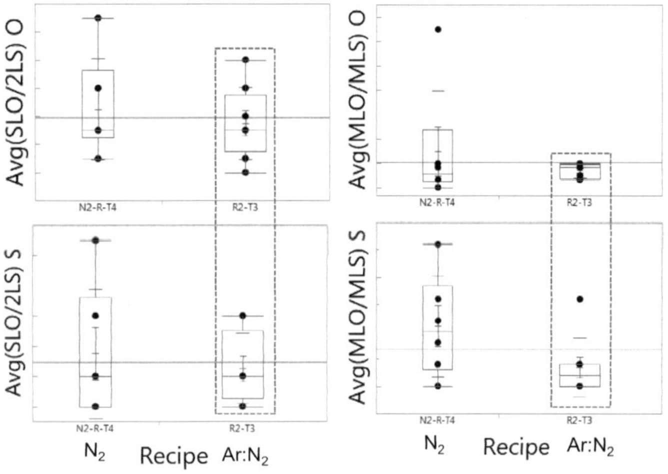

D_0 Reduction (%) by Ar addition			
SLO/2LS		MLO/MLS	
O	S	O	S
15	28	52	68

Ar ($Ar:N_2$) aerosol cleaning shows better opens and shorts performance than N_2 aerosol cleaning

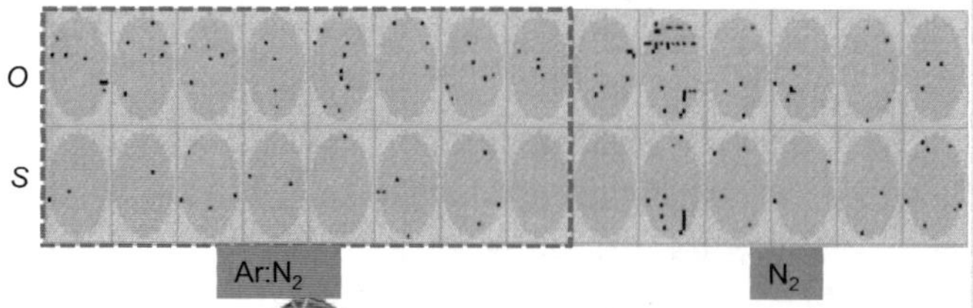

Summary:

- **$Ar:N_2$ aerosol cleaning shows reduced D_0 than N_2**

April 14, 2016

GLOBALFOUNDRIES

6

Effect of Process Gas on Performance

28 nm CFM Defectivity, and Yield

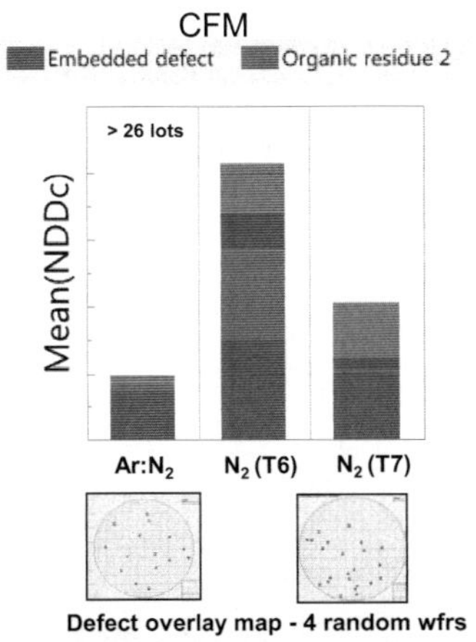

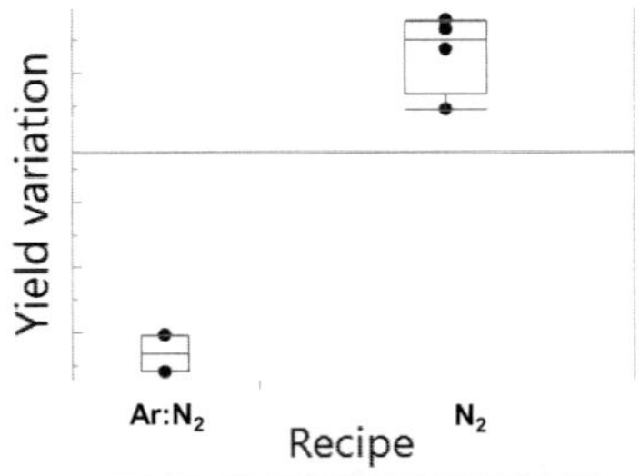

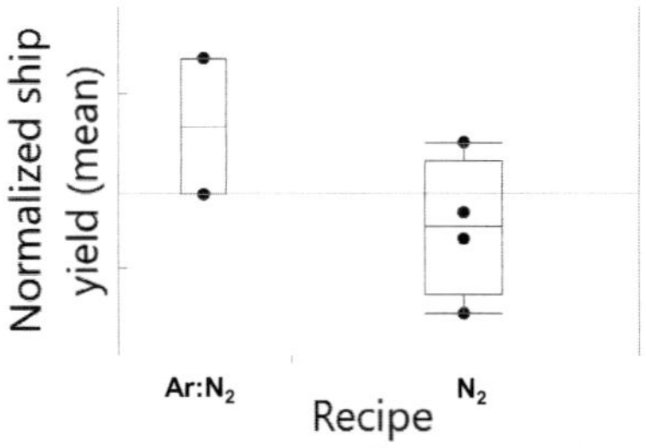

Yield variation ⇩ 74%

0.7 % ⇧ in HVM Yield

NDDc / normalized defect density counts ⇩ > 55 %

Summary:

Ar:N_2 aerosol cleaning show improved inline performance

- **NDDc reduction > 55%**
- **Yield improvement by ~ 0.7%**
- **Yield variation reduced by ~74%**

GLOBALFOUNDRIES

Outline

Background

Particle Removal in Cryogenic Aerosol Cleaning

- Effect of process gas on particles (Sub 28 nm node)

Methods

Throughput Improvements on Ar Aerosol Cleans

- **Recipe parameters tuning**
  - Ar flow usage reduction, process time & throughput

Results

Throughput Improvements on Aerosol clean @ GLOBALFOUNDRIES

- UPH gain and saving (14 nm)

GLOBALFOUNDRIES

Throughput Improvements on Aerosol Cleans:

Recipe parameters tuning

Engineering Process Parameters

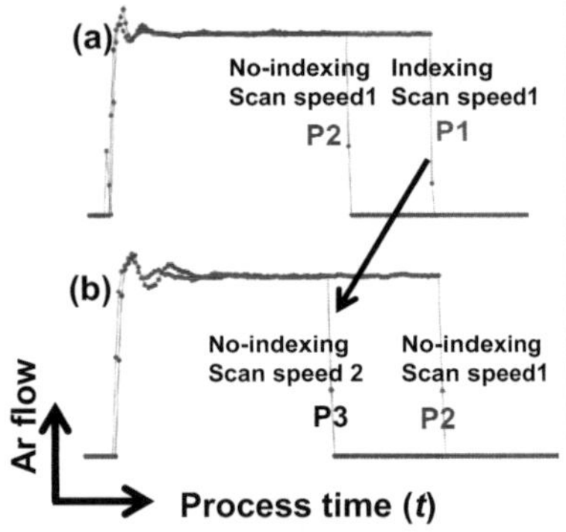

Process	Recipe parameters	T
P1	Indexing, scan speed ***S1***	*t1*
P2	No-indexing, scan speed ***S1***	*t2*
P3	No-indexing, scan speed ***S2***	*t3*

t1 > t2 > t3
Scan speed S2 > Scan speed S1

P1 → P2 (~32% process time reduction)

P2 → P3 (~28% process time reduction)

Tuning parameters

- Process time
- Particle removal efficiency
- Throughput

Summary:

- **P1 → P3, process time reduction by ~ 60%**
- **Throughput gain**
- **Cost of ownership**
- **Ar flow usage reduction**

Throughput Improvements on Aerosol Cleans:

P1, P2, P3: Comparison of True Adders

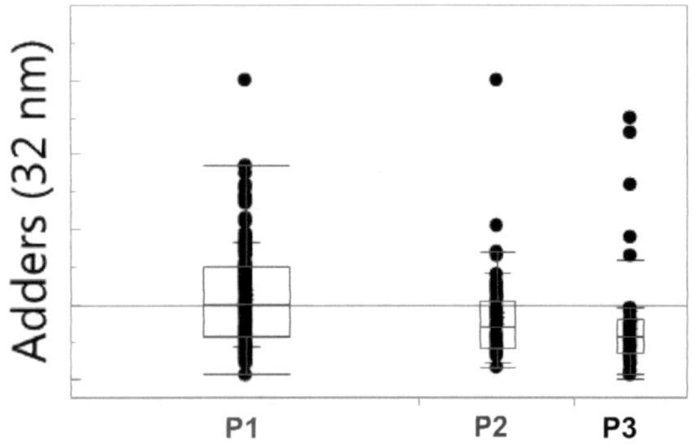

Summary:

- **True adders in, P1 to P2 to P3, are comparable**

Throughput Improvements on Aerosol Cleans

Adders (@ 19 nm) for 14 nm HVM

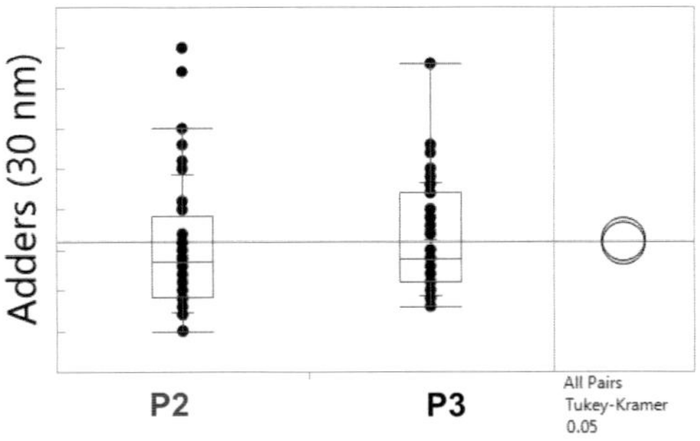

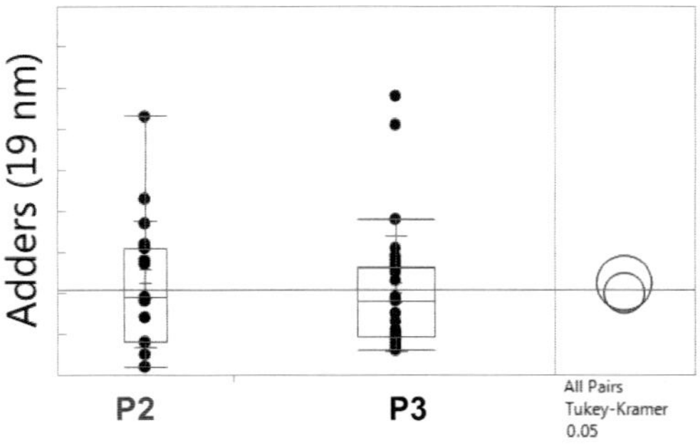

Summary:

- Comparable adders performance @ 19 nm

GLOBALFOUNDRIES

Throughput Improvements on Aerosol Cleans:

28 nm product: Via Opens and Shorts

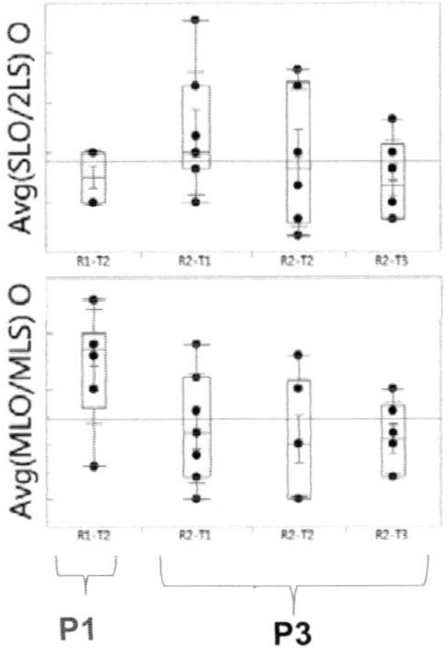

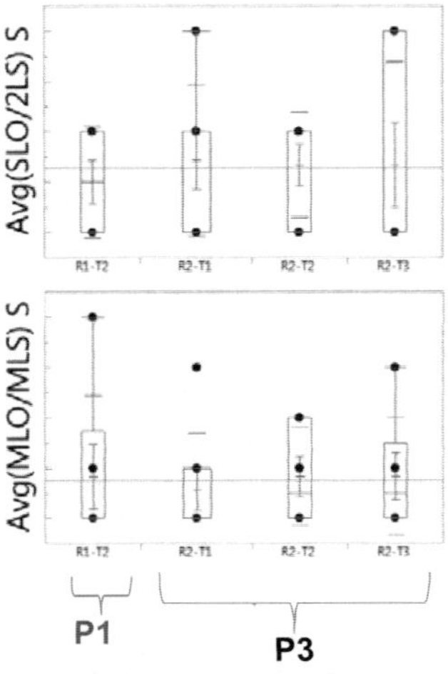

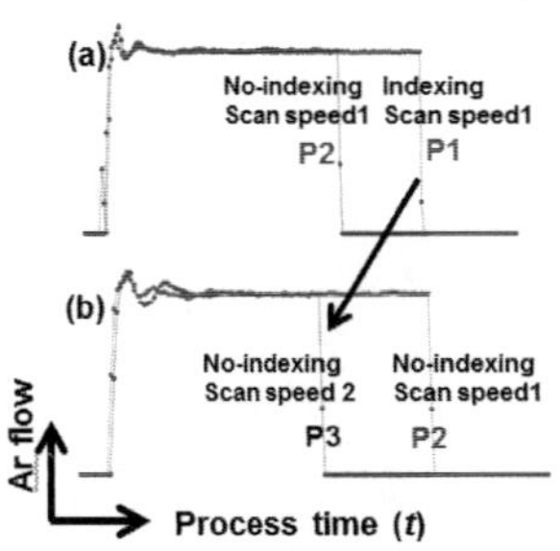

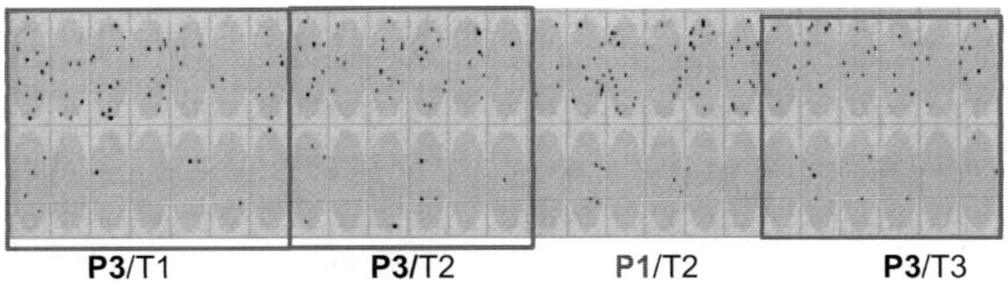

Summary:

P1 and P3 show comparable D_0 (opens and shorts) performance of 28 nm HVM

GLOBALFOUNDRIES

Outline

Background

Particle Removal in Cryogenic Aerosol Cleaning

Effect of process gas on particles (Sub 28 nm node)

Methods

Throughput Improvements on Ar Aerosol Cleans

- **Recipe parameters tuning**
  - Process time, throughput, Ar flow usage reduction

Results

Throughput Improvements on Aerosol clean @ GLOBALFOUNDRIES UPH gain and saving (14 nm)

GLOBALFOUNDRIES

Throughput Improvements Aerosol Cleans:

UPH Gain and Cost Saving

Engineering Process Parameters

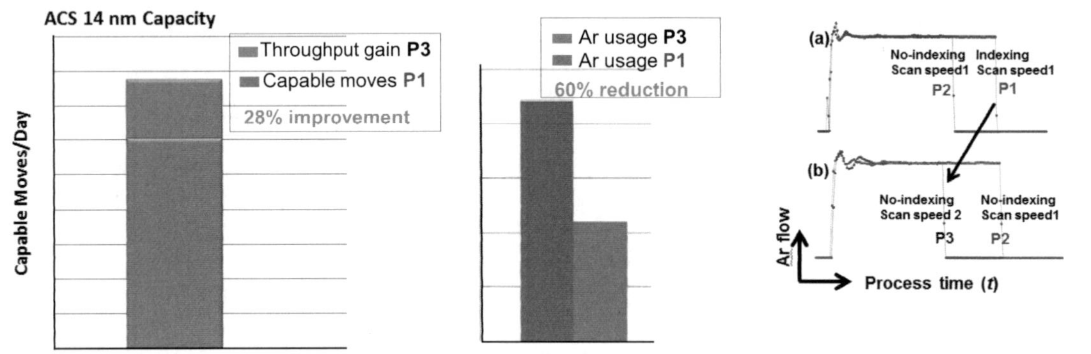

Summary:

P3 implementation at GLOBALFOUNDRIES

- faster throughput
- capable moves
- lowering CoO
- saving on Ar

CapEx saving to date (%)	CoO saving on Ar usage
28%	60%

GLOBALFOUNDRIES

Throughput Improvements Aerosol Cleans

14 nm product: Via Opens and Shorts

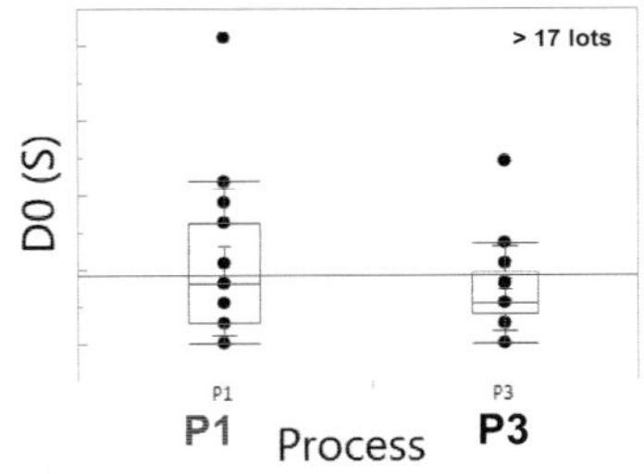

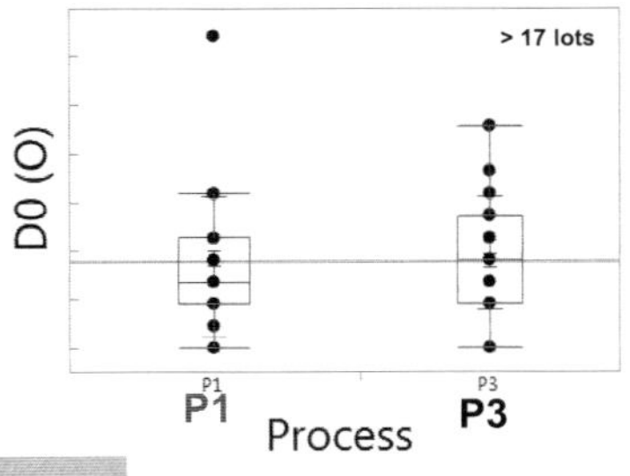

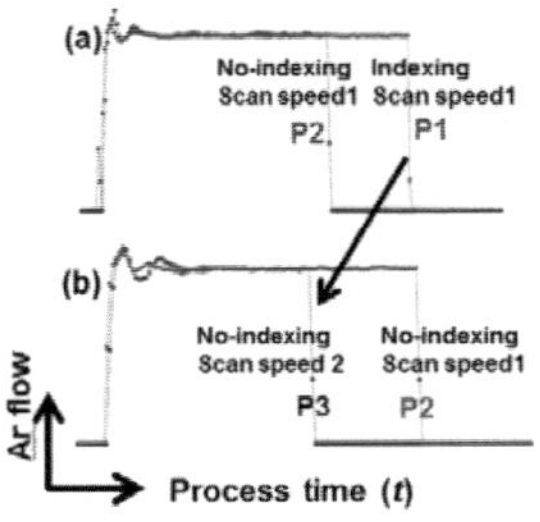

D_0 via opens and shorts are comparable for P1 and P2 process

D_0-Opens/shorts wafer maps

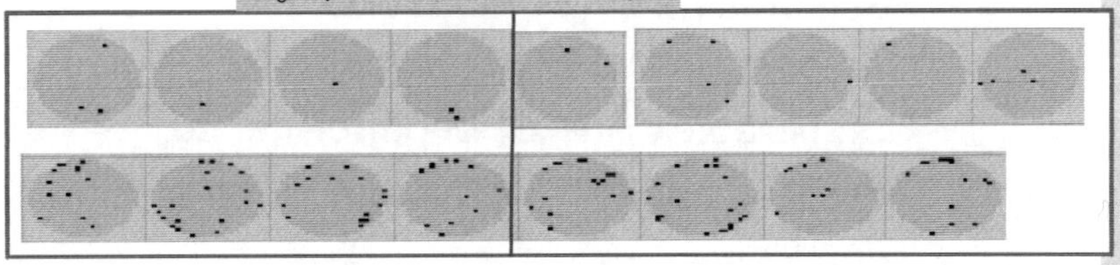

P1 P3

Summary:

- **P1 and P3 show** comparable D_0(opens and shorts) performance on product
- Faster throughput recipe cleaning performance could be maintained at BEOL cleaning process steps

GLOBALFOUNDRIES

April 14, 2016

15

Throughput Improvements Aerosol Cleans

Impact 14 nm CFM Defectivity

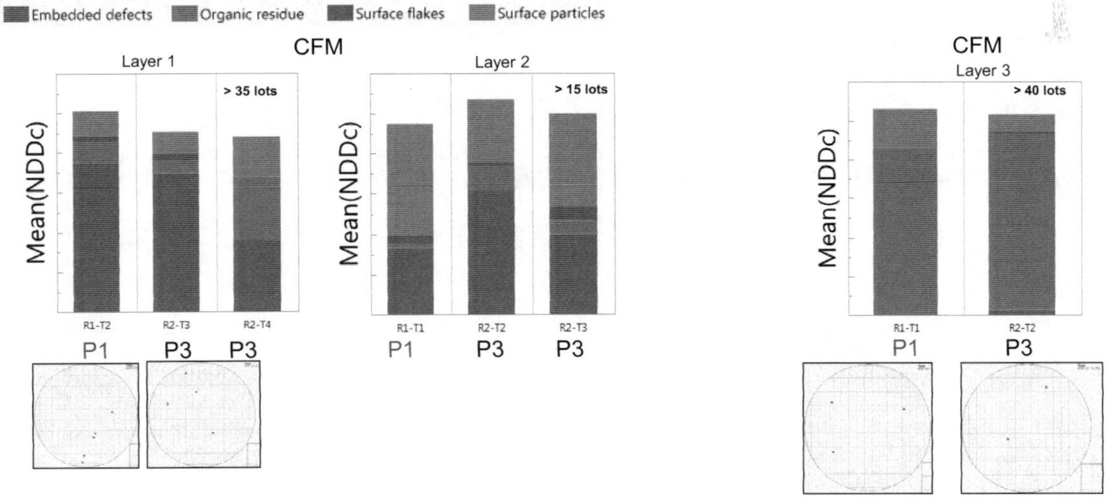

Defect overlay map - 4 random wfrs

NDDc (defects) comparable

Summary:

P1 and P3 show comparable inline CFM performance on 14 nm HVM

GLOBALFOUNDRIES

April 14, 2016

16

Summary and Conclusions

Ar (Ar:N_2) aerosol

- Shows reduced D_0 than N_2
- NDDc improvement > 55%
- HVM Yield improvement by ~ 0.7%

Throughput improvement, P3 implementation at GLOBALFOUNDRIES

- Faster recipe on 14 nm HVM
- ✓ Comparable D_0 (opens and shorts)
- ✓ Comparable inline CFM performance
- Increased capacity with same number of tools
- CoO ⇩ by ~ 28%
- Ar gas consumption ⇩ ~ 60%

April 14, 2016 GLOBALFOUNDRIES 17

Acknowledgements

Grateful thanks to:

- Wet Cleans Colleagues in GLOBALFOUNDRIES

- Process Integration Colleagues in GLOBALFOUNDRIES

April 14, 2016 GLOBALFOUNDRIES 18

Questions?

GLOBALFOUNDRIES®

SPCC 2016 Panel discussion

Jeff Hemphill, INTEL CORPORATION

April 20, 2016

Where are the greatest challenges facing formulated cleans?

Purity

- Raw material control is critical → Deep technical understanding of supply chain
- Vertical integration or custom purification. Consistency of "new" components.
- Extensive filtration required for yield.
- Robust inline SPC; paradigm shift in chemical industry

Cost

- Volume growth drives major price pressure
- Formulation "boundary condition" / Prefer highly aqueous cleans

Selectivity

- Increasing list of process materials → Complex Requirements
- Atomic Layer Control; Surface-termination; Grain structure; Patterned wfr sensitivity

Global Supply Management | 2

What's the best approach to creating a novel formulated clean?

Novel Solution to Existing Problem?	Anticipate New Problems – Proactive Solutions
• Changes are disruptive & resource-intensive • "Crowded" market • Motivation must be large • Major breakthrough in cost, efficiency or performance	• Track industry trends & research • Build a chemistry "Library" • "Shotgun" R&D approach • Early characterization is critical to HVM success

Global Supply Management — intel 3

SPCC Clean Panel Discussion

Cuong Tran - Director, Post-CMP Cleans

April 20, 2016

Entegris
creating a material advantage

Topics of Discussion

Where do the challenges lay in the formulated clean world?

Is collaboration the key to success?

Entegris

Challenges for Formulated Clean

#1 Technical

- Fragmentation
  - Each customer is different, each node is different
- Increasing complexity:
  - More materials, increased selectivity, lower defects
- Lack of relevant test vehicles and metrology
  - Only customer wafers tell the story
- Supply chain, quality and change control requirements tightening

#2 Commercial

- Dilutions are increasing
- ASP Pressure intensifies
- Volume per solution significantly reduced

#3 Competition

- Commodity chemicals
- Traditional competitors
- Customers
  - Increased activity in developing formulations and tolling
- OEMs
  - Driving down chemical consumption to improve CoO for tool sales
- Fast followers
  - Increasing numbers of local companies attacking N-2 business (28nm)

CONFIDENTIAL | 3

Real World Example for PCMP Cleans

	Formulation A	Formulation B
First Sold	2000	2015
Customers	5	1
Fabs	13	2
Tech Nodes	130 – 28nm	10nm
CoA Parameters	~10	>45
Complexity	-	5x
ASP Expectation	-	-20%
Dilution	-	2x
Volumes		

CONFIDENTIAL | 4

Success Depends on Increased Collaboration

Opportunity to "put the pieces together" in the materials space

- Advanced precursors, advanced surface preparation, novel doping
- Materials handling, sensing, and delivery components

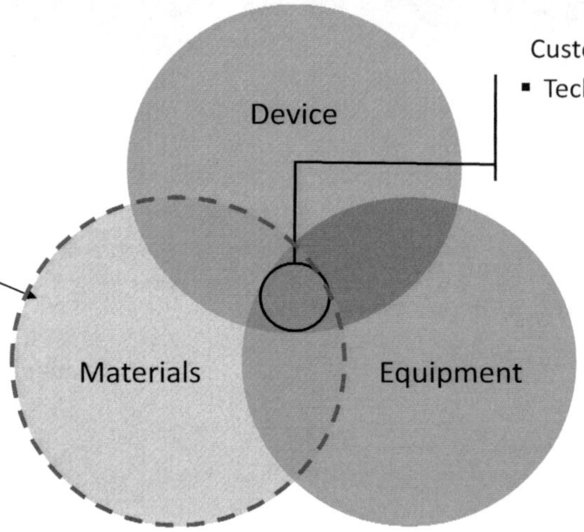

Customers Care About:

- Technology **Performance**
- Process **Yield**
- Device **Cost**

Significant overlap within the Semiconductor Ecosystem

- Targeted collaborations are the key to profitable solutions for all parties

Entegris

Accelerating Development Cycles and Improving Yields

- Combine capabilities to unlock new value
  - Technical capability for rapid innovation
  - Breadth of technologies
  - Global reach, local support
- Enable new paradigms within the semiconductor ecosystem
  - Wet chemistries and wetted surfaces
  - Gases and gas-wetted surfaces
  - Solids and solid material delivery
- Create differentiated solutions to accelerate customer yield ramps

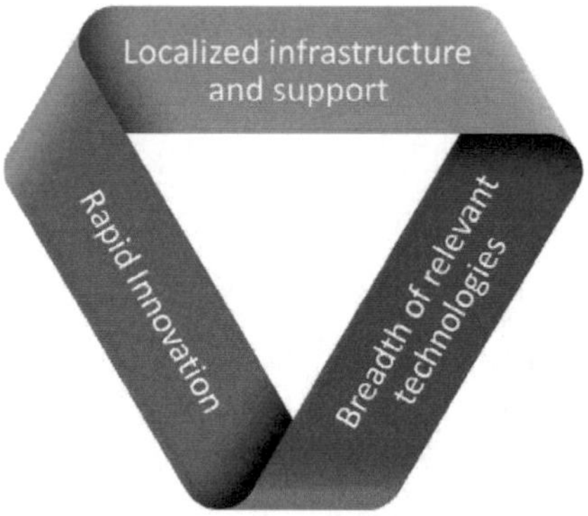

Enable the customers to focus on process integration and product performance

Entegris

Gary Dailey

Vice President, Global Marketing - Electronic Materials

Avantor Performance Materials

gary.dailey@avantormaterials.com

Challenges Facing Formulated Cleans

Technical Challenges

- Device Complexity
- Increased Purity Requirements
- Formulation Complexity
- Speed to Ramp

Commercial Challenges

- Product customization
- Post ramp support
- Supply Chain Management
- Economics are challenging future investments

2

Collaboration Is the Key

- Chemistry can get us where we need to go
- Collaboration is the clear path to success
- Need to consider the differing needs of:
  - Device Manufacturer
  - Formulator
  - Tool Manufacturer

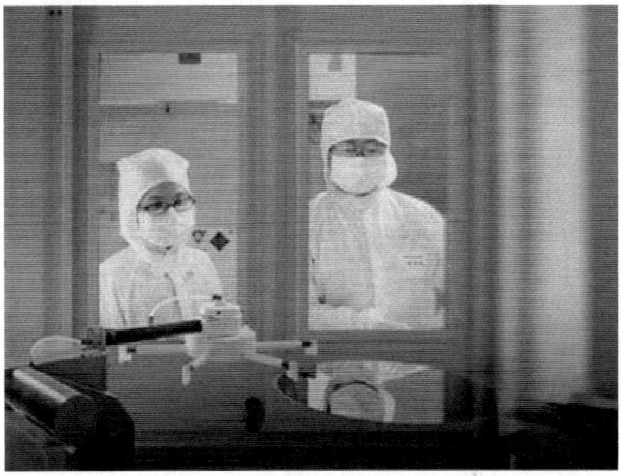

J.T.Baker | AVANTOR Electronic Materials

3

Advancing Cleans

SPCC 2016
Surface Preparation and Cleaning Conference

The Future of Formulated Cleans

Panel Discussion

Tianniu Rick Chen, Ph.D.
Global Business
Marketing Manager

AIR PRODUCTS

Advancing Cleans

Challenges facing formulated cleans

1. Getting closer to each other
   - OEMs/IDMs
   - RMs (raw materials) suppliers
   - Academia/Consortia

2. Adding values
   - vs. high purity bulk solvents/acids
   - vs. less environmentally-benign products
   - Integrated offerings

AIR PRODUCTS

Our approaches

Advancing | Cleans

Getting closer to partners/customers

Global infrastructure/Local support

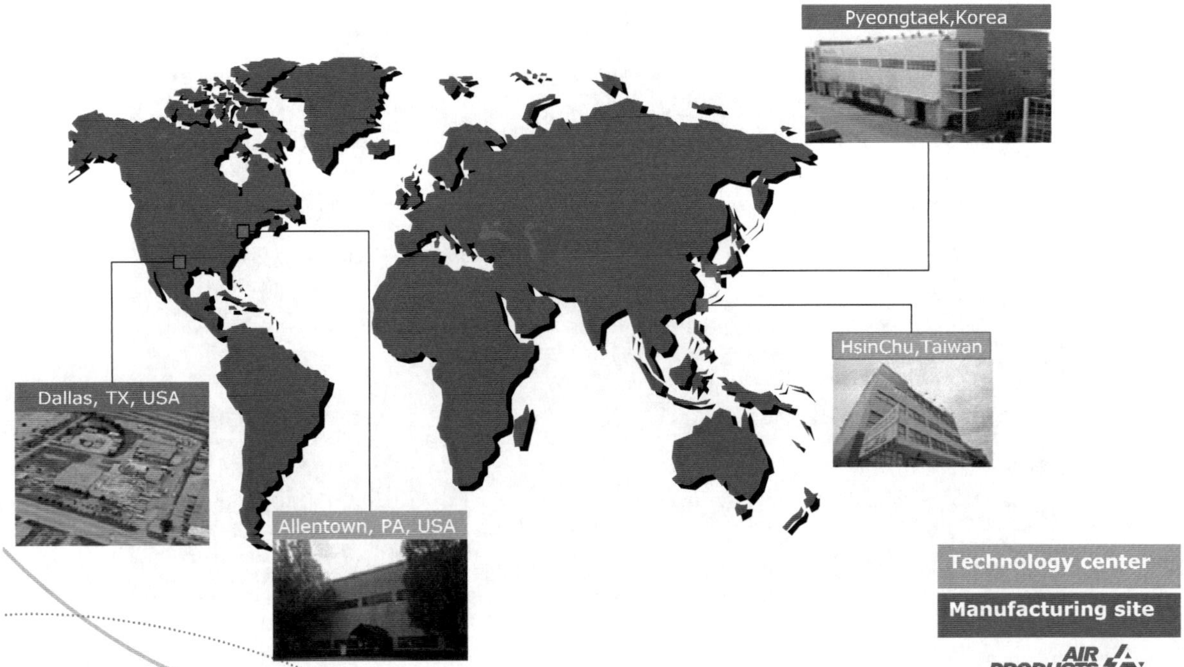

AIR PRODUCTS

3

Our approaches (continued)

Advancing | Cleans

Adding values

Integrated Offerings

Air Products Electronic Materials Division:
The Materials Partner of Choice of the Semiconductor Industry

- Focus on Semiconductor (IC) Materials where materials provide low-cost in use/high value in use
- Leverage technology leadership, global scale, quality and reliability capabilities, and partnership with customers and OEMs to develop and commercialize the next generation technologies which will advance the industry
- Expand into adjacent segments within IC

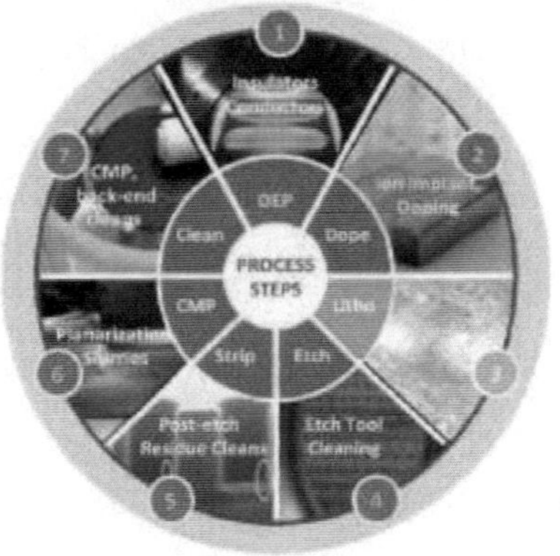

AIR PRODUCTS

4

Our approaches (continued)

Advancing Cleans

Adding values

Integrated Offerings

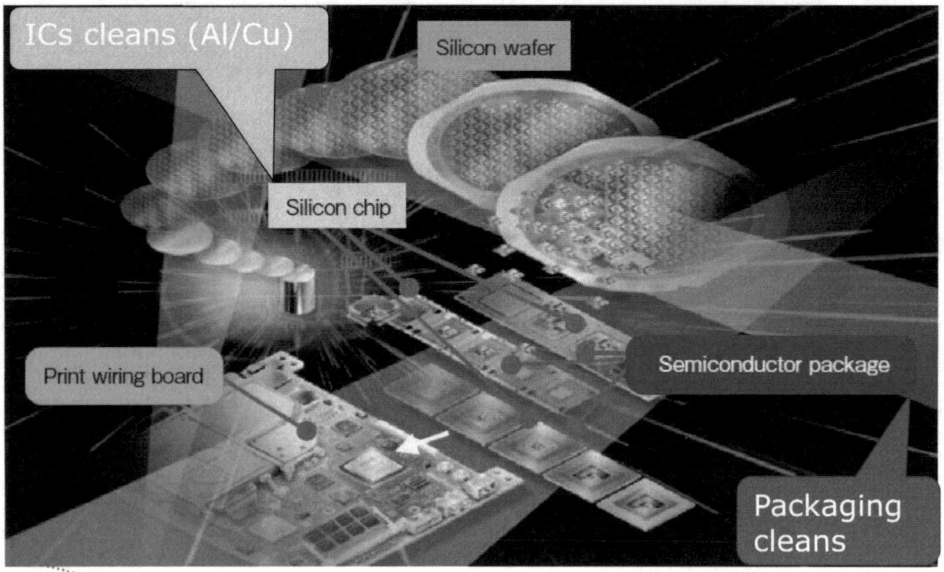

* Nakamura, H.; Katogi, S. *Hitachi chemical technical report*, *55*, pp 24-29.

AIR PRODUCTS

5

What can we help you advance?

airproducts.com/advancing

CHENT7@airproducts.com

AIR PRODUCTS

6

Study of particle attachment on silicon wafers during rinsing

Kurita Water Industries Ltd.
Takeo Fukui, ○Koji Nakata

KURITA

Outline

KURITA

Introduction : Requirement for Ultrapure water quality

(ITRS2013 Yield Enhancement Table YE3)

	2013	2014	2015	2016	2017	2018	2019
Critical particle size[nm]	**20**	**17.9**	**15.9**	**14.2**	**12.6**	**11.3**	**10**
Number of particles[#/L]	**1000**	**1000**	**1000**	**1000**	**1000**	**1000**	**1000**

According to the ITRS, the number of particles should be reduced less than 1000 #/L for the size of at 10 nm particles by the year of 2019.

KURITA

Introduction : Particle monitoring 1990s to 2019

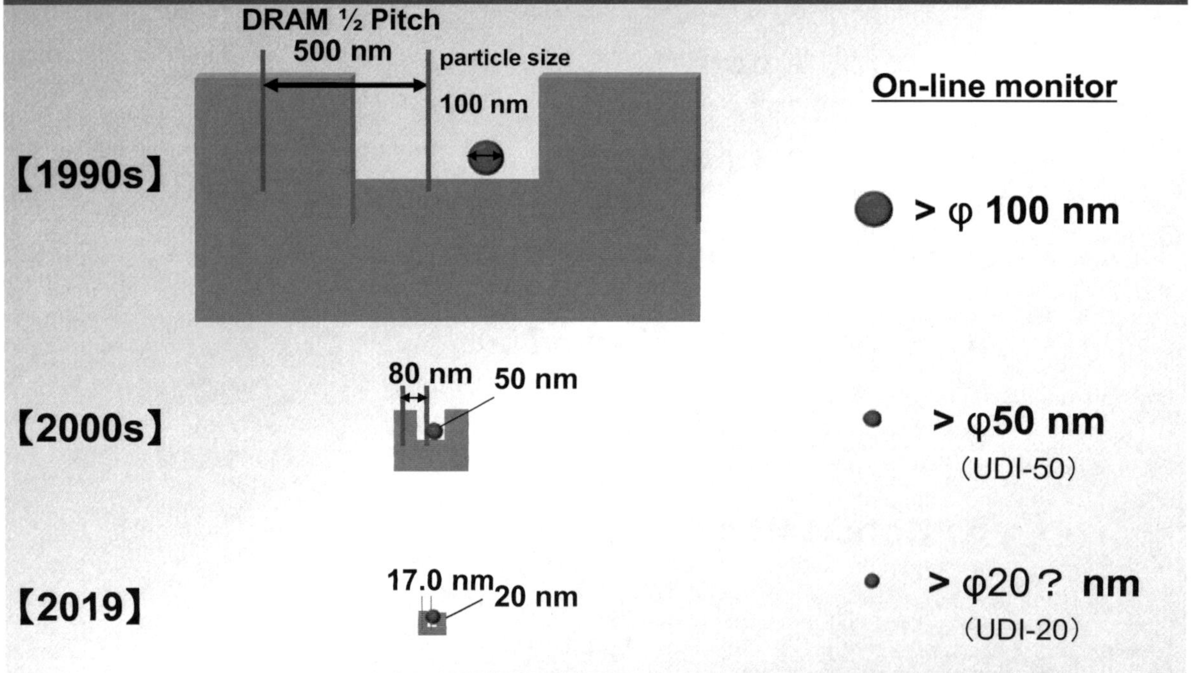

It has been reaching the size detection limit of LPC

KURITA

Introduction : Requirement for wafer cleanliness by the ITRS

（ITRS2013 Front End Process Table FEP11）

	2013	2014	2015	2016	2017	2018	2019
Critical particle size[nm]	14.2	12.0	11.3	10	8.9	8.0	7.1
Number of particles[#/wafer]	12.6	34.2	34.2	34.2	34.2	34.2	34.2

According to the ITRS, the number of particles on the wafer should be controlled very severely.

KURITA

Introduction : Purpose of this study

<Needs for particle in UPW>

- **How should we detect the smaller particle ?**
- **How should we reduce the particle in UPW?**
- **What level should we remove particle in UPW to? ①**
- **What impurity increases particle adhesion on wafer ②**

Focus of this study

Purpose of this study

To clarify how the particle in UPW adsorb onto wafer surface

KURITA

Experimental Method① : Experimental condition and flow

<Experimental condition>

- Wafer : 150 mm SiO_2/Si(100)
- Injected particle : SiO_2
- Injected particle size : around 100 nm
- Particle concentration in UPW : 20,000 #/mL
- Pre-cleaning : O_3 (30ppm) water rinse in 180 sec

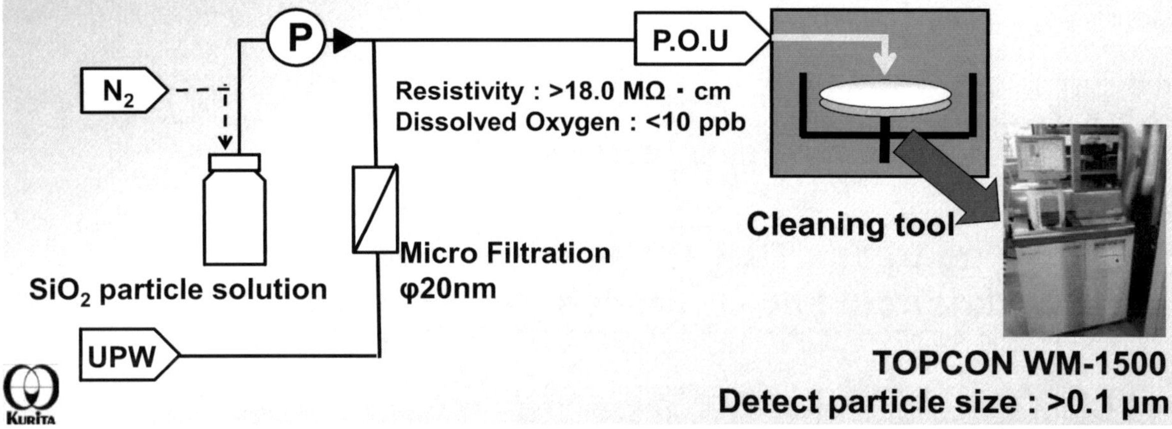

Experimental Method ① : Rinsing recipe

< Rinsing recipe >

Operation No.	1	2	3	4	5	6
Rotation speed [rpm]	500	300	200	ω 1 (1000)	ω 2 (1000)	200
Time [sec]	2	5	5	t_1 (20)	t_2 (40)	5

Rinsing process (Operations 2–4)

Drying process (without N2 gas) (Operations 5–6)

The test conditions(ω and t) were changed in order to clarify the behavior of particle adsorption on wafer surface.

Result ① : Effect by the rinsing time on particle adsorption

Operation No.	1	2	3	4	5	6
Rotation speed [rpm]	500	300	200	1000	1000	200
Time [sec]	2	5	5	t_1	40	5

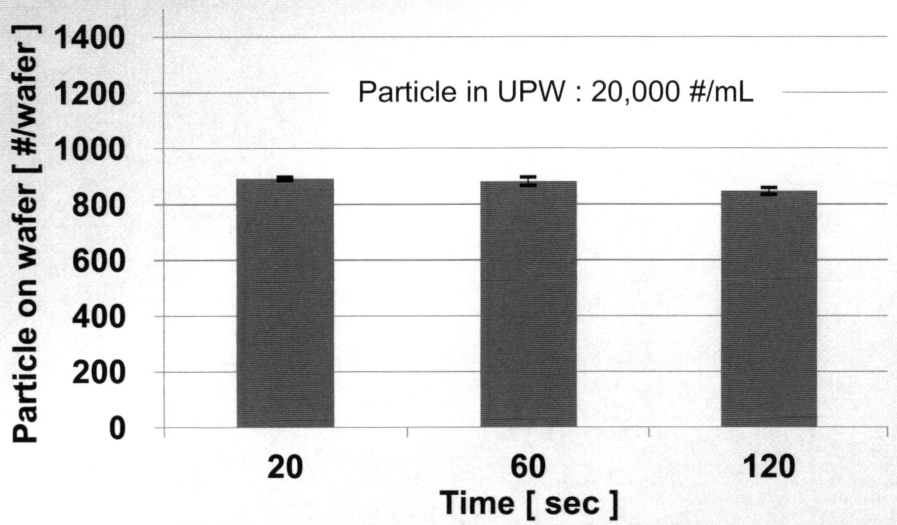

Rinsing time does not affect the number of particle on wafer.

Result ① : Effect by the rotation speed in rinsing process on particle adsorption

Operation No.	1	2	3	4	5	6
Rotation speed [rpm]	500	300	200	ω_1	1000	200
Time [sec]	2	5	5	20	40	5

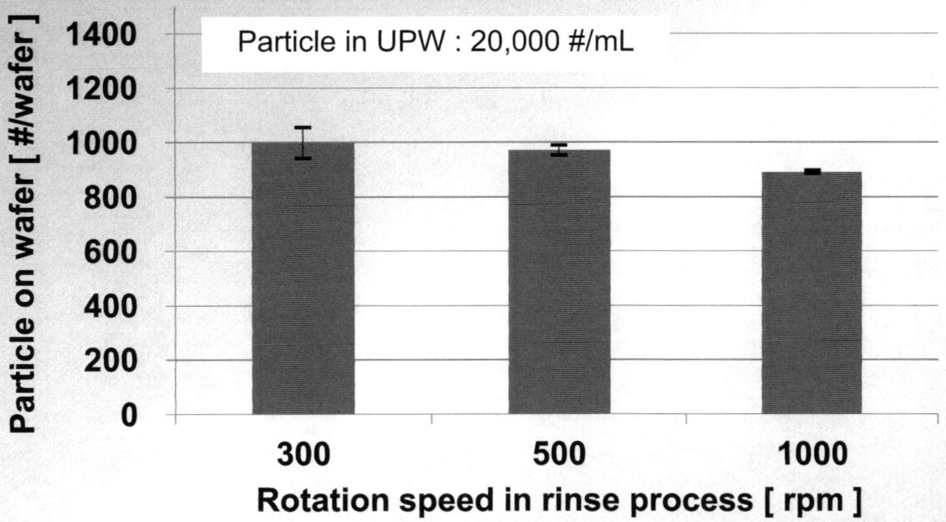

Rotation speed in rinsing process does not affect the number of particle on wafer.

Result ① : Effect by the drying time on particle adsorption

Operation No.	1	2	3	4	5	6
Rotation speed [rpm]	500	300	200	1000	1000	200
Time [sec]	2	5	5	20	t_2	5

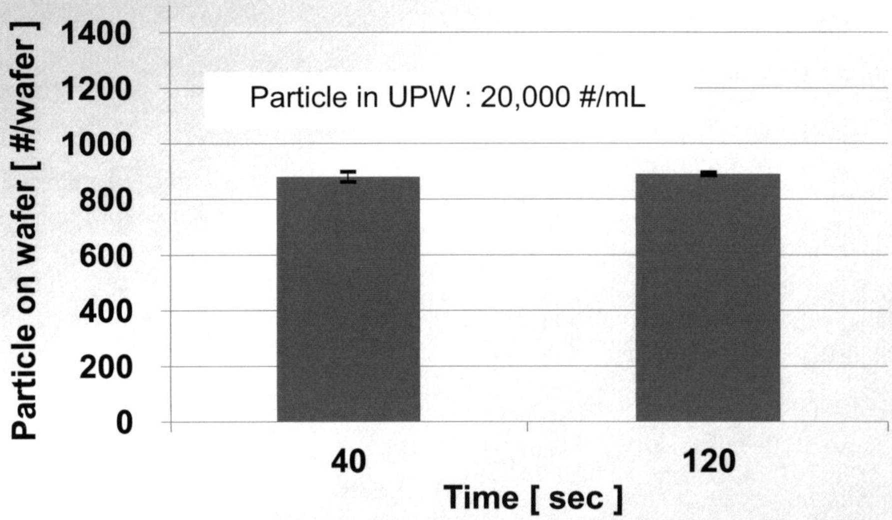

Drying time does not affect the number of particles on wafer.

KURITA

Result ① : Effect by the rotation speed in drying process on particle adsorption

Operation No.	1	2	3	4	5	6
Rotation speed [rpm]	500	300	200	1000	ω_2	200
Time [sec]	2	5	5	20	40	5

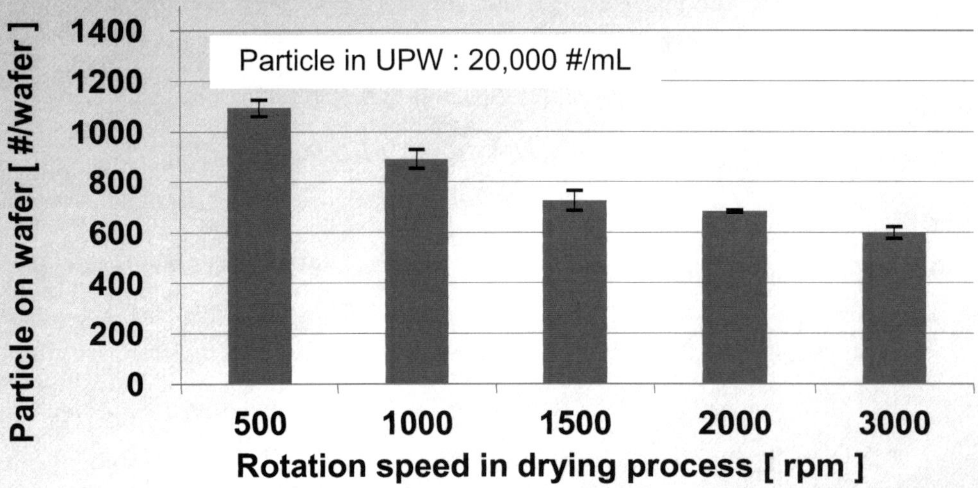

Particle on wafer decreases as rotation speed in drying process increases.

KURITA

Discussion ① : The model of particle adsorption onto wafer surface

We divided two process of particle adsorption onto wafer surface. One is rinsing process and another is drying process. We evaluated each of effect.

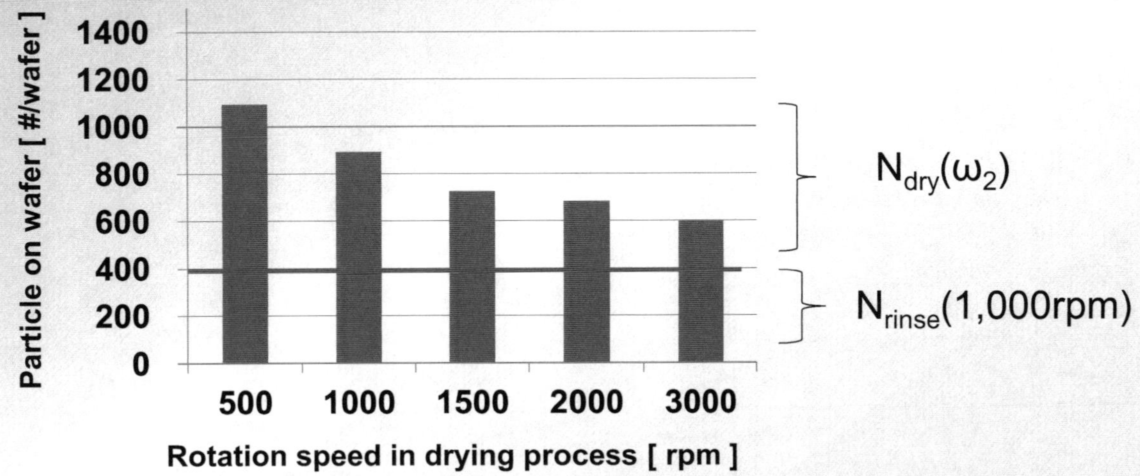

$$N = N_{rinse}(\omega_1) + N_{dry}(\omega_2)$$

KURITA

Discussion ① : The model of particle adsorption onto wafer surface

1.In the rinsing process

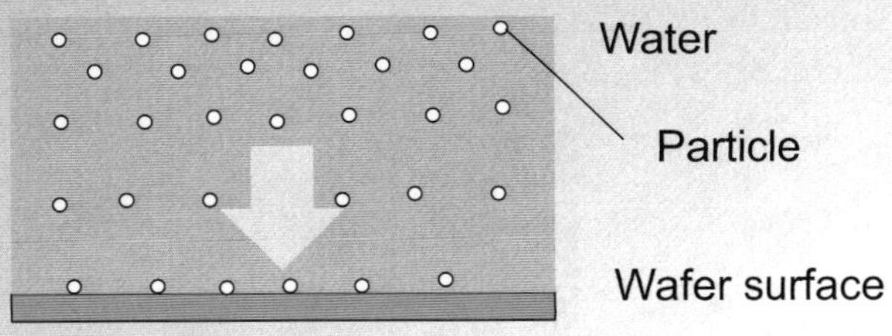

➢Particle adsorption to wafer surface by equilibrium reaction

2.In the drying process

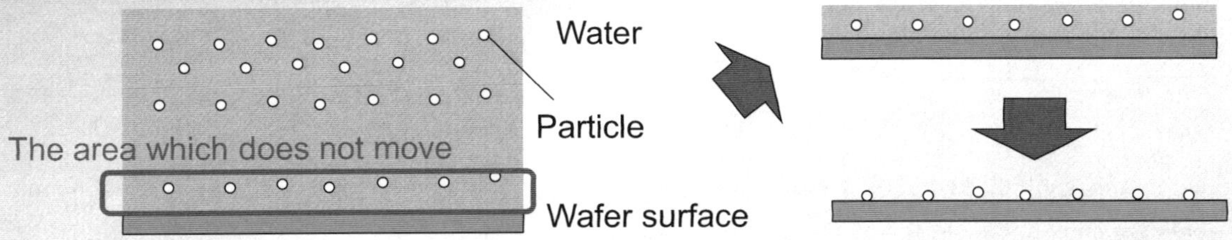

➢Particles in the boundary layer accumulate on the wafer surface by water evaporation.

KURITA

Discussion ① : How to calculate water property

The water thickness and the average velocity of water are shown as the equation (1) and (2).

⬇

The boundary thickness (3) can be calculated by the results of equation (1) and (2).

⬇

The number of particles adsorbed on the wafer is calculated.

$$h = e\frac{r}{(r+l)}\frac{v_0}{v_{av}} \quad (1)$$

$$v_{av} = \sqrt{\frac{2(E-\rho ge)}{\rho}} \quad (2)$$

$$E = \frac{1}{2}\rho v_{av}^2 + \rho ge = \frac{1}{2}\rho v_0^2 + \rho ge + \int_r^{r+l} dE$$

$$dE = (C_f - F_f)dl = \left(\rho(r+l)\omega^2 - \frac{3\mu v_{av}}{h^2}\right)dl$$

$$v(z) = -\frac{3v_{av}}{2h^2}(z^2 - 2hz) \quad (3)$$

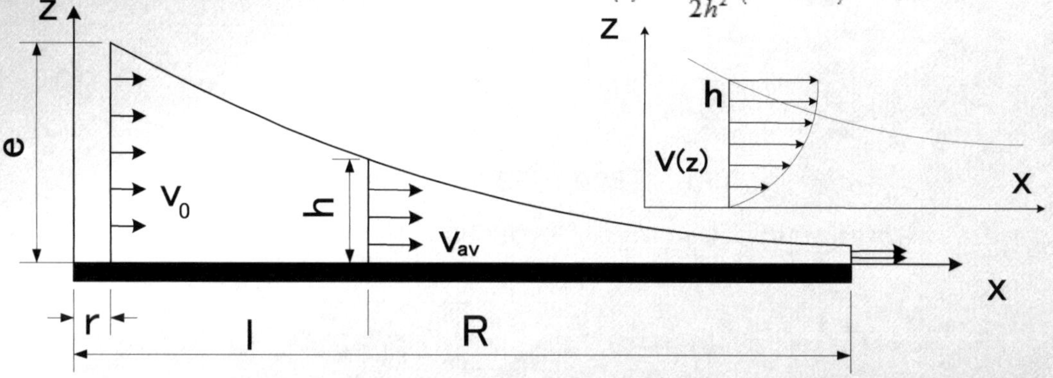

Discussion ① : Model calculation of spin cleaning

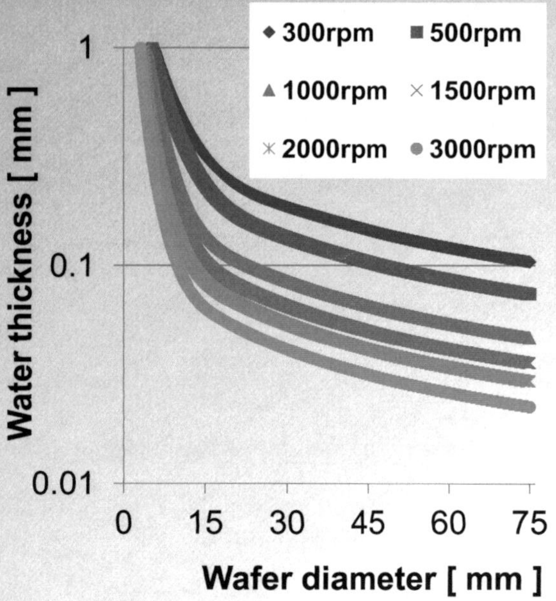

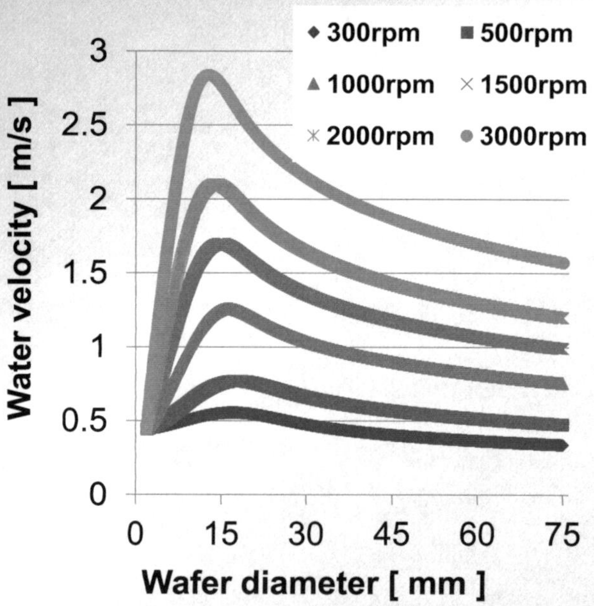

Water thickness and water velocity were calculated as functions of distance from the center of wafer.

Discussion ① : Boundary layer calculation

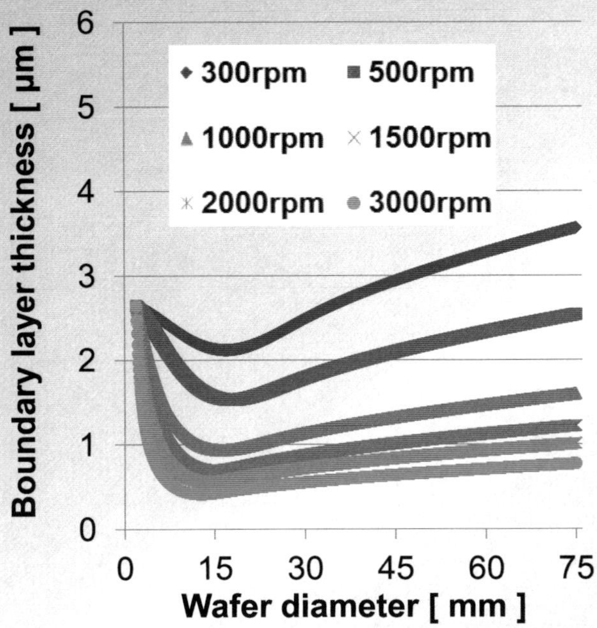

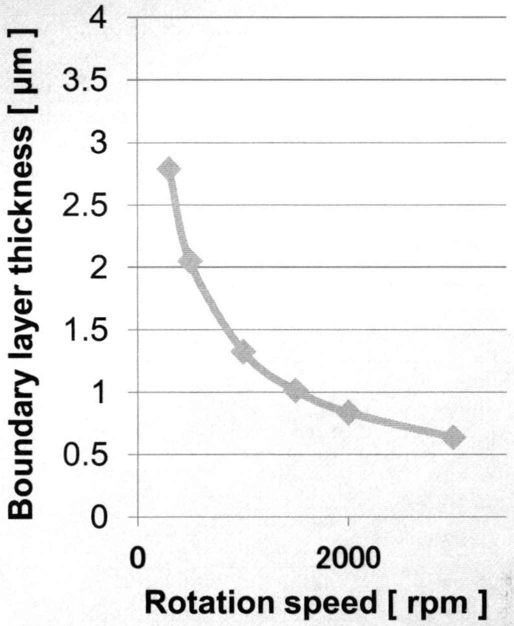

Boundary layer was calculated as a function of distance from the center of wafer.

KURITA

Discussion ① : Comparison between the model and experimental result

$$N = N_{rinse}(\omega_1) + N_{dry}(\omega_2)$$

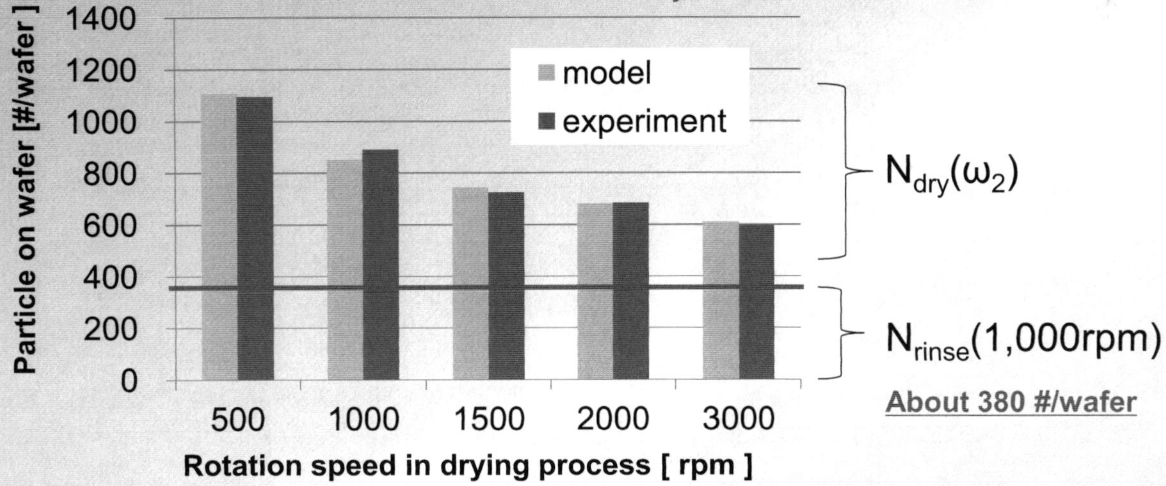

- The calculation result well fitted with experimental result.
- The impact of rinsing process is roughly the same degree as that of drying process and cannot be ignored.

⇒ To avoid adhering particles to wafer on UPW cleaning process, the number of particle in UPW should be decreased.

KURITA

Experimental Method ②: Experimental condition and flow

<Experimental condition>

- Wafer : 300 mm SiO_2/Si and SiGe(Ge:50%)/Si
- Injected particle : SiO_2
- Injected particle size : 50 nm (for SiO_2/Si), 150 nm (for SiGe/Si)

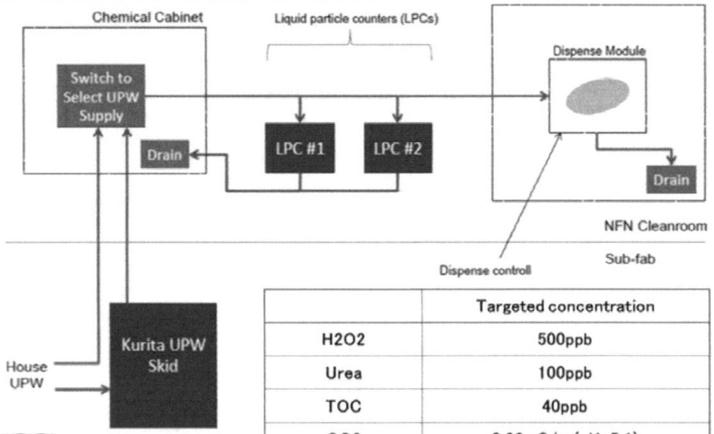

	Targeted concentration
H2O2	500ppb
Urea	100ppb
TOC	40ppb
CO2	0.26mS/m (pH=5.1)
O2	900ppb

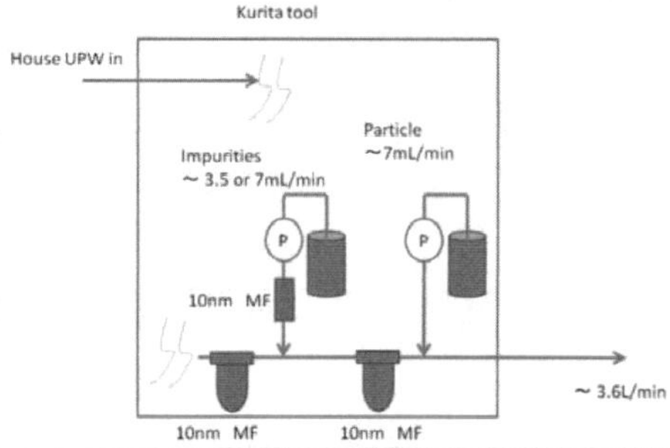

Kurita's water quality control unit

Experimental Method ② : Kurita's water quality control unit

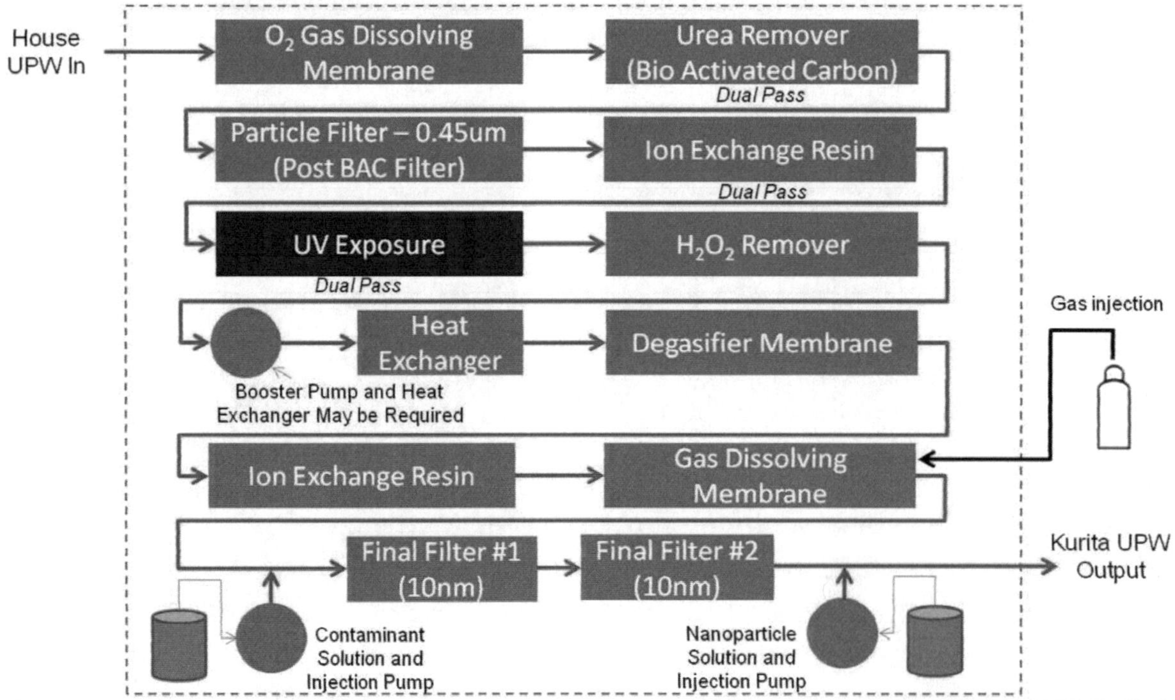

Result & Discussion ②: Effect of chemicals on SiO_2/Si wafer and SiO_2 particle

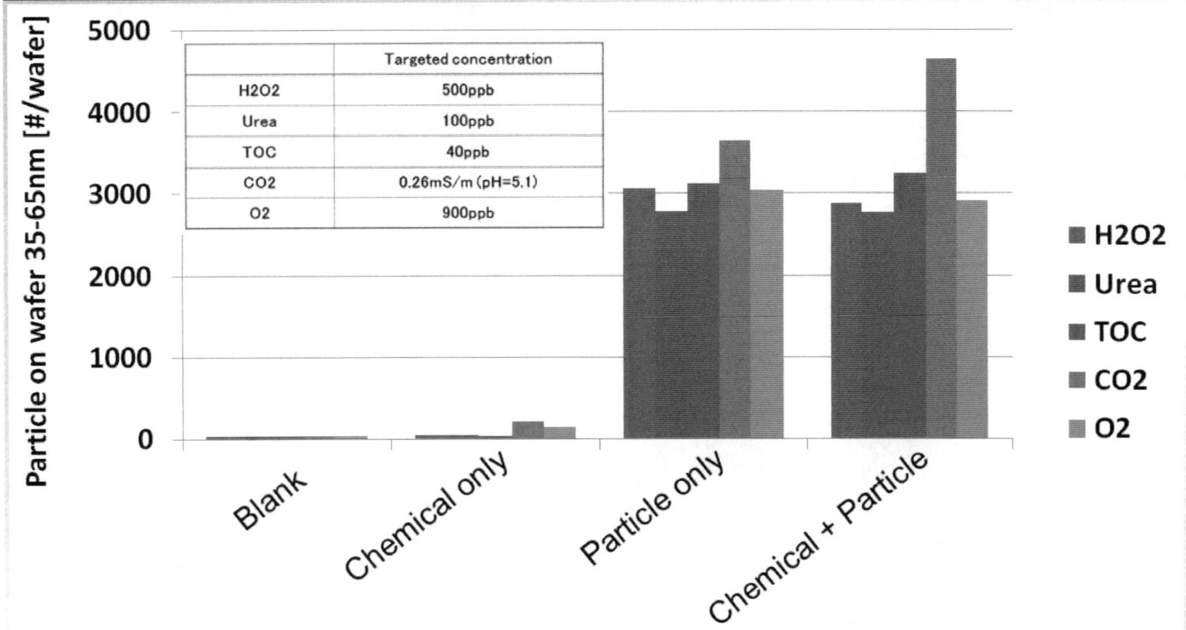

CO_2 has the influence on particle adhesion to SiO_2 surface.

KURITA

Result & Discussion ②: Effect of chemicals on SiGe/Si wafer and SiO_2 particle

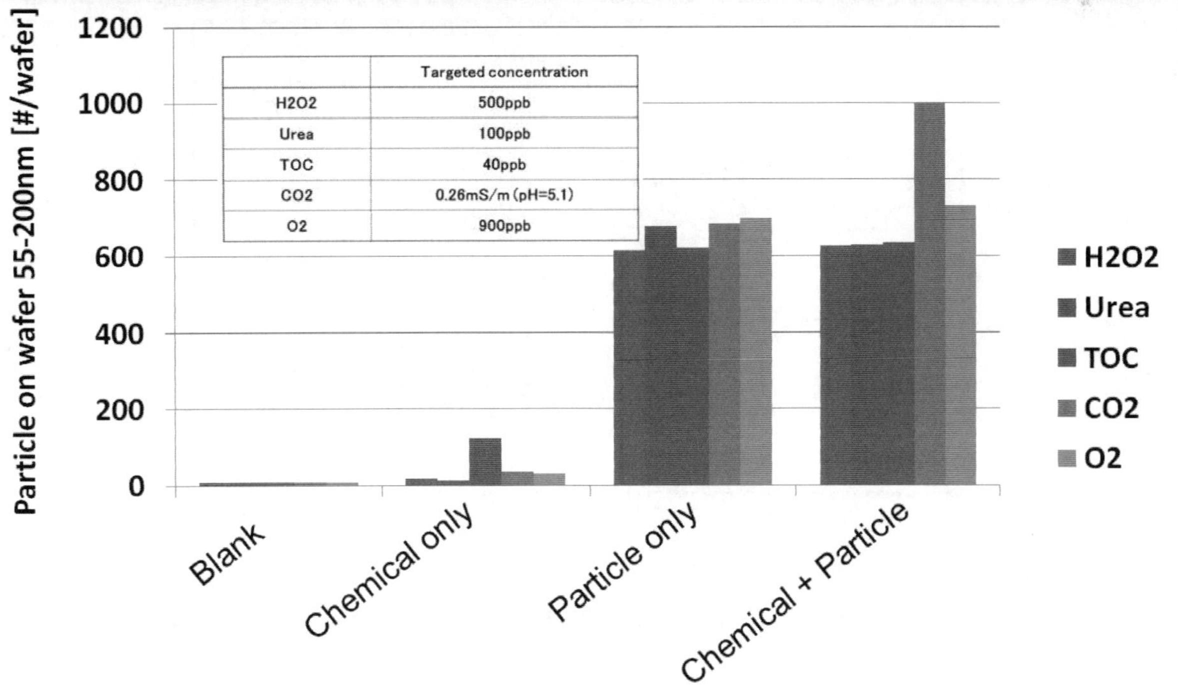

CO_2 has the influence on particle adhesion to SiGe surface.

KURITA

Result & Discussion ②: Effect of CO_2 water on SiO_2/Si wafer

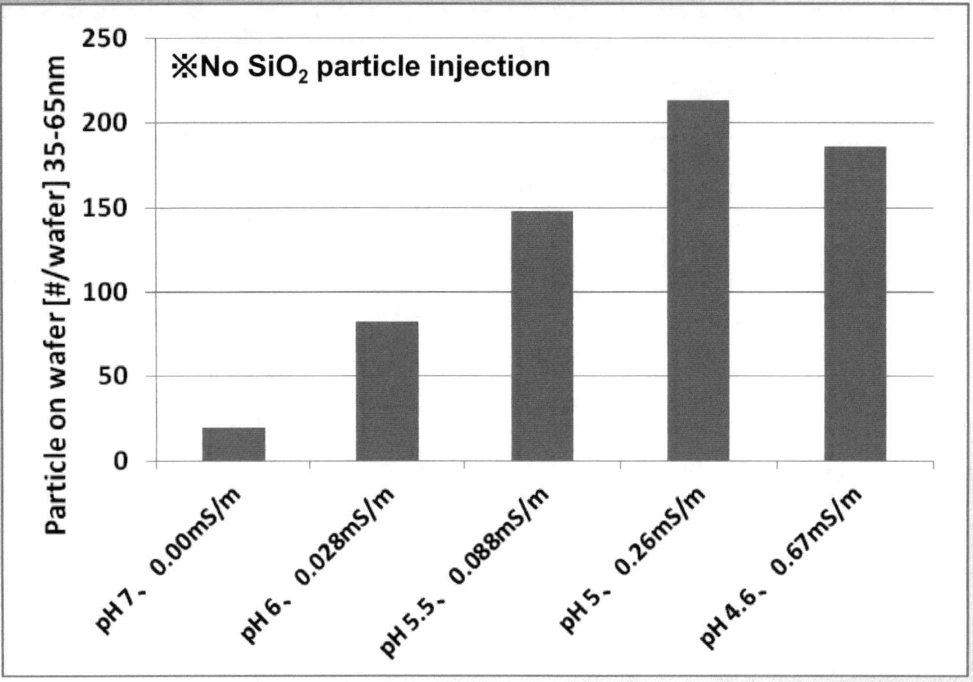

CO_2 has a strong influence on the number of particle on wafer.

KURITA

Result & Discussion②: Relationship between zeta potential and pH

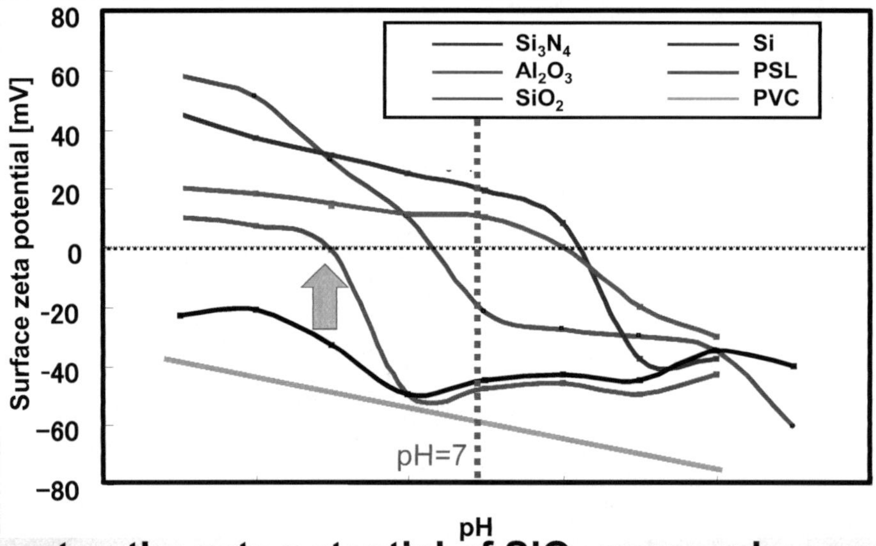

In CO_2 water, the zeta potential of SiO_2 approaches zero , resulting in the drop of the repulsive force between the wafer and particles.
⇒ We would like to discuss the effect of the surface charge by changing particle species in the next study.

KURITA

Conclusion

We studied about the particle adhesion to wafer and the impact of impurity.

(1) The rotation speed of drying process is one of the most effective parameters in the particle adsorption.

(2) The number of particles adsorbed at the rinsing process is large and almost constant.

(3) CO_2 has the influence on particle adhesion to SiGe and SiO_2 surface to wafer.

Next, we would like to investigate the effect of the surface charge and the particle size in detail.

KURITA

We can support quality stabilization of a semiconductor products.

Not just water, Creativity

Thank you for your attention

KURITA WATER INDUSTRIES LTD.

Al Corrosion-free Photoresist Stripping and Etch Residue Removal Process with Dilute Halide Solutions at Room Temperature

[1]Steve Sangwook Ryu
Email: steve.ryu@avantormaterials.com
Phone: +82-10-2784-9515

[2]Janglae Park, Miyeon Oh, Chia Cheng Yang, Ph.D.
Chien-Pin Sherman Hsu, Ph.D.
1 Senior Application Manager in Korea, 2 Application Engineer in Korea

J.T.Baker®

AVANTOR™ Electronic Materials

Objectives

- Demonstrate removal of photoresist (PR) and etch residue while maintaining high compatibility with Al/TiN/SiO2/SiN materials
  - Minimize corrosion defects post plasma etch process
  - Prevent galvanic corrosion during cleaning process

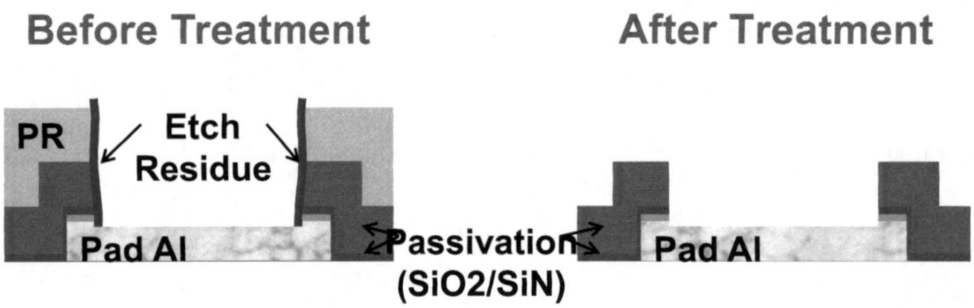

2

Challenges

- Removal of PR and PERR on Al surfaces using halide-containing materials which typically cause corrosion of the Al bond pads

J.T.Baker | AVANTOR Electronic Materials

3

Challenges

Al Corrosion by Fluorocarbon Pathway

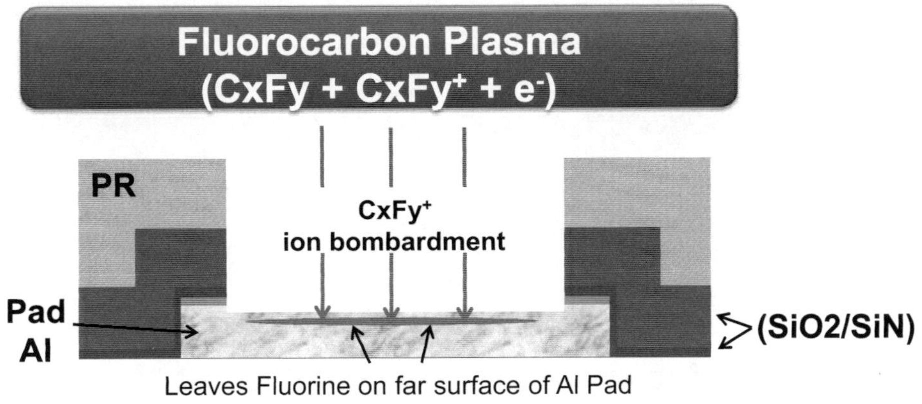

<u>Pad Al Corrosion by AlFx and AlxOyFz Formation</u>

$[AlF_x]^{(x-3)-} + 3e- \Leftrightarrow Al + 6F-$ (X=3 or 6)

$[AlF_x]^{(x-3)-} + Al_2O_3 \rightarrow Al_xO_yF_z$

J.T.Baker | AVANTOR Electronic Materials

4

Challenges

Al Corrosion by Formation of Galvanic Cell

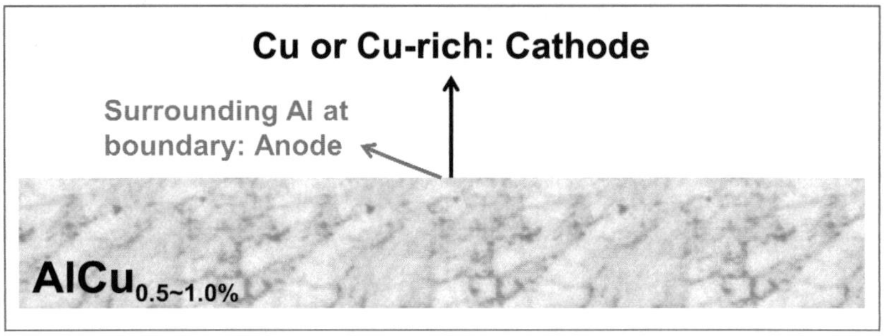

Anode: Al ⇔ Al^{3+} + 3e : - 1.662 V
Cathode: Cu^{2+} + 2e ⇔ Cu : + 0.337 V

O_2 + $2H_2O$ + 4e ⇔ 4 OH^-
4Al + $3O_2$ + $6H_2O$ ⇔ $4Al^{3+}$ + $12OH^-$ ⇔ $4Al(OH)_3$

J.T.Baker | AVANTOR Electronic Materials

5

Testing Plan

Goals:

- Demonstrate that halide-containing product (XM-426) does not remain on the surface following cleaning
    - XM-426 removes photo resist from the wafer surface
    - XM-426 cleans post-etch residue

Method:

- Expose blanket and pattern wafers to CxFy etch processing
- Perform cleaning by XM-426

Analysis:

- **Time of Flight SIMS (TOF-SIMS):** Depth profile analysis for halide contamination remaining on Al bond pad
- **Auger Electron Spectroscopy (AES):** Grazing angle analysis for halide remaining on the Al surface
- **SEM Image Analysis** for corrosion analysis and cleaning performance
- **Optical Imaging Analysis** for corrosion analysis

J.T.Baker | AVANTOR Electronic Materials

6

Test Parameters

Sample ID	Test Condition	Optical Analysis	SEM Analysis	TOF-SIMS	AES	Purpose
S1	Before chemical treatment	X	X	X	X	Measure residual fluoride before cleaning
S2	Chemical treatment without DI rinsing (air dried)	X	X	X	X	Reference for DI rinsing effect
S3	Normal chemical treatment with DI rinsing	X	X	X	X	DI rinsing effect
S4	Chemical treatment with multiple concentration (3x) of halide component	X	X	X	X	Examine corrosion by increased halide concentration

J.T.Baker | AVANTOR Electronic Materials

7

Results – Corrosion Prevention of Al Pad after Cleaning

Optical Microscope Inspection of Al Bond Pads

(S1) Before Cleaning | (S2) Cleaning w/o DI Rinsing | (S3) Cleaning w/ DI Rinsing | (S4) Cleaning with 3X of Halide

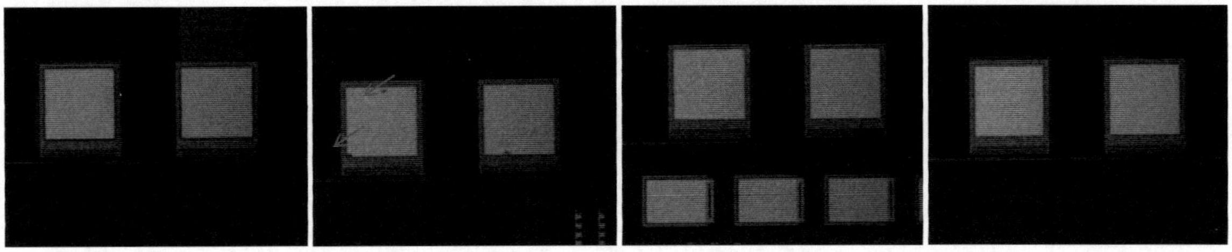

- Sample S2 shows corrosion on Al pad after chemical treatment without DI rinsing
- Samples (S1, S3, S4) show no corrosion following processing

J.T.Baker | AVANTOR Electronic Materials

8

Results – Corrosion Prevention of Al Pad after Cleaning

SEM Inspection of Al Bond Pad

(S1) Before Cleaning	(S2) Cleaning w/o DI Rinsing	(S3) Cleaning w/ DI Rinsing	(S4) Cleaning with 3X of Halide

- Samples S1 and S2 shows residue on Al Pad
- Samples S3 and S4 show a clean surface free of corrosion

J.T.Baker | AVANTOR Electronic Materials

9

Results – Corrosion Prevention of Al Pad after Cleaning

TOF-SIMS Analysis

Total Count of Halogen: Without DI Rinse

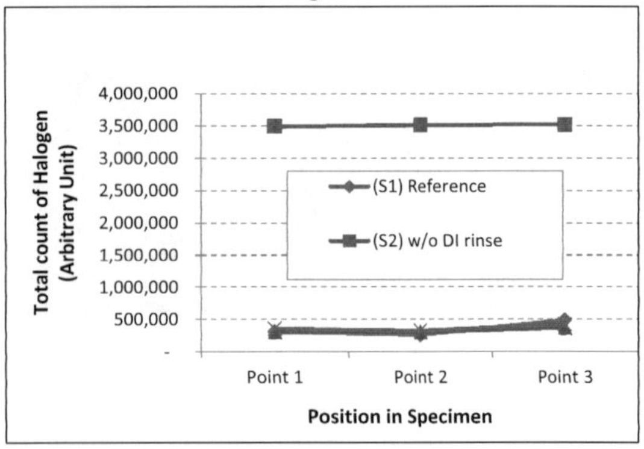

- Dominant existence of halide component was observed in S2

Total Count of Halogen: After DI Rinse

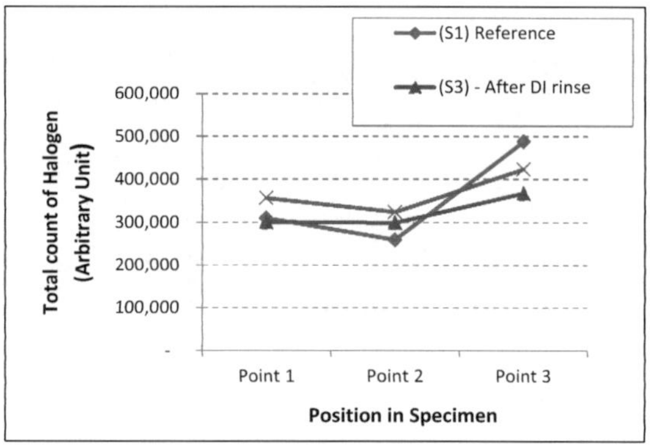

- Following the clean and rinse process, the halide concentration is within range of the control sample

J.T.Baker | AVANTOR Electronic Materials

10

Results – Corrosion Prevention of Al Pad after Cleaning

AES Analysis: Spectra of S1, S3 and S4

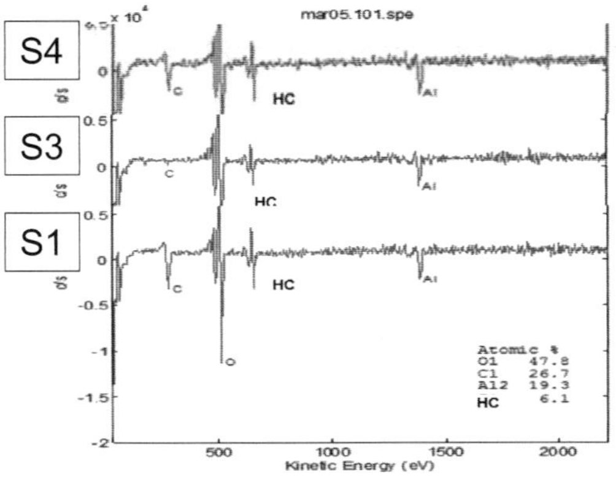

Sample	Al	C	O	HC
S1	19.3	26.7	47.8	6.1
S2	8.0	82.6	6.7	2.5
S3	23.6	N/A	67.8	4.1
S4	22.3	20.7	51.7	5.4

- S1 and S4 show similar HC peak/Al peak ratio
- S3 peak ratio shows HC reduced compared to S1 and S4
- S2 appears to be masked by residue

11

Results – Corrosion Prevention of Al Pad after Cleaning

Summary of Analysis

Variable	Optical Analysis	SEM Analysis	TOF-SIMS	AES
(S1) Before chemical treatment	No corrosion	Residue visible	Minimum halide signature	Baseline
(S2) Chemical treatment without DI rinsing (air dried)	Residue/ corrosion visible	Residue visible	Halide signature	Halide masked by residue
(S3) Normal chemical treatment with DI rinsing	No corrosion	No residue/ corrosion	Minimum halide signature	Halide signature reduced
(S4) Chemical treatment with multiple concentration (3X) of halide component	No corrosion	No residue/ corrosion	Minimum halide signature	Halide signature similar to S1

12

Results – Cleaning and Stripping Performance

Photoresist Cleaning and Material Compatibility

60 seconds @ 23 °C	Positive Photoresist	Al	TiN	TEOS	SiN
Target ER (Å/min)	> 4,000	< 50	< 50	< 50	< 50
XM-426 etch rate (Å/min)	> 20,000	1.2	0	2.8	0.3
Non-halide containing Al PERR	> 10,000	3.0	< 0.1	< 0.5	N/A
Non-halide/NMP containing Al PERR	> 10,000	1.8	< 0.1	0.1	0.3

- XM-426 demonstrates similar performance to industry-accepted products but contains halides

13

Results – Cleaning and Stripping Performance

Performance of XM-426

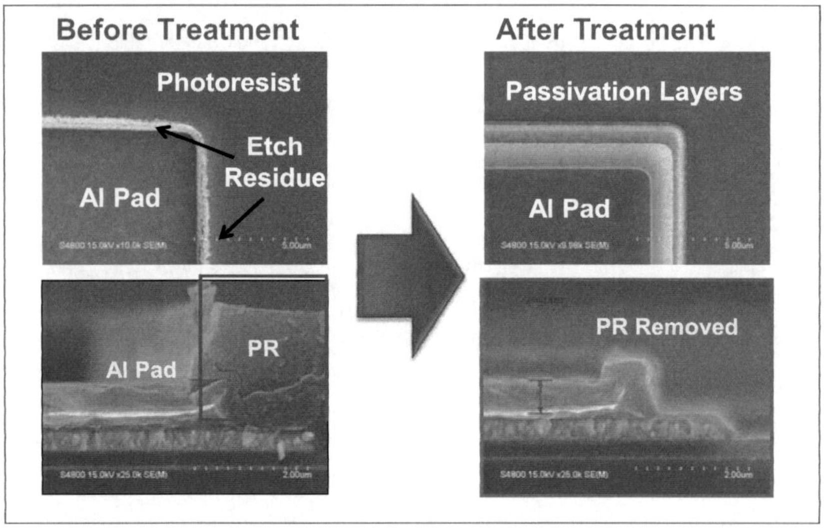

- XM-426 demonstrates ability to remove photoresist and clean post-etch residue without damaging the Al bond pad

14

Conclusion

- Al pad corrosion can occur when fluorocarbon gas is used during the plasma etch process for opening of passivation layers
- Avantor's study tested a J.T.Baker® corrosion-free photoresist stripping and etch residue removal cleaning solution
- Results were confirmed by AES and SIMS which showed complete removal of the photoresist and post-etch residue without corrosion to the Al bond pad

J.T.Baker® | AVANTOR Electronic Materials

15

Finite Element Analysis of CVD Stripping Kinetics for *In Silico* Optimization of Semiconductor Manufacturing Parts for Sub-20 nm Technology Nodes

Osama Khalil†‡, Ardeshir Sidhwa Ph.D. †, John Deem†, David Zuck†, Bin Mu Ph.D.‡, Erica Forzani Ph.D.‡
† Quantum Global Technologies, LLC, Quakertown, PA
‡ Arizona State University, Tempe, AZ

Introduction

Quantum Global Technologies® (QGT: Quantum Clean® & ChemTrace®), a leader in the semiconductor parts cleaning and analytical business through their Atomically Clean Surface (ACS™) process, has developed new technologies aimed at exceeding OEM performance and meeting stringent defectivity requirements for sub-20 nanometer technologies. This is achieved by highly optimized parts cleaning and coating processes. This paper outlines recent advances in the ACS™ parts cleaning process by focusing on particle removal methods. The focus of the present work is to explain how finite element analysis is used to simulate the selective wet etching of CVD with a cleaning solution that was optimized by QuantumClean's Innovation and Technology group. Optimizing cleaning processes *in silico* for a given part surface allows QuantumClean to better predict outcomes of real surface reactions to adapt a multitude of chamber parts to the ASC™ process.

The present work considers a solid sample that is subjected to wet etch immersion clean that selectively removes layers of silicon nitride deposition from the coupon surface. The kinetic expressions associated with this chemical process for the purpose of the simulation in the present work describe the rates of adsorption and desorption of dilute species X and dilute product C onto the surface deposition B and the rate of depletion of species B through a surface reaction.

Reactant sorption: $X + B \leftrightarrow XB$

Surface reaction: $XB \leftrightarrow CB$

Product sorption: $CB \leftrightarrow C + B$

Finite element analysis allows for the facile scaling of physical phenomena such as the cleaning process of the present work. This study considers a small, metallic part with seven honeycomb-shaped ribs that is coated in silicon nitride. The part is shown in Figure 1.1 bare of any silicon nitride deposition. This part is digitally constructed arbitrarily and is not representative of any real industrial machine component. During the timespan of simulation, the part is immersed in an 800-mL cylindrical space (100 mm radius) containing upward-flowing aqueous Solution X (Figure 1.2).

Bare Part Ray-Traced Rendering

Fig. 1.1

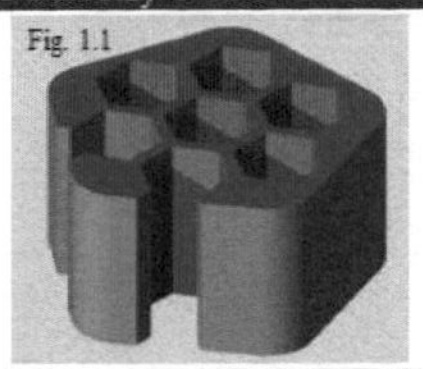

Simulation Space and Meshing

Fig. 1.2

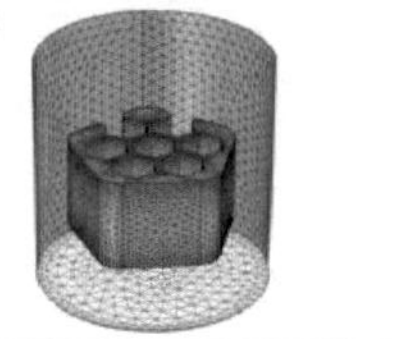

This study takes advantage of the strong relationship between the thickness of thick silicon nitride layers and their observable color (Figure 2.1) to closely approximate the true color of the deposition (Figure 2.2) as it is removed throughout the simulation. Real time lapse imagery and color observation were used to estimate film removal progress without the regular use of more expensive metrology, such as refractometry or ellipsometry.

Simulation Results and Conclusions

Silicon Nitride Thickness Color Legend[1] (Fig 2.1) and Real Part (Fig. 2.2)

Fig. 2.1

Fig. 2.2

False-Color Thickness Profile

Fig. 3

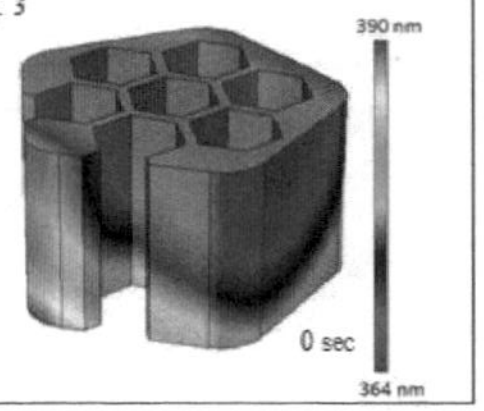

Simulation Time Lapse

Fig. 4.0

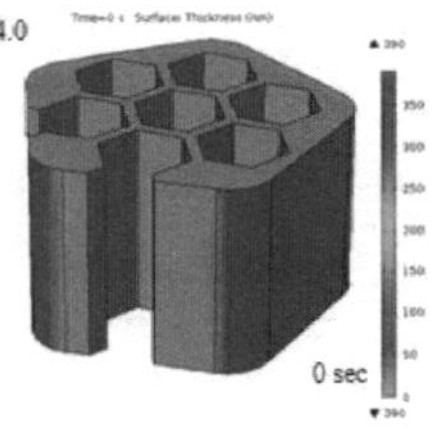

Fig. 4.1

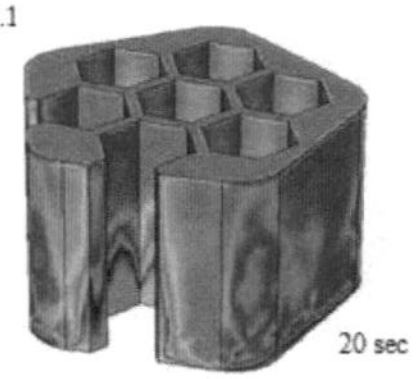

Fig. 4.2

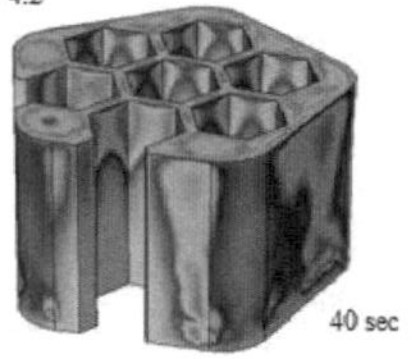

Fig. 4.3

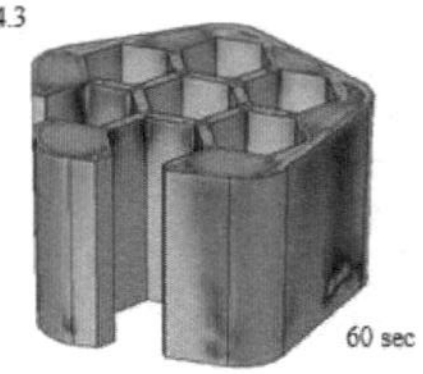

Fig. 4.4

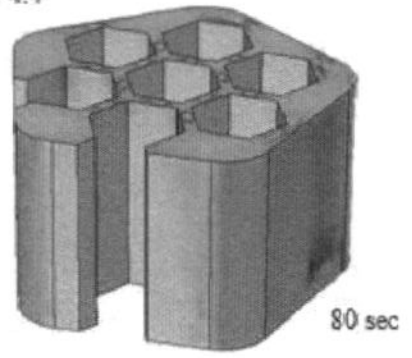

Cut-Line Thickness Profile Time Lapse

Fig. 5.1

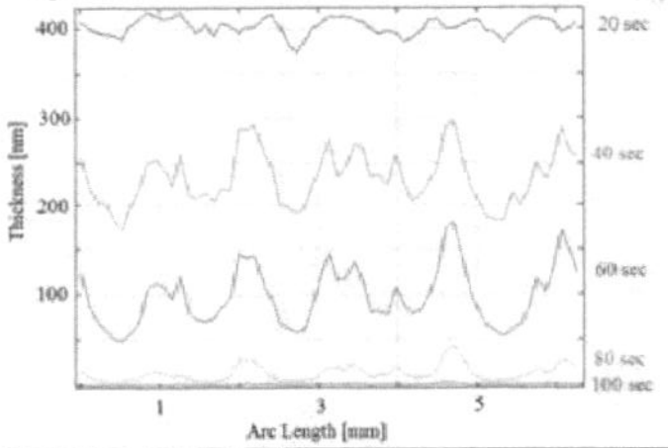

Fig. 5.2

Silicon nitride is not uniformly distributed over the part surface, but an approximate range of thickness is known, as shown in false color in Figure 3. To attempt to capture the uncertainty in the observation of the initial condition of the part surface, an arbitrary thickness (h_g) distribution function,

$$h_g(x, y, z) = a(y^2 + z^2) + b$$

is applied to the simulation volume and describes an infinite paraboloid that is used to simulate the minute degree of variance in thickness across the part surface. Figure 3 is a surface plot of silicon nitride thickness shown in false color on the part at the start of the simulation (t = 0 s). However, on the scale of the complete simulation, which encompasses the entire thickness range from 0 nm to 390 nm, variance shown in Figure 3 is indiscernible, as shown in Figure 4.0.

Figures 4.0 through 4.4 present the results of the simulation at 20-second intervals, with the color legend shown in Figure 4.0. Figure 5.1 is an overlay plot showing the thin film thickness profile along the 6.2-mm cutline shown in Figure 5.2.

This simulation was built from the observable color of the deposition on parts at different stages in the cleaning process and is effective in clearly demonstrating the kinetics of deposition removal by wet chemical etch. The model results show the impact finite element analysis can have on reducing the analytical costs of determining spatially-dependent etch selectivity on any number of parts using similar kinetics models.

References

[1] Henrie, J., S., K., S.M., S., & A., H. (2004). Electronic color charts for dielectric films on. Optical Society of America, 12 (7), 1464-1469.

Contact

Osama Khalil
osama.khalil@quantumclean.com
Quantum Global Technologies, LLC
1900 AM Drive, Suite 200
Quakertown, PA 18951

QUANTUM GLOBAL TECHNOLOGIES, LLC

SPCC 2016
Surface Preparation and Cleaning Conference

Acoustic Characterization of a Photomask Cleaning System

X. Chen [1], M. Zhao [3], P. Yam [1], C. Zanelli [1], O. Nagaya [2], H. Arai [2], K. Kutsuna [2], M. Zhao [1], and M. Keswani [2, a]

[1] Onda Corporation, 1290 Hammerwood Ave., Sunnyvale, CA, USA
[2] R&D Division, Honda Electronics Co., Ltd., Toyohashi, Japan
[3] Department of Materials Science and Engineering, University of Arizona, 1235 E James E. Rogers Way, Tucson, AZ, USA
[a] manishk@email.arizona.edu

INTRODUCTION

Although megasonic technology is widely used to clean photomasks, the acoustic performance is not well understood. Of all the process parameters that influence cleaning (e.g., temperature, flow, pH, gas concentration, mechanical translation, etc.) the characterization of the ultrasonic field remains elusive. The shift to EUV lithography processes elevates this issue further since the risk of yield loss is even higher in the absence of a pellicle. This study aims to achieve a deeper understanding of the complex acoustic behavior by presenting results from three independent measurement techniques.

METHODS

Hydrophone and Mask Sensor Scanning Setup

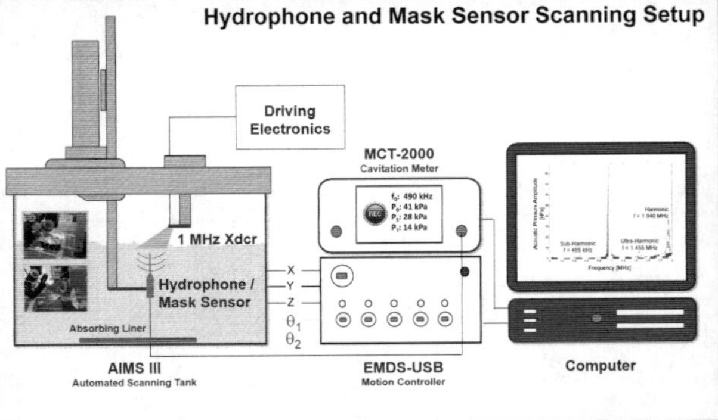

Schlieren Imaging Setup

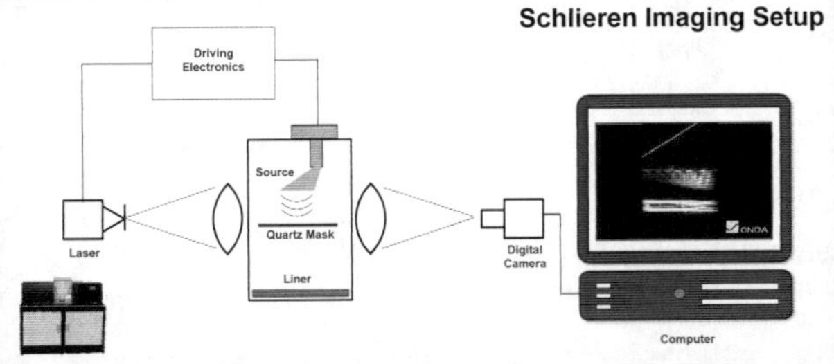

RESULTS & DISCUSSION

Schlieren Imaging

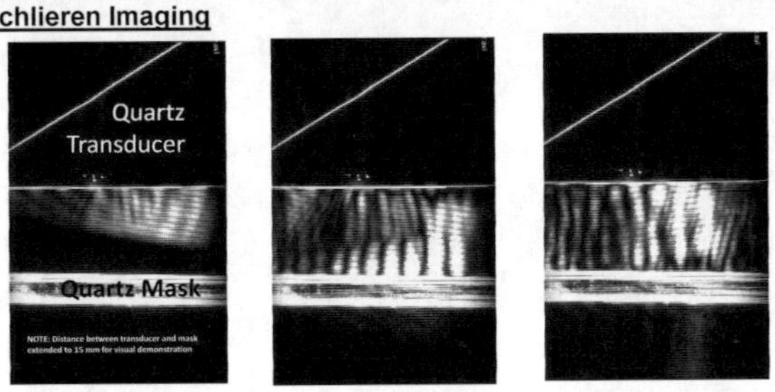

Wave propagates at an offset angle from transducer

Incident wave disturbed by reflected wave from both top and bottom surface of quartz; some waves transmit through quartz mask

The resultant sound field reveal a complex pattern from multiple reflections

RESULTS & DISCUSSION

Total Pressure Uniformity

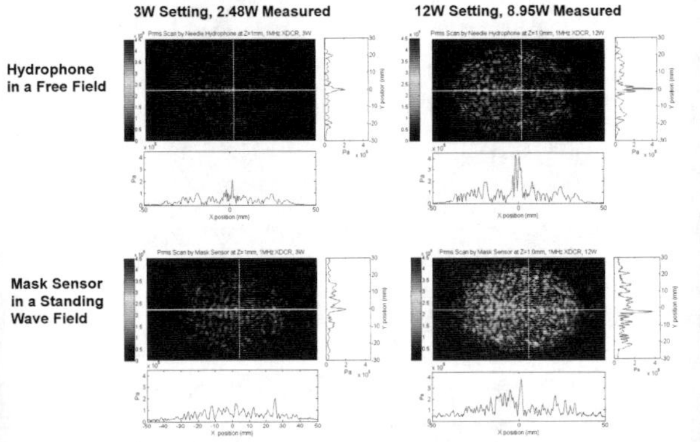

- 2D plots indicate variability in pressure uniformity for different power settings
- Mask sensor measurements yield a more complex acoustic field from reflections off quartz mask
- Scanning away from transducer highlight standing wave effects.

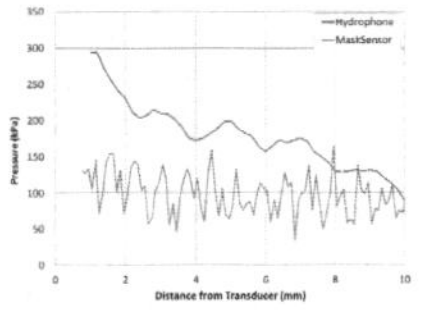

Direct Field and Cavitation Pressure

Cavitation vs. Power

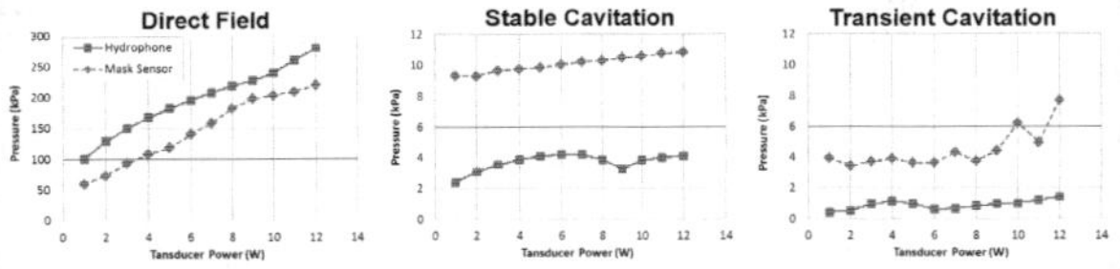

Cavitation vs. Position

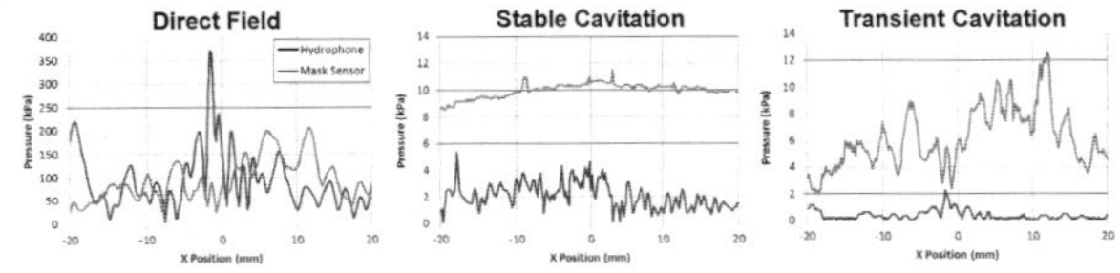

- Direct field pressure trend as expected with power, namely Power ∝ Pressure^2
- Low levels of stable and transient cavitation detected, even at high power levels
- Cavitation level increased in the presence of a photomask.

CONCLUSIONS

Through the use of different measurement techniques, a deeper understanding of the acoustic performance for a mask cleaning system has been achieved. This helps bridge the gap in connecting the ultrasound with cleaning performance.

Advanced Metrology for Post Etch Residue Removal

Chuannan Bai, Eugene Shalyt, Guang Liang, Jingjing Wang, Michael MacEwan and Helen Lu
ECI Technology, 60 Gordon Drive, Totowa, New Jersey 07512, USA

Introduction

Cleaning processes in sub-10 nm node have an increase challenge due to smaller geometry, increased number of wetter materials, and increased functionality/selectivity demands where some components have to be removed at a fast rate with a minimum etch rate for other components. As a result, most modern formulations include at least 6 subcomponents with a special emphasize on corrosion inhibitors and promoters. As an example, fluoride is a potent etching agent for W, Al and SiO_2 while BTA and similar compounds are effective corrosion inhibitors for Cu.

• Measuring Inhibitors by HPLC

Chromatogram below shows an unique inhibitor peak, even in a full complex matrix. HPLC can measure inhibitor at ppm level without dilution

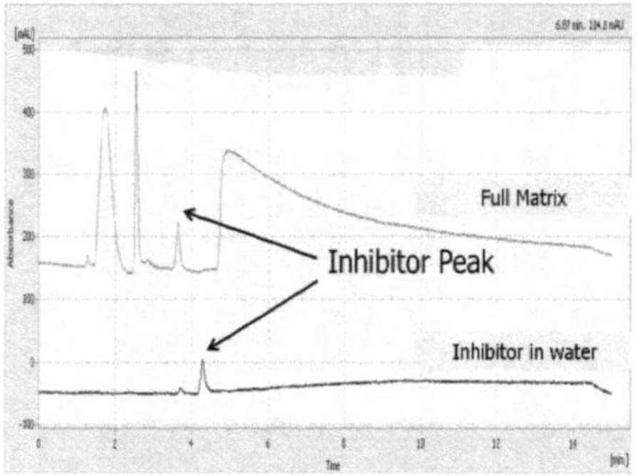

• Measuring Inhibitors by UV-VIS spectroscopy

Cu inhibitors are commonly used to passivate previously plated Cu layers. Organic groups in inhibitor compounds usually have good UV-VIS signals. Such inhibitors can be measured by UV-VIS spectroscopy. In case multiple inhibitors are used, spectral features can overlap, multi variate analysis (MVA) is needed. The inhibitor concentration is expressed as:

$$c_{inhibitor} = c_{\lambda_1} \times A_{\lambda_1} + \cdots c_{\lambda_i} \times A_{\lambda_i} + \cdots c_{\lambda_n} \times A_{\lambda_n} \quad (1)$$

where c_{λ_i} is the coefficient at wavelength λ_i and A_{λ_i} is the absorbance at wavelength λ_i. The coefficient set is predetermined by a large amount of calibration data and multivariate analysis.

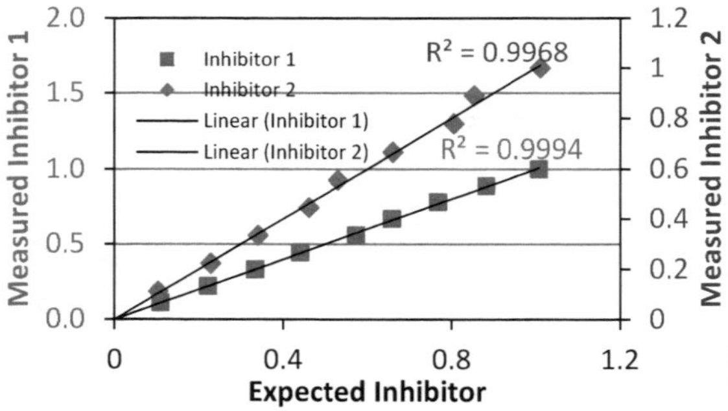

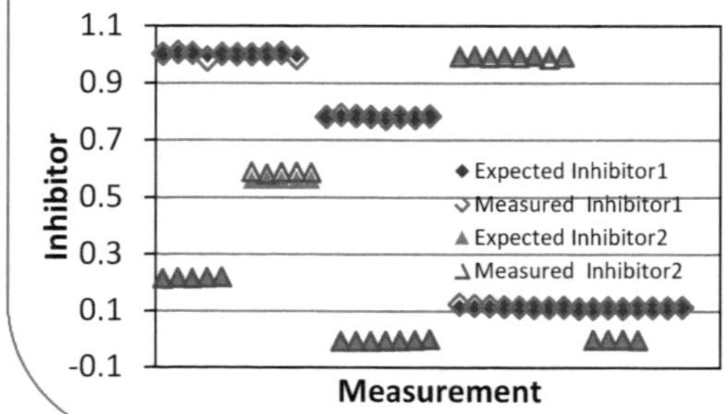

• Measuring Fluoride by ISE

Low concentration of Fluoride is measured by Fluoride Ion Selective Electrode (FISE)method. Under ideal conditions, the potential (E) of a FISE is given by the well-known Nernst equation:

$$E = Eo - (2.303\ RT/nF) \log [F^-] \quad (2)$$

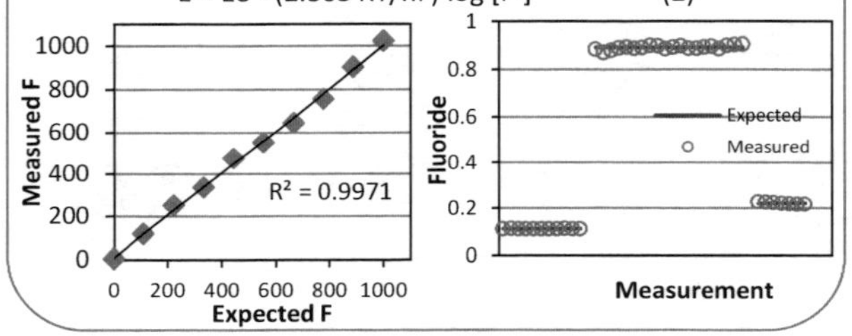

• Measuring contaminant Ti in TiN removal

Titanium is a contaminant from the TiN removal. Monitoring of Ti level is needed for the process control. Figure below shows an excellent correlation of spectroscopic signal to the Ti concentration at ppm level.

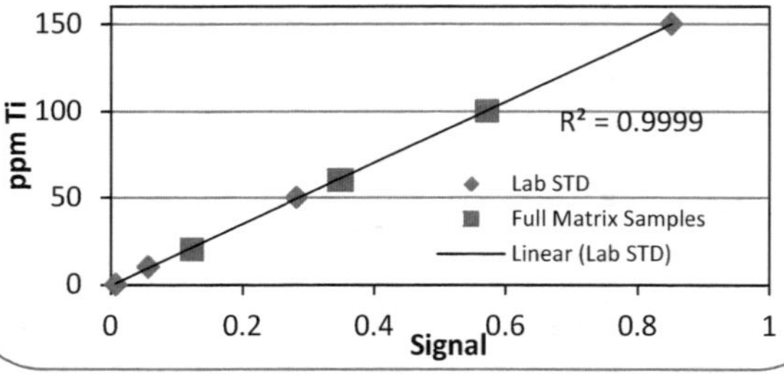

Conclusion

With the combination of various metrology techniques (spectroscopy, ISE, HPLC, etc.), multiple components in the cleaning solutions, including contaminant can be measured with fast, accurate, and repeatable results

References:
Eugene Shalyt, Guang Liang, Jingjing Wang, Michael Pavlov, Vladimir Dozortsev, Chuannan Bai and Peter Bratin *ECS Trans. 2013 volume 58, issue 6, 159-167*
J. J. Wang, E. Shalyt, C. N. Bai, G. Liang, M. MacEwan, V. Parekh, "Advanced Monitoring of TMAH Solution", *Solid State Phenomena, Vol. 219, pp. 81-84, Sep. 2014*
Handbook of Cleaning for Semiconductor Manufacturing: Fundamentals and Applications by Richard F. Reidy; Karen A. Reinhardt Published by John Wiley & Sons, 2011

SPCC 2016
Surface Preparation and Cleaning Conference

Rapid Cleaning Using Novel Processes With Coatings

Alex Brewer and John Moore

Daetec, LLC

1227 Flynn Rd., Unit 310 Camarillo CA 93012

www.daetec.com jmoore@daetec.com

Surface Preparation and Cleaning Conference
April 2016

DAETEC — 2016 SPCC Santa Clara California — 1

Washable Coatings

- **Coatings that Clean**
  - **CoatsCleans™ – emulsify PR**
  - **Washable Primers – lift off/ protect from metal X-linking**

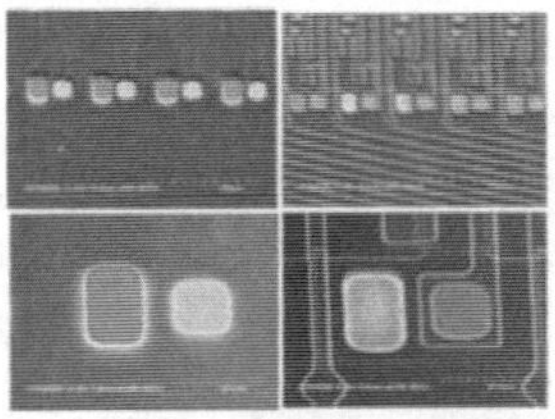

- **Washable Coatings/Adhesives**
  - Laser processing
  - Temporary bonding
  - Dicing and polishing
  - Planarizing

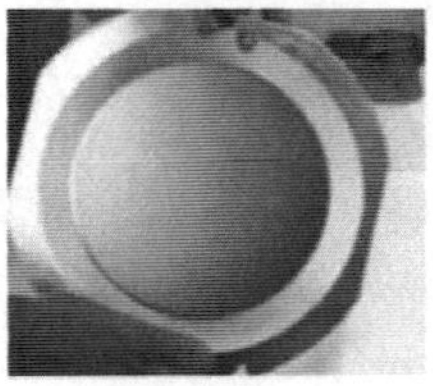

FF Tape
Washable Safe

DAETEC — 2016 SPCC Santa Clara California — 2

CoatsCleans™ vs. Immersion

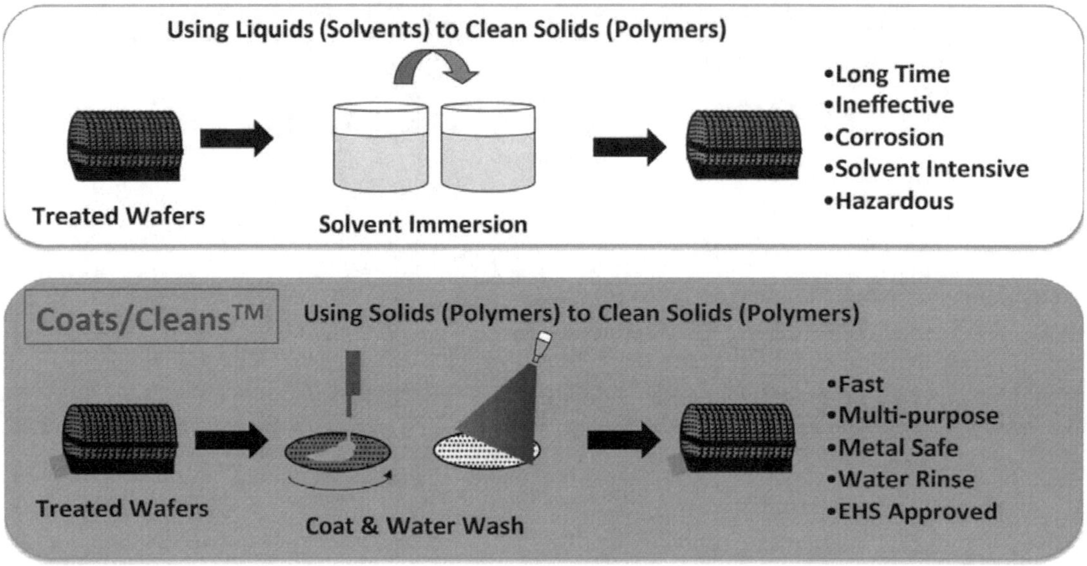

Stripping Neg-Acrylic PR (Bumping)

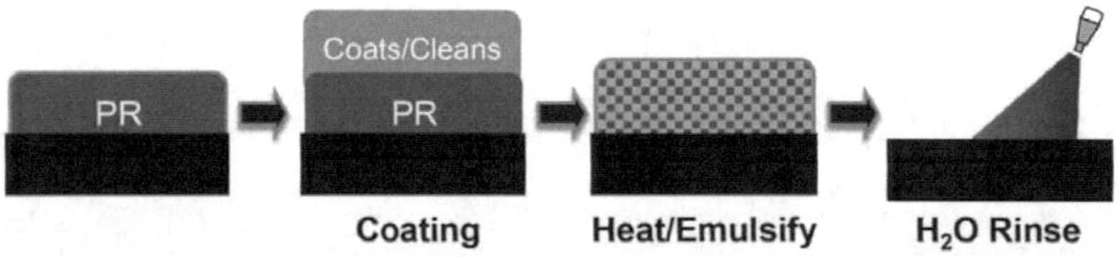

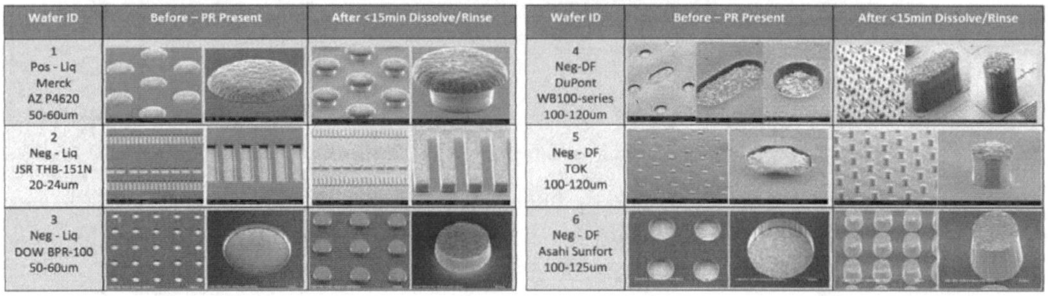

DAETEC 2016 SPCC Santa Clara California

Washable Primer - Eliminates Residue

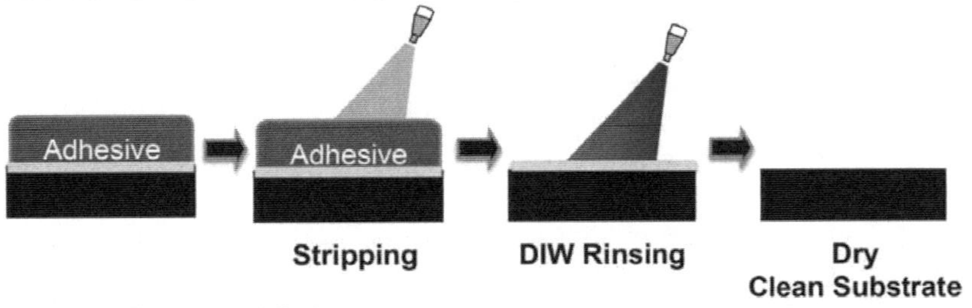

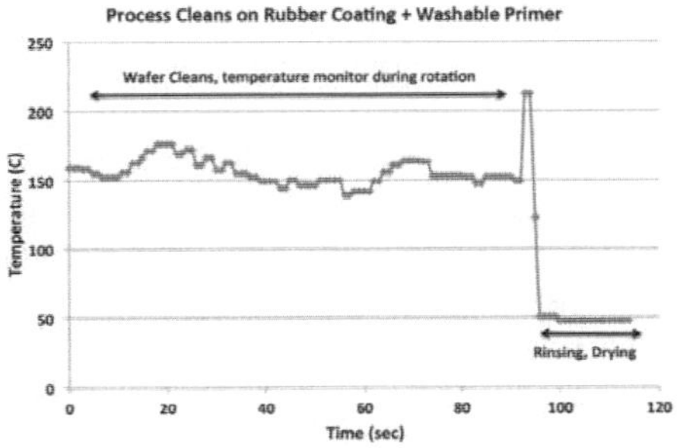

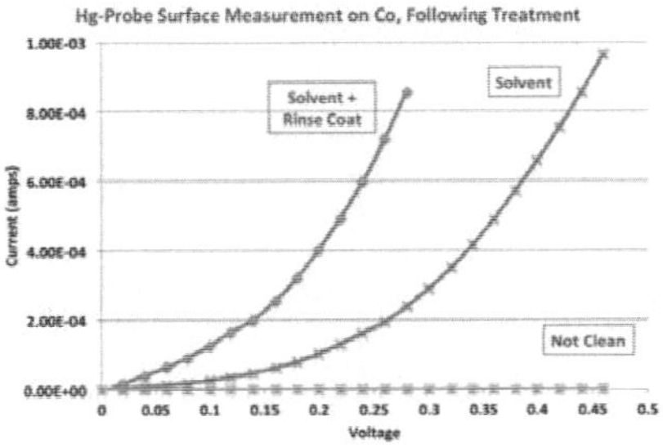

Improved Thermal Resistance

Thermal Resistance

- Chemical functionality
  - Phenyl
  - Polyester

DaeCoat™ Systems

- Phenyl silicones
- Polyphenylsulfones
- Salt conjugates

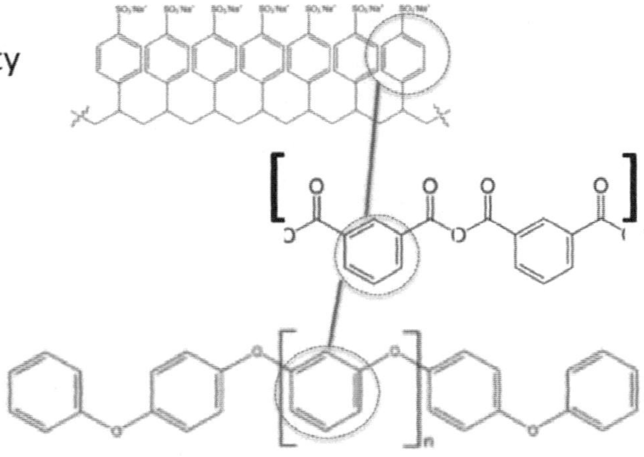

Thermal Resistant Washable Coatings

Thermal Exposure — After RT Water Rinse

PVA PVP DaeCoat™ — PVA PVP DaeCoat™

200C

250C

300C

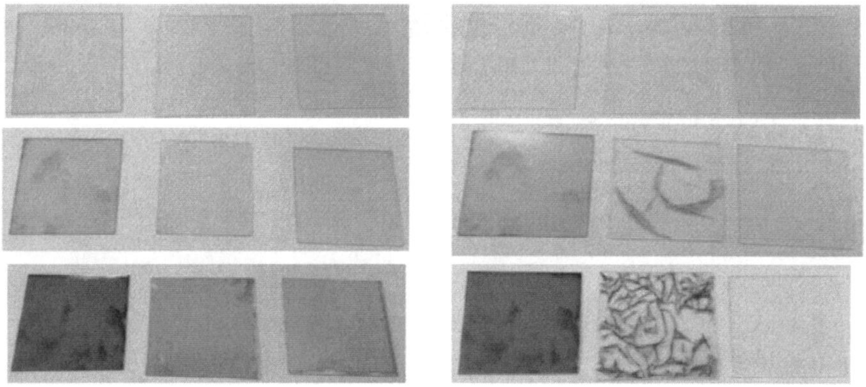

Ex.: Wafer Temporary Bonding

Process Demand

- **Objective:** Wafer thinning, backside processing
- **Mechanical (e.g. grind):** Yes
- **Thermal resistance:** <300C
- **Process/chemicals:** Yes
- **Uniformity:** ~2um

Recommendation

- **DaeCoat™ 355**
  - Green solvent washable, DaeClean™ 300
  - Broad chemical resistance
  - Thermal resistance: >300C
- **Carrier:** Solid, due to small die, simple release/cleans
  - chemical diffusion
  - recycled

Thinning, Processing, Release

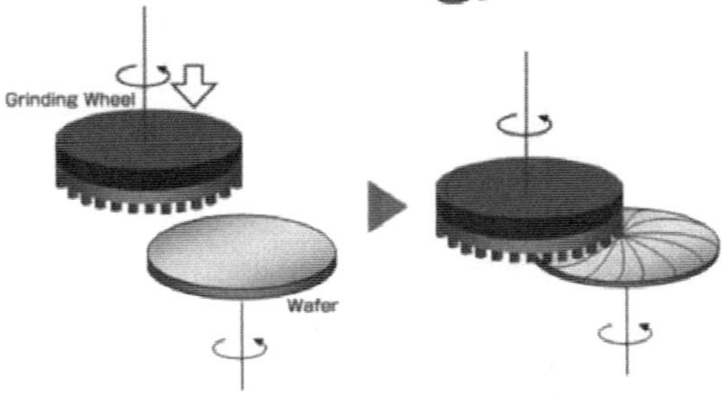

Grinding, backside processing, singulation

Singulation offers 1-2mm channel between devices to enable simple debond & wash

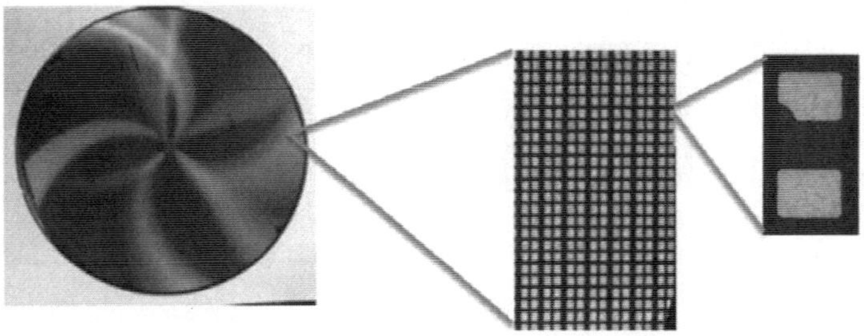

Green Solvent Wash Adhesive

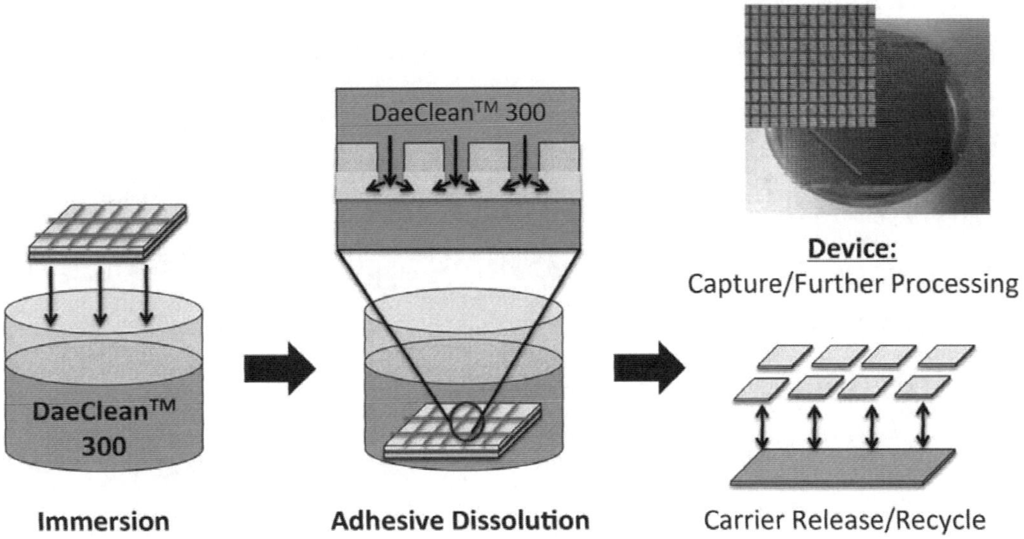

Ex.: Device Temporary Bonding

Process Demand	Recommendation
• **Objective:** LTCC flip-chip bond & encapsulate • **Mechanical (e.g. grind):** No • **Thermal resistance:** ~275C • **Process/chemicals:** limited, RT flux cleaner • **Uniformity:** <10%	• **DaeCoat™ 535** – Hot DIW washable – RT chemical resistance – Thermal resistance: >300C • **Carrier:** Porous – chemical diffusion – recycled

LTCC/HTCC

- Microelectronics on a ceramic substrate
- Multi-layer packaging
- MEMS, military, RF, wireless
- Thickness <50um to >250um
- Commonly 100-150um
- Green tape – several suppliers
- Extremely fragile – handling challenge!

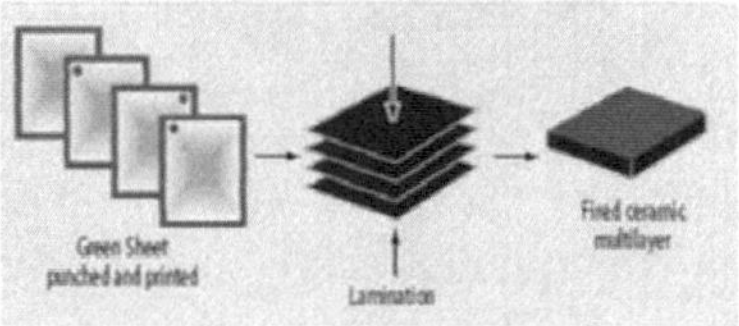

DIW Wash Adhesive (LTCC)

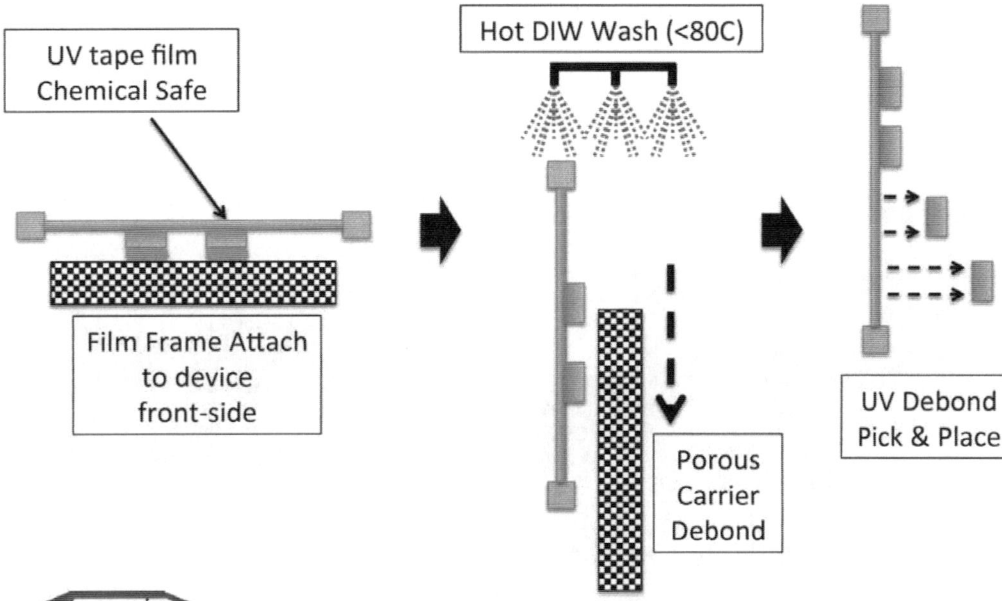

Ex.: Wafer Planarization

Process Demand	Recommendation
• **Objective:** Wafer planarizing coating for backside processing • **Mechanical (e.g. grind):** No • **Thermal resistance:** <300C • **Process/chemicals:** Yes • **Uniformity:** <5% • **Special:** Desire to finish on FF tape	• **DaeCoat™ 357** – Green solvent washable, DaeClean™ 300 – Broad chemical resistance – Thermal resistance: >300C • **Carrier:** desire FF tape – Safe for DaeClean™ 300

Washable Planarization Coating

Sputtering Test

- Sputter deposition of 200nm Ti:W + 300nm Copper on 250μm thick DaeCoat™ 357 using LLS802 multi target tool

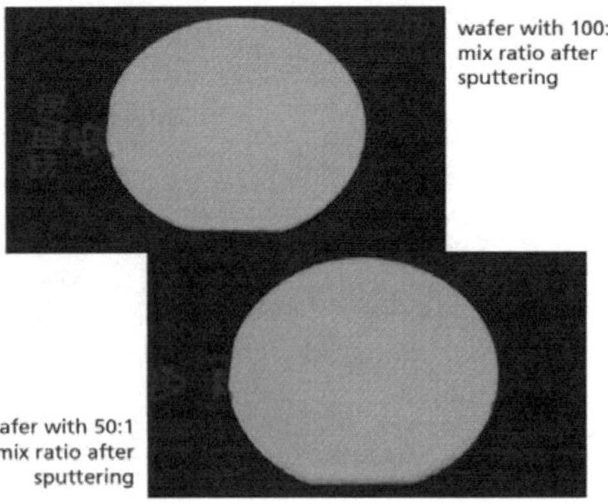

wafer with 100:1 mix ratio after sputtering

wafer with 50:1 mix ratio after sputtering

Chamber Capability:
24 x 4" - 6" wafers per batch
8 x 8" wafers per batch
4 x 300 mm wafers per batch

Parameter	DaeCoat™ 355	DaeCoat™ 357	DaeCoat™ 515	DaeCoat™ 535	DaeCoat™ 615
Coating Thickness	<5-100 um	<5-250 um	<5-100 um	<5-60 um	<5-60 um
Cure	UV/Thermal	UV/Thermal	Thermal	Thermal	Thermal
Max temp	~300C	~300C	~300C	~300C	~200C
Application	Temp Bonding or Coating	Temp Planarizing Coating	Laser Processing	Temp Bonding or Coating	Temp Bonding or Coating
Resists RT DIW*	✔	✔	✖	✔	✔
Resists Acids*	✔	✔	✖	✖	✔
Resists Litho Stripper Chemistries*	✔	✔	✖	✖	✖
Clean Conditions	DaeClean™ 300 (Safe Solvent)	DaeClean™ 300 (Safe Solvent)	RT DIW	80C, DIW	DaeClean™ 150 (Detergent)

Surface Contamination Control through Final Surface Finish (FSF™) Processing for Semiconductor Equipment Parts for Sub-16 Nanometer Nodes

Ardeshir "Ardy" J. Sidhwa Ph.D., Osama Khalil, John Deem, David Zuck, Matthew House
Innovation and Technology (R&D)
1900 AM Drive, Suite 200
Quantum Global Technologies, LLC
Quakertown, PA, USA

Contact info: ardy.sidhwa@quantumclean.com, Mobile: 602-499-6510

Abstract

- Meeting defectivity requirements are challenging as technology nodes continue to scale down. Defect variability such as particles and residues create killer defects at the component-level and is expected to severely impact product yields at sub-16 nm technology nodes. Additional complexity, in the form of new processes and materials, will further challenge component-level design and performance.
- Quantum Global Technologies® (QGT: Quantum Clean® & ChemTrace®), a leader in the semiconductor parts cleaning and analytical business through their Final Surface Finish (FSF™) processes, has developed new technologies aimed at exceeding OEM performance and meeting stringent defectivity requirements for sub-16 nm technologies.
- This work addresses some of the major challenges and solutions towards achieving contamination-free manufacturing (CFM) in a cost-effective manner.

Experimental Setup for FSF™

- Final Surface Finish (FSSF™) was developed to clean different substrates precisely. Substrate material can vary from aluminum, stainless steel, titanium, ceramic, quartz, ceramic coating, etc.
- To further complicate the surface condition, the substrate can be textured or non-textured material. However, textured material introduces additional complexity to the trace metal analysis due to increased surface area from surface morphology. During the parts cleaning process each substrate is exposed to multiple process parameters.
- Figure-1 depicts the above scenario in graphical format. The ultimate goal is to yield the best possible Final Surface Finish (FSF™) for critical substrates.

Graphical Presentation Format for FSF™

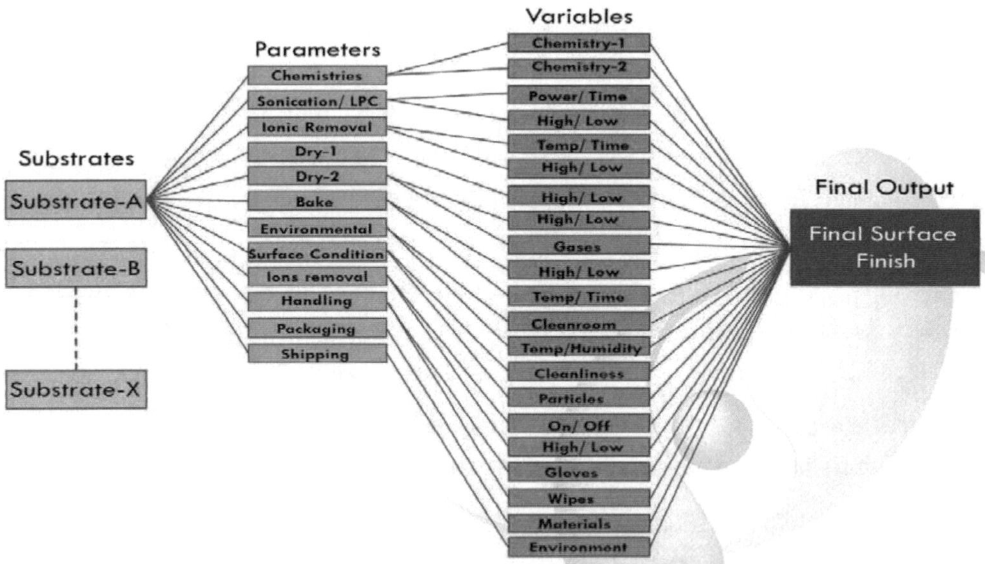

Figure-1:Experimental pathway for developing FSF™ recipe builder

Introduction to Final Surface Finish (FSF™)

- In order to confront the issue of adapting parts cleaning technology to shrinking critical dimensions of integrated circuitry, a pathway to achieving atomically clean surfaces through experimentation was defined. Key process parameters that affect surface cleanliness metrics that were identified were first evaluated within an experimental parametric space that would serve as a baseline for improvements made to the final surface treatment process. It is imperative that OPC facilities unify analogous processes so they are consistent and copied exactly to reduce variability and maintain the quality that is expected by the customer, regardless of the cleaning site.
- For QuantumClean, this meant that research was to be targeted at comparing and selecting the best chemical treatments, surface passivation, texturing, and coating techniques, and cleanroom protocol to be developed and carried out at all sites.

QUANTUM GLOBAL TECHNOLOGIES, LLC

SPCC 2016
Surface Preparation and Cleaning Conference

Prepared by Ardy Sidhwa, PhD Slide 5

Final Surface Finish (FSF™)- cont'd

- OEMs and customers create the specifications that define their surface cleanliness acceptance criteria. For example, a specification for surface trace metals defines the maximum count of leachable metals for a part acceptable for installation after a clean.
- To screen the effects of altering process steps on surface trace metals, a "progressive" experimental design was implemented, in which process steps are optimized in phases, then optima from each phase are transferred to the next phase until all factors and process steps are optimized. In the present work, each phase is a full-factorial experiment centered upon a single process step. Table I presents the phase structure of the FSF experiments. The output to these tests were a combination of 30 element trace metal analysis by inductively-coupled plasma mass spectrometry (ICP-MS), and ion chromatography by deionized water extraction.

QUANTUM GLOBAL TECHNOLOGIES, LLC

SPCC 2016
Surface Preparation and Cleaning Conference

Prepared by Ardy Sidhwa, PhD Slide 6

Final Surface Finish (FSF™)- cont'd

- From these data, linear regressions were generated relating each experimental factor to model-predicted outcomes of each trace metal or ionic contaminant. For any given configuration of process parameters within the experimental parametric space, a set of regression models can predict the counts of each element or ionic compound. The measure of goodness for a particular configuration of factor levels, or desirability, is a function used to describe the minimization of contaminants on a scale from zero to one. Desirability functions were designed such that maximum desirability (lower target limit) is set to the lower detection limit of the measurement instrument, while the upper target limit is set to the average metal or ion count within the experiment.
- Figure 2 shows an example of how process parameters applied power and immersion time were tuned so to maximize the desirability function for a particular phase of quartz experimentation. Only two elements are shown in Figure 4 (zinc and zirconium), but all 30 trace metals analyzed by ICP-MS are accounted into each desirability function.

Cleaning Process Step	Process Parameters Used By Phase				
	Phase 1	*Phase 2*	*Phase 3*	*Phase 4*	*Phase 5*
Final Chemical Treatment		P1 Optimum Chemistry	P1 Opt. Chem.	P1 Opt. Chem.	P1 Opt. Chem.
Cleanroom Step 1	POR Step 1		P2 Opt. CR Step 1	P2 Opt.CR Step 1	P2 Opt.CR Step 1
Cleanroom Step 2	POR Step 2	POR Step 2		P3 Opt. CR Step 2	P3 Opt. CR Step 2
Cleanroom Step 3	POR Step 3	POR Step 3	POR Step 3		P4 Opt. CR Step 3
Packaging	POR Pkg	POR Pkg	POR Pkg	POR Pkg	

TABLE-1: PROGRESSIVE OPTIMIZATION EXPERIMENT DESIGN

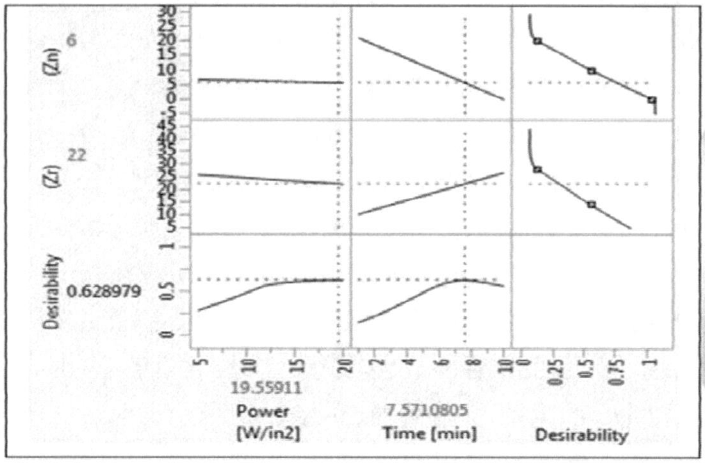

Figure-2: A section of the array of regression plots and desirability functions (top two rows, showing regressions for zinc and zirconium). Dotted crosshairs indicate the values of each element and desirability function when the factors power and time are set to maximize desirability.

Surface Texture Compensation

- When considering trace metallic or ionic contaminants, it is important to consider that the test method for acquiring these data involves an approximation of the macroscopic surface area from which the leachable metals are extracted, in order to report data in units of contaminants per unit area. However, the microscopic surface area of a textured surface can be much greater than that of a non-textured or polished surface, therefore subjecting textured parts to the same specification for contaminants per macroscopic surface area leaves those parts at a disadvantage. To address this issue, models were formulated to approximate microscopic surface areas of materials subjected to grit blasting from a given grit size or R_a. Samples of aluminum, titanium, stainless steel and quartz textured to roughnesses prevalent in QuantumClean grit blast procedures were analyzed by optical profilometry (Zemetrics ZeScope) and validated by 408-nanometer laser scanning microscope (Keyence VK-X250).

- Figure 3 shows Zemetrics ZeScope optical profilometer outputs z values on an x,y array, so to numerically compute surface area, {*x,y,z*} values on the captured array are used to construct a polyhedral surface constituent of polygons of calculable area *A*, according to Equation 1, where a half parallelogram is constructed from data points $p_{i,j}(x_{i,j}, y_{i,j}, z_{i,j})$, $p_{i+1,j}(x_{i+1,j}, y_{i+1,j}, z_{i+1,j})$, and $p_{i,j+1}(x_{i,j+1}, y_{i,j+1}, z_{i,j+1})$.

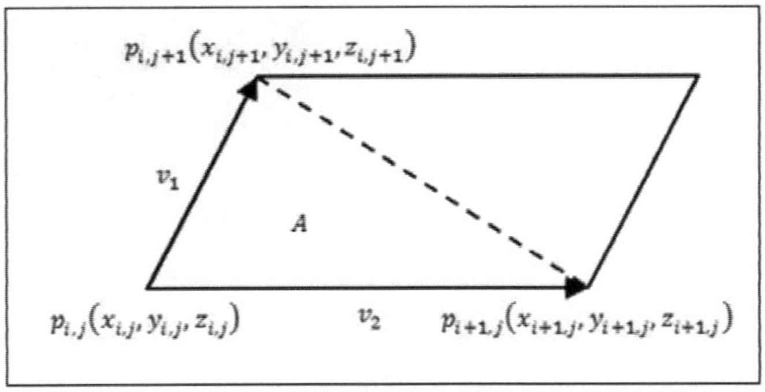

$$A = \frac{1}{2}|\mathbf{v}_1 \times \mathbf{v}_2|$$

Where $v_1 = p_{i,j+1} - p_{i,j}$ and $v_2 = p_{i+1,j} - p_{i,j}$.

Figure 3: A parallelogram formed by the two vectors v1 and v2 used to calculate area A.

- Total area is the summation of all values of A within the assay array. The model resulting from performing this calculation on any variety of part surfaces adds a major component to the FSF™ Recipe Builder that now accounts for microscopic morphology that is intrinsic to part surfaces, yet affects contamination measurements and parts cleaning first-pass yield. Figure 4 depicts the Surface morphology maps of non-textured, lightly textured and heavily textured quartz material. Figure-5 depicts normalized computed microscopic surface area of non-textured versus textured stainless steel.

Figure 4: Surface morphology maps of non-textured, lightly textured, and heavily textured quartz captured by Zemetrics ZeScope optical profilometry.

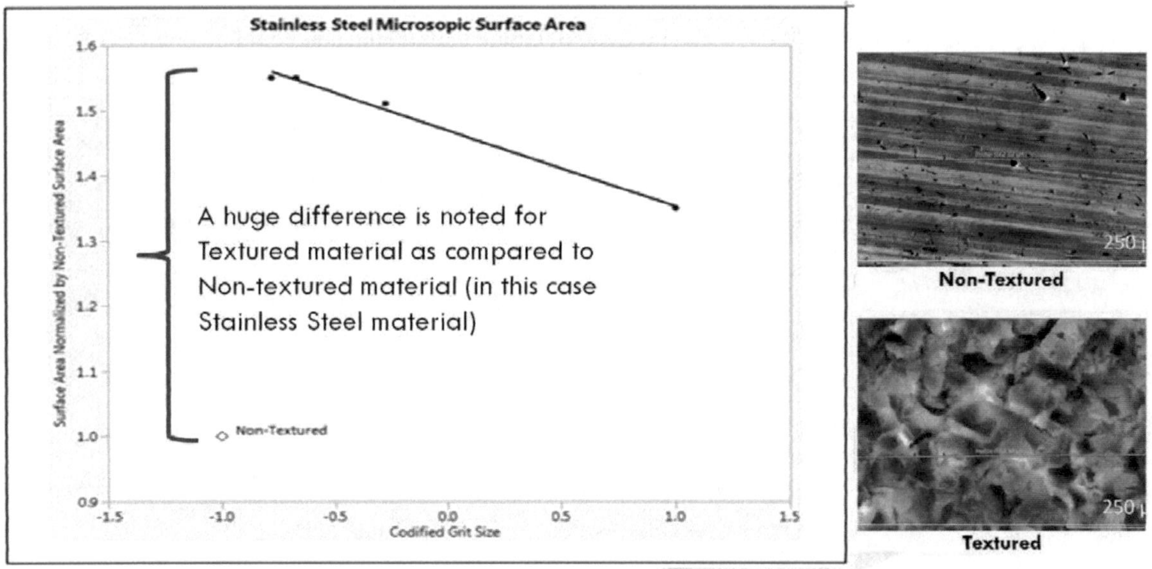

Figure 5: Computed microscopic surface area of textured stainless steel, normalized by non-textured surface area and measured by Zemetrics ZeScope optical profilometry related to codified grit sizes ranging from -1 to +1

Results and Discussion

- Table 2 presents the codified results of the Final Surface Finish progressive optimization experimental process (actual process parameter values converted to -1 to 1 scale). These values have been used as a reference datum for subsequent efforts to mitigate trace metallic and ionic contamination. The chemical composition column of Table 2 shows a majority of substrates performing best under Chemistry B. This may be indicative of an opportunity to unify cleaning processes, which can drive down the cost of consumables associated with each of those substrates.
- Figure 6 presents the trace metal improvement with Chemistry B from the process baseline. Given the sensitivity of sub-20 nanometer processes and that of surface cleanliness measurement systems, process development for sub-20 nanometer surface cleaning requires this kind of baseline reference in order to confidently determine whether a new process makes measureable improvements. Every time the FSF™ processes are improved by a new finding, baseline defectivity falls to a new low. Figure 7 shows actual results of FSF™ process

	Chemistry		Cleanroom Step 1		Cleanroom Step 2		Cleanroom Step 3	Package
	Comp.	*Time*	*Set*	*Time*	*Set*	*Time*	*Type*	*Procedure*
Levels	*Chem-A* *Chem-B* *Chem-C*	*-1* *0* *+1*	*-1* *+1*	*-1* *+1*	*-1* *+1*	*-1* *+1*	*A* *B*	*Bag-A* *Bag-B* *Bag-C*
Textured Aluminum	Chem-B	+0.65	-1.00	-0.08	+0.75	+0.96	A	Bag-B
Non-Textured Aluminum	Chem-A	+0.66	+0.21	-0.55	+0.94	+0.86	A	Bag-B
Textured Titanium	Chem-B	+0.35	+1.00	+1.00	-0.96	+1.00	A	Bag-B
Non-Textured Titanium	Chem-A	+0.44	-1.00	+1.00	+0.44	-1.00	A	Bag-B
Textured SST	Chem-B	+1.00	-0.15	+1.00	-1.00	+1.00	A	Bag-B
Non-Textured SST	Chem-B	+1.00	+1.00	+1.00	+1.00	+1.00	A	Bag-B
Textured Quartz	Chem-B	+1.00	+1.00	+1.00	-1.00	-1.00	A	Bag-B
Non-Textured Quartz	Chem-B	+1.00	+0.97	+0.32	+1.00	+0.88	A	Bag-B
Anodized Aluminum	Chem-B	-1.00	+1.00	-1.00	+1.00	+0.91	A	Bag-B
Alumina (Ceramic)	Chem-B	+0.03	+1.00	+1.00	-1.00	+1.00	A	Bag-B
Bulk Yttria	Chem-C	+0.02	-0.98	+1.00	-1.00	-1.00	A	Bag-B

TABLE-2: OPTIMIZED PROCESS PARAMETER LEVELS BY SUBSTRATE

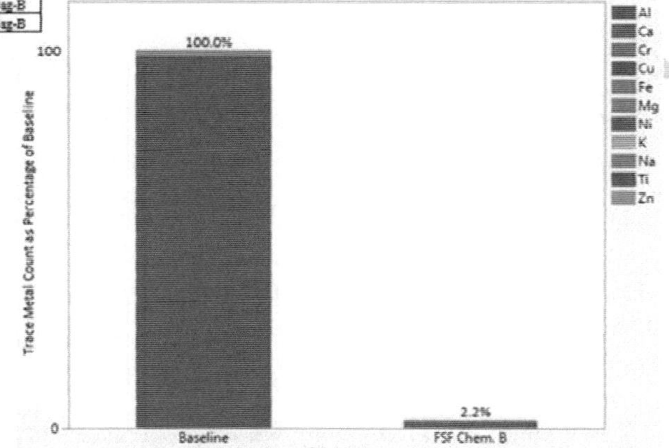

Figure-6: Trace metal results comparing new test chemistries to FSF baseline for non-textured aluminum cleaning

Final Surface Finish (FSF™) for Aluminum Substrate													
	OEM-1 Spec			OEM-2 Spec					QC-Historical			
10^{10} atom/cm^2	Non-Textured 6061 Al Part-A	Non-Textured 6061 Al Part-B	Critical Aluminum Part (AL6061-T6) Part-C	Non-Textured 6061 Al Spec PART-AB	FSF Validation-1	FSF Validation-2	FSF Revalidation Multiple Samples	Predicted Value	Part-A	Part-B	Part-C	Part-AB
Element	Part-A	Part-B	Part-C	Part-AB	Batch-1	Batch-2	Multiple-Batches	Values from Simulation	Control Data	Control Data	Control Data	Control Data
Al	100,000	100,000	220,000		9,900	4,200	6,500	12,466	96,418	39,571	219,222	
Ca	1,500	1,500	1,500	5,000	85	140	400	157	516	1,300	674	584
Cr	200	200	250		20	20	20	20	69	20	135	
Co				100	5	5	5	5				87
Cu	5,000	5,000	5,000		98	110	360	224	2,458	66	38	
Fe	5,000	5,000	5,000		21	20	53	38	4,366	182	572	
Pb					3	3	3	3				
Li				1,000	10	20	20	10				795
Mg	5,000	5,000	5,000		660	300	330	1,566	4,882	1,223	2,305	
Mn					6	5	5	1				
Mo				100	2	2	2	2				79
Ni	100	500	100	2,000	11	10	10	10	25	97	238	146
K	400	400	400	5,000	50	150	50	50	105	367	107	345
Na	2,000	2,000	2,000	10,000	50	210	170	94	1,690	1,040	409	1,469
Sn					5	11	5	5				
Ti	25	100	250		20	20	20	20	33	28	125	
Zn	120	350	500		27	20	20	40	378	596	212	

20nm, 16nm and 14nm OEM Specifications — Final Surface Finish Data — Actual 20nm, 16nm and 14nm Parts Data

Figure-7: Shows actual FSF™ data compared to current specification requirements

Summary and Conclusion

- Parts cleaning requirements have changed significantly since they were originally outsourced to specialized third parties. These changes are driven through increasing numbers and combinations of materials and processes that allow improvements in device performance and footprint reduction.

- Trace metal, ionic and particulate contamination levels that are near or even beyond current metrology capability are now found to be major impediments to chamber process performance and device yield - "Things that didn't matter before now do".

- The challenges moving forward include the development of increasingly complex selective stripping technologies, improved handling and packaging solutions and continued improvements in real-time metrology that can detect ever smaller contamination levels and allow predictable chamber performance to help drive the industry forward.

Static control in Ultrapure DI water application

Eric Guo[1], Crystal Wang[1], Sandy Qian[1], Mars Wang[1], Harry Zhang[1], Keanu Wu[1], Fei Xu[2], Jian Shen[2], Quanyuan Li[2]

[1]Semiconductor Manufacturing International Corporation
[2]Changzhou Ruize Microelectronics Co., Ltd.

Surface Preparation and Cleaning Conference
April 19~20, 2016 Santa Clara, California

Outline

- Introduction
- System Design
- Results

Surface Preparation and Cleaning Conference
April 19~20, 2016 Santa Clara, California

Introduction

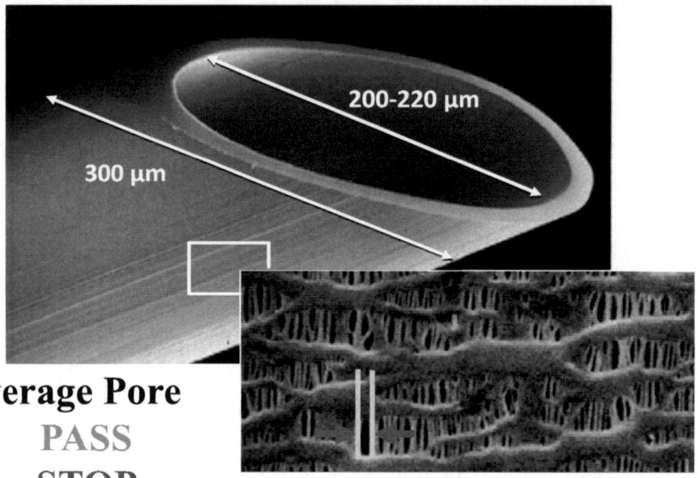

0.03 µm Average Pore
CO_2 gas PASS
H_2O liquid STOP

$$CO_2 + H_2O \rightarrow H^+ + HCO_3^-$$

Surface Preparation and Cleaning Conference
April 19~20, 2016 Santa Clara, California

Industrial Applications

General surface cleaning
Wafer dicing process
Single substrate spin cleaning process

Biggest issue for $DICO_2$ process

How to keep the conductivity set point with wide flowrate span

Surface Preparation and Cleaning Conference
April 19~20, 2016 Santa Clara, California

SMIC REC

System Design

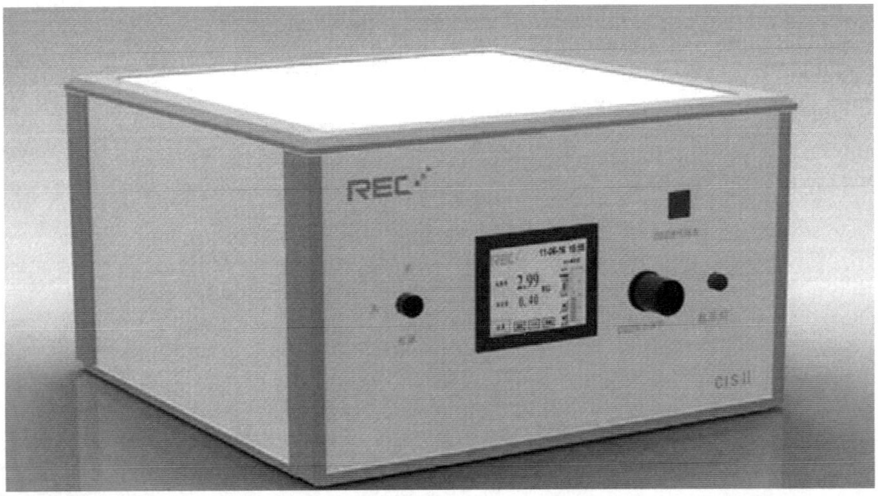

Surface Preparation and Cleaning Conference

April 19~20, 2016 Santa Clara, California

SMIC REC

CO_2 & DI mixing

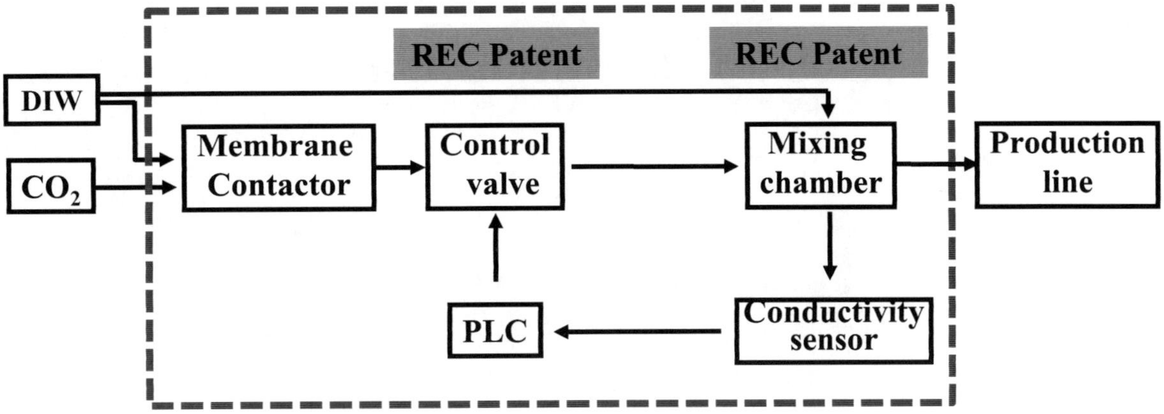

Control Mechanism:

Step motor driven needle valve in the control valve modulates the flow rate of highly concentrated CO_2 water entering the 2nd mixing chamber based on the signal from conductivity sensor

Surface Preparation and Cleaning Conference

April 19~20, 2016 Santa Clara, California

【Top view】

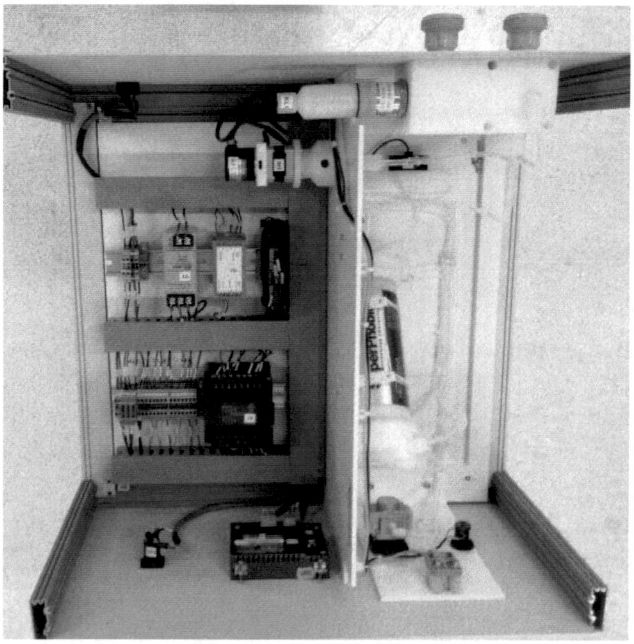

Surface Preparation and Cleaning Conference

April 19~20, 2016 Santa Clara, California

【Product key technique I】

Gas adding membrane

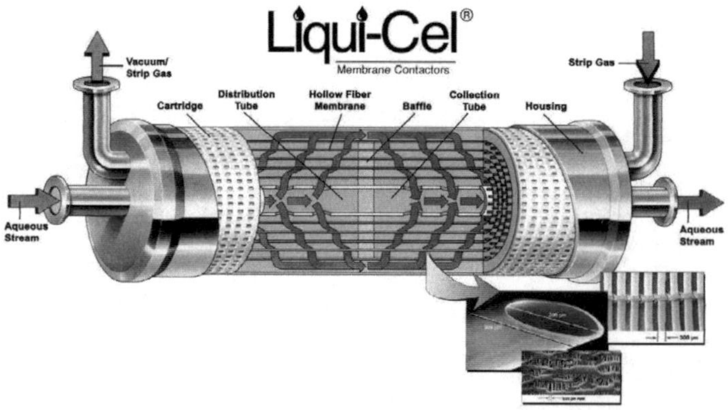

Large surface area membrane produces highly concentrated CO_2 re-ionized water

Surface Preparation and Cleaning Conference

April 19~20, 2016 Santa Clara, California

【Product key technique II】

Control valve

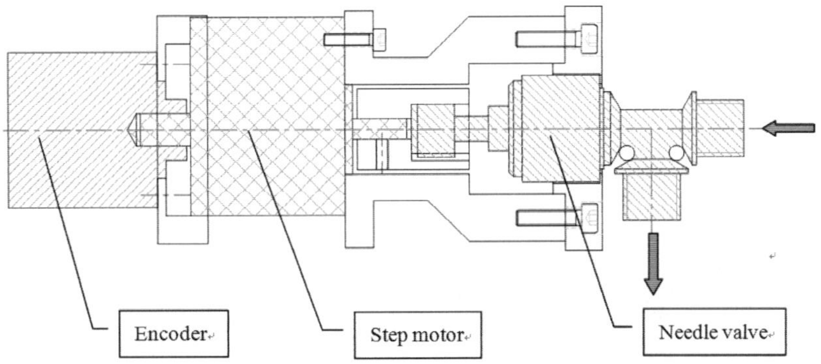

Step motor driven mechanism precisely control the volume of concentrated CO_2 re-ionized water for 2nd tier mixing

Surface Preparation and Cleaning Conference
April 19~20, 2016 Santa Clara, California

【Product key technique III】

2nd tie mixing chamber

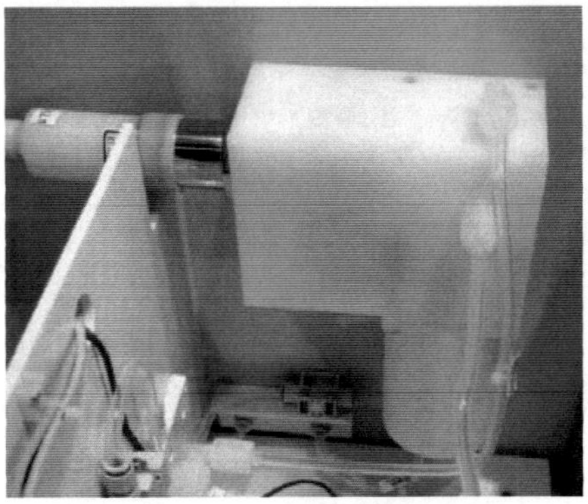

Internal check valve design to prevent down stream water backflow causing contamination on membrane and DI water supply

Surface Preparation and Cleaning Conference
April 19~20, 2016 Santa Clara, California

Key techniques

① *Large surface area membrane*
② *REC patent feedback loop control mechanism*
③ *REC patent 2nd tier mixing chamber design*

↓

Major advantages

① *Wide range for $DICO_2$ flow rate*
② *Precise control for resistivity set point with wide flow rate span*
③ *Fast response for resistivity set point change*

Surface Preparation and Cleaning Conference
April 19~20, 2016 Santa Clara, California

Results

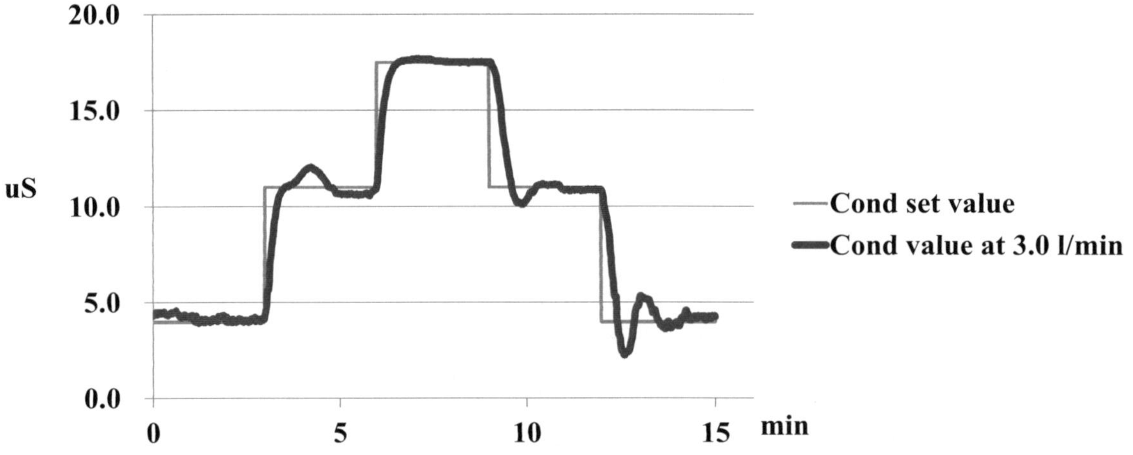

Time to Set Value (+/- 1uS): 15sec~30sec

Surface Preparation and Cleaning Conference
April 19~20, 2016 Santa Clara, California

SMIC REC

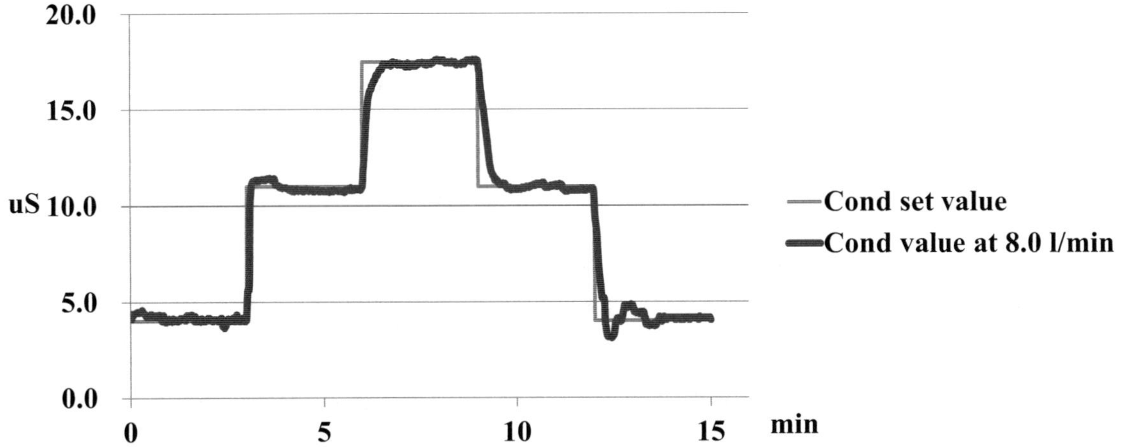

Time to Set Value (+/- 1uS): 6sec~18sec

Surface Preparation and Cleaning Conference

April 19~20, 2016 Santa Clara, California

SMIC REC

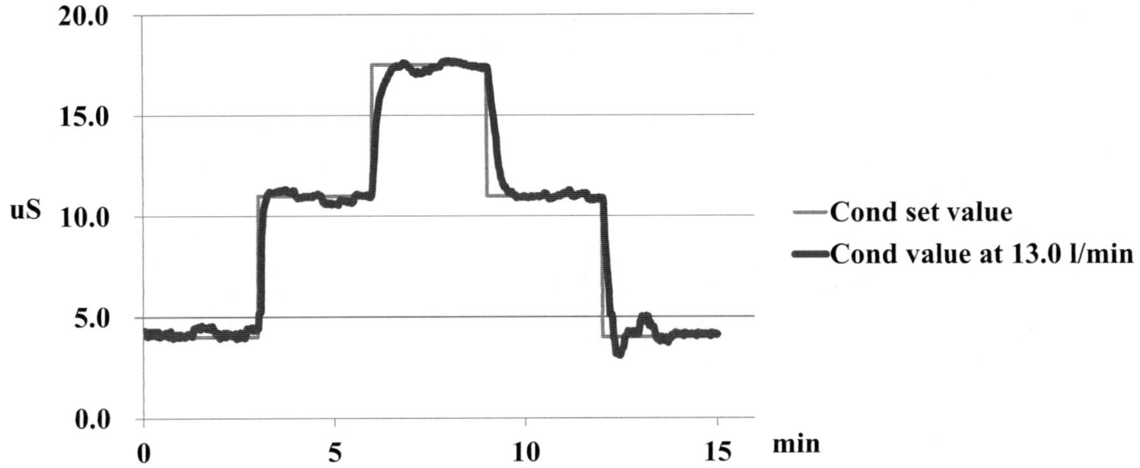

Time to Set Value (+/- 1uS): 8sec~21sec

Surface Preparation and Cleaning Conference

April 19~20, 2016 Santa Clara, California

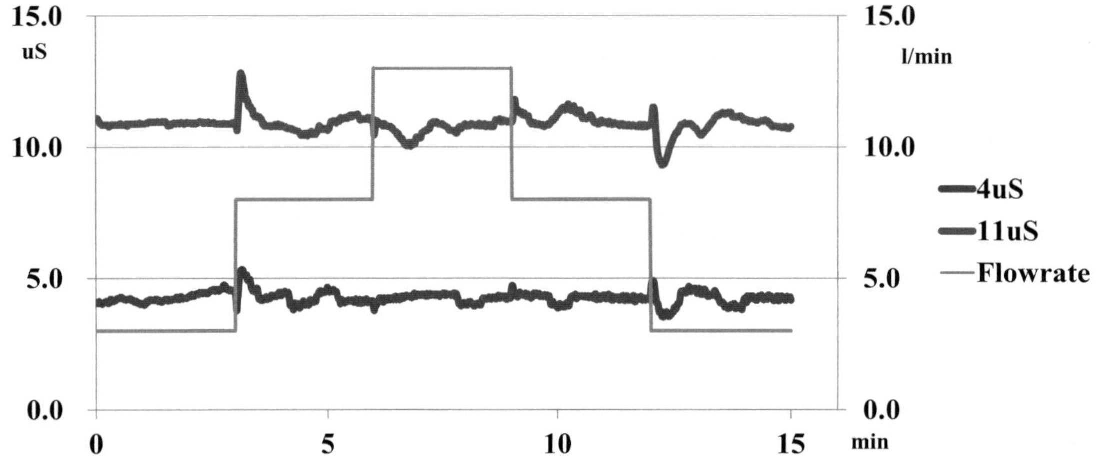

Time to Set Value (+/- 1uS): 13sec~15sec
Total range: Set value +/- 1.5uS

Surface Preparation and Cleaning Conference
April 19~20, 2016 Santa Clara, California

Acknowledgements

1. Sponsored by Program of Shanghai Subject Chief Scientist.
2. Management and technical support from Semiconductor Manufacturing International Corporation and Changzhou Ruize Microelectronics Co., Ltd. .

Thank You!

Surface Preparation and Cleaning Conference
April 19~20, 2016 Santa Clara, California

BALAZS™ nanoanalysis

AIR LIQUIDE™

Contamination Control of Manufacturing Practices Using Critical Wipers

Balazs NanoAnalysis, 46409 Landing Parkway – Fremont CA 94538

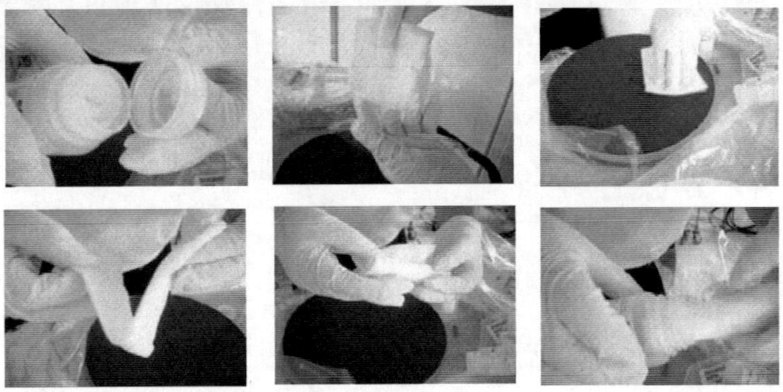

Agenda

Authors: Victor K.F. Chia and Jennifer Jew
Contact: victor.chia@airliquide.com

- Introduction to Clean Manufacturing
- Contamination Sources
- Value of Critical Wipe Test
- Critical Wipe Test Procedure
- Case Studies
- Contamination Mapping
- Contamination Risk Assessment
- Summary

Why is Clean Manufacturing Important?

- Clean Manufacturing (Operation) is a **prerequisite** to successful manufacturing
- It is often **overlooked** because it is a culture and it is considered a hindrance to the production process and schedule
- But it is important because it can **impact**:
  - Product line down time
  - Quality and delivery issues
  - Not sharing of Best Practices
  - Planning delays
  - Overtime cost

Clean Manufacturing is a culture of doing it right when no one is looking or checking

What is Needed for Clean Manufacturing?

- **Analytics** provide insight to cleanroom operation and manufacturing that are affecting product quality

- **Monitor and understand the sources of contamination** during cleanroom operations, procedures and activities
- **Contamination control systems** for effective control of physical and environmental parameters and process-to-process interactions
- **Effective wipe down** to improve surface cleanliness by removing contaminants that can originate from
  - People
  - Environment
  - Parts and equipment

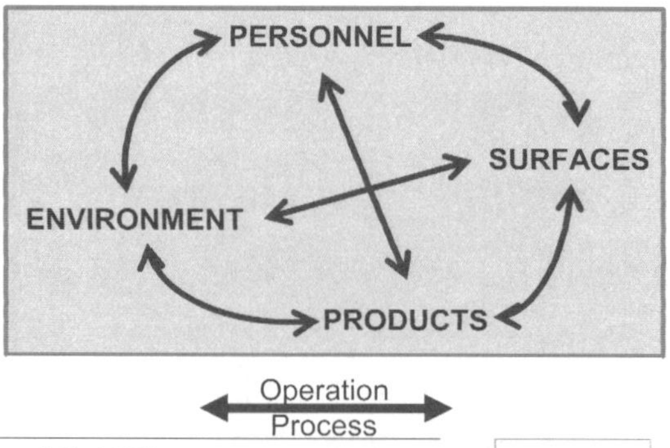

Primary Contamination Sources

- **Facility** - gas, water, chemicals and infrastucture
  - **Cleanroom environment** - cleanroom, laminar flow hood
  - **Cleanroom consumables** – garments, gloves, bags, wipes, packaging
- **People** - contamination generator and mechanism for transfer
- **Manufacturing materials** – lubricants, detergents, polishing agents.
- **Supplier** – in addition, any process steps completed by an external supplier must also be evaluated

5 | Contamination Control using Critical Wipes | BALAZS™ The MicroContamination Experts | AIR LIQUIDE

Value of Critical Wipe Test

- IC devices vary greatly in complexity and application
- When processing advanced IC devices it is important to know the cleanliness state of normal cleanroom operation, before and after facility upgrade and process tool PM
- Critical wipe monitoring can provide valuable insight to the cause of a contamination event affecting the product quality; contamination sources have unique signatures

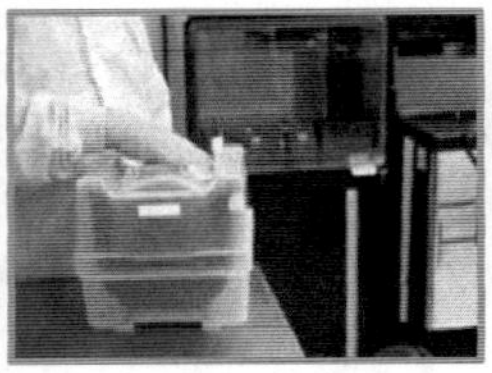

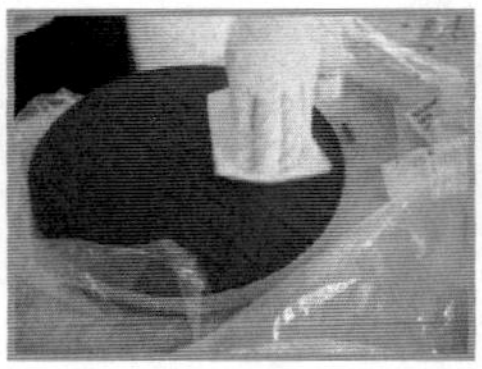

6 | Contamination Control using Critical Wipes | BALAZS™ The MicroContamination Experts | AIR LIQUIDE

Value of Critical Wipe Test

The metals monitored by Critical Wipe Test can be associated with People, Process Equipment and the condition of the Facility

People

Element	RL
■ Sodium (Na)	10 Total ng/wipe
■ Potassium (K)	50 Total ng/wipe

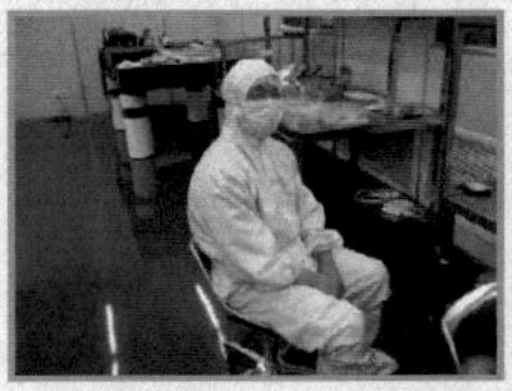

Process/Equipment

Element	RL
■ Nickel (Ni)	10 Total ng/wipe
■ Iron (Fe)	50 Total ng/wipe
■ Copper (Cu)	10 Total ng/wipe
■ Chromium (Cr)	10 Total ng/wipe
■ Aluminum (Al)	10 Total ng/wipe

Facility

Element	RL
■ Magnesium (Mg)	10 Total ng/wipe
■ Calcium (Ca)	50 Total ng/wipe

Contaminants

Galvanized steel, Fe, Ni, Zn

People, Na

Gypsum = $CaSO_4$, Ca/$MgCO_3$. Na/KCl, FeS_2

Dirt = SiO_2 (quartz), $CaCO_3$,(limestone), Na, K, Mg, P, S, Fe, and Mn

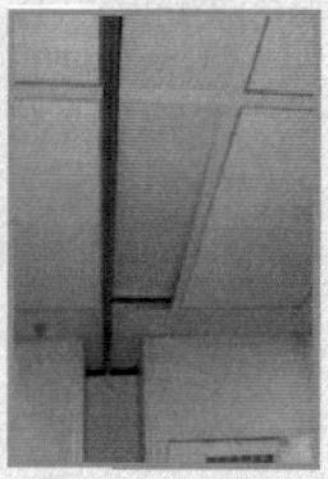

Critical Wipe Test Procedure

- **Balazs pre-cleaned wipers are used for critical wipe test**
- A surface should be swiped two (2) times in the same location using a fresh wipe surface
- Target surface area sampled is 16 cm^2

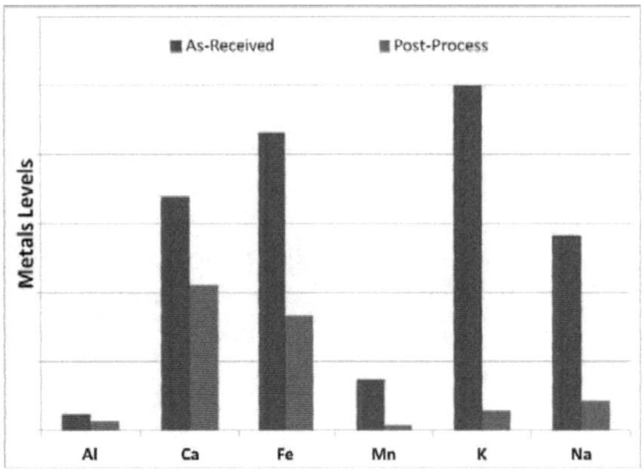

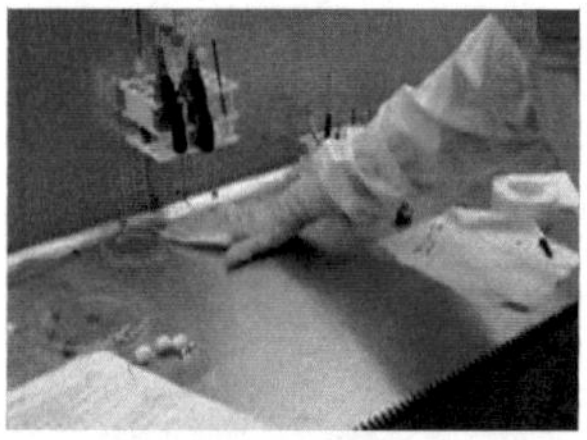

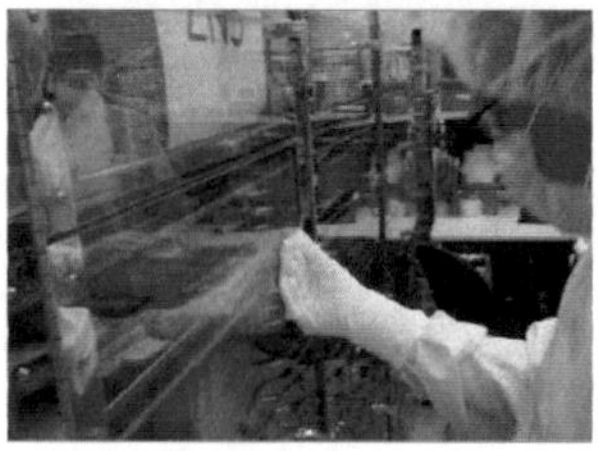

Critical Wipe Test Procedure

Double glove. Clean outer glove using IPA.

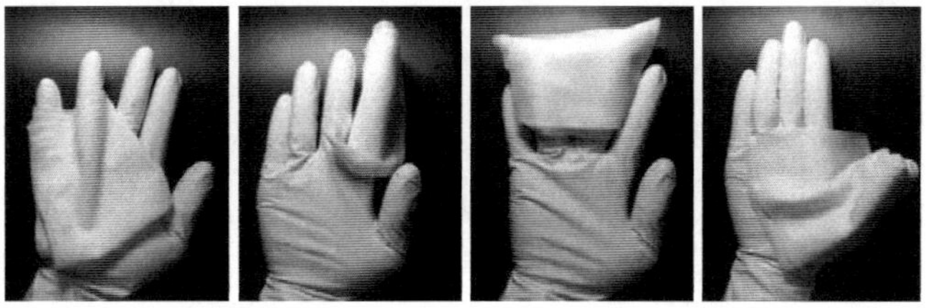

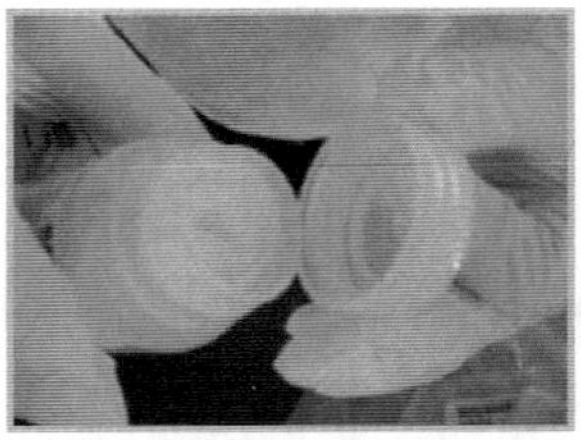

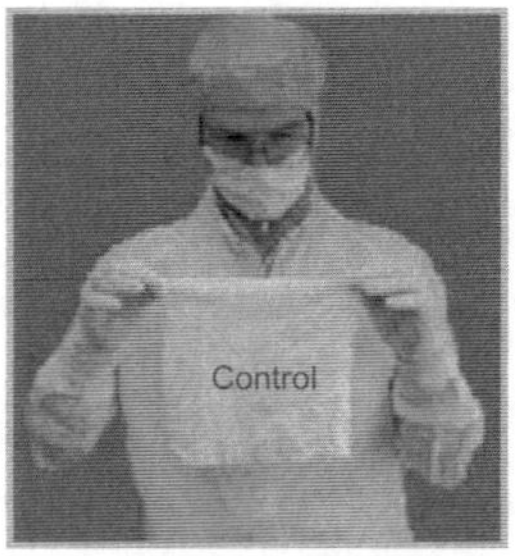

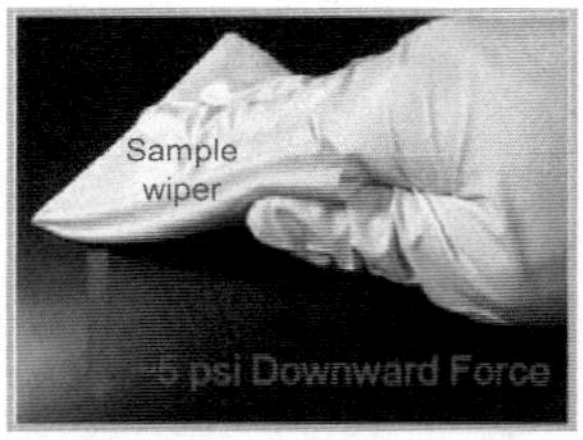

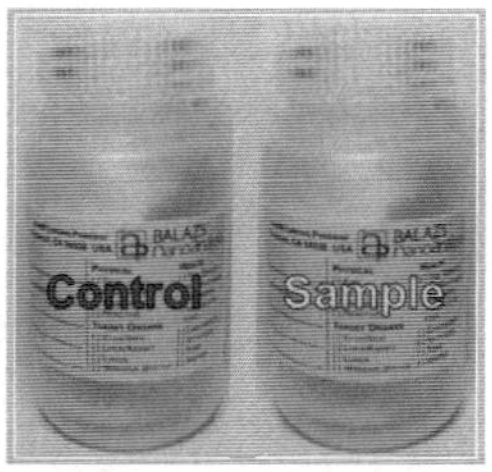

Critical wiper bottle containing wiper for Control and Sampling

i. **Control wiper to sample cleanroom environment**
ii. **Sample surfaces using the Sample Wipers**
iii. **Return wipers into their containers**

Wipers ready for ICP-MS analysis

Critical Wipe Test Results from the Wall

- Based on this critical wipe test result, one can establish usage and activity levels; adherence of protocol, e.g. if frequency of wipe down was performed; and track contamination carry over

11/13	12/13	Diff.	
1500	2300	800	
80	81	1	
33000	*75000*	42000	Ca
100	90	-10	
490	*900*	410	
2500	*3800*	1300	Fe
120	100	-20	
8400	16000	7600	
7700	12000	4300	Mg
*	50	50	
170	200	30	
8900	17000	8100	K
46000	*74000*	28000	Na
640	740	100	
170	290	120	
2200	3200	1000	
111970	205751	93781	

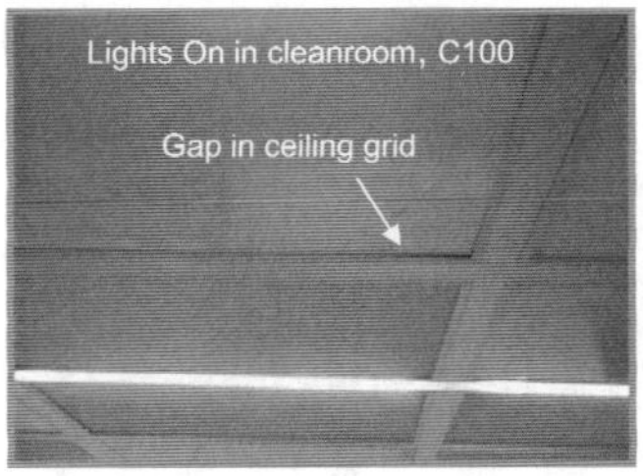

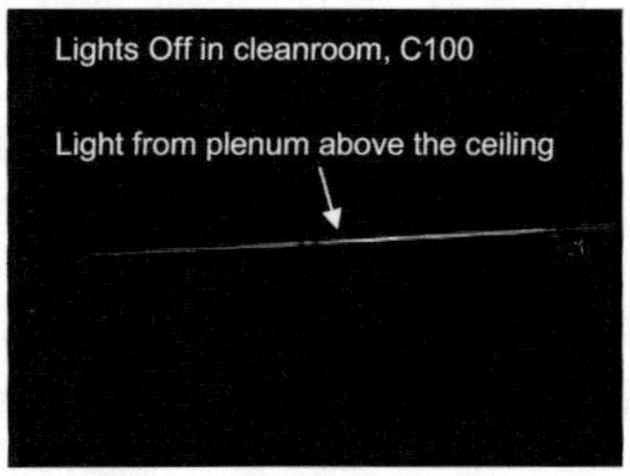

Gypsum = $CaSO_4$, $Ca/MgCO_3$. Na/KCl, FeS_2

Critical Wipe Test Results from Equipment

- The critical wipe test result of inspection scopes can establish usage and activity levels; adherence of protocol, e.g. if frequency of wipe down was performed; and contamination carry over

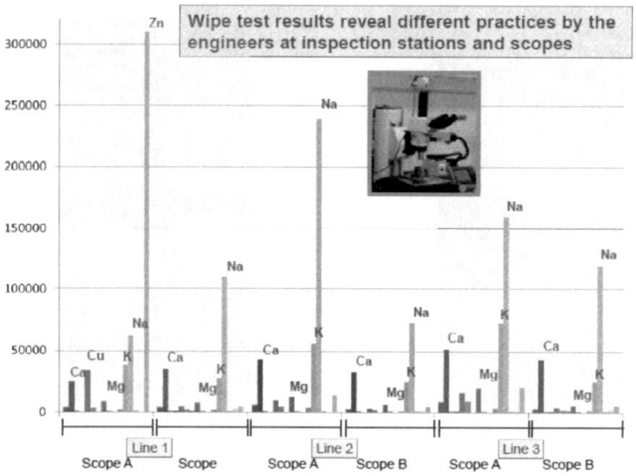

	Start of Shift	Mid-Shift	End of Shift
Working Surfaces	Yes	After lunch	Yes
Equipment Tools	Yes	Prior to use for each step	Yes
Microscopes	Yes	Not required	Yes
Fixtures	Prior to use	Not required	Not required

Critical Wipe Test Results from Process Tool

- After PM, critical wipe testing revealed that locations F and J had the highest levels of Sodium and Potassium, while location E had the lowest metals

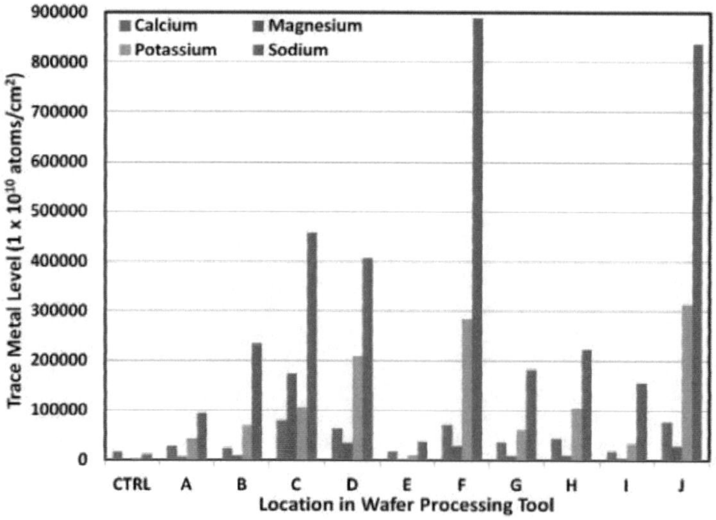

Critical Wipe Test Results of Process Tool Assembly

- Critical wipe testing identified key areas of operations contributing to cross-contamination

Critical Surface Wipe Testing can be used to determine the CAUSE during Chamber Assembly

Steps of Chamber Integration

Parts Receiving ➡ Chamber Assembly ➡ Chamber Testing ➡ Chamber Shipment

Check For Contamination Sources During Chamber Assembly

- Poor cleanroom protocols: gowning, contact with chamber
- Contaminated tools: dirty wrenches, tools to lift chamber
- Contaminated environment: walls, workbenches, hoods
- Airborne molecular contamination
- Poor Clean Manufacturing practices

GAP ANALYSIS

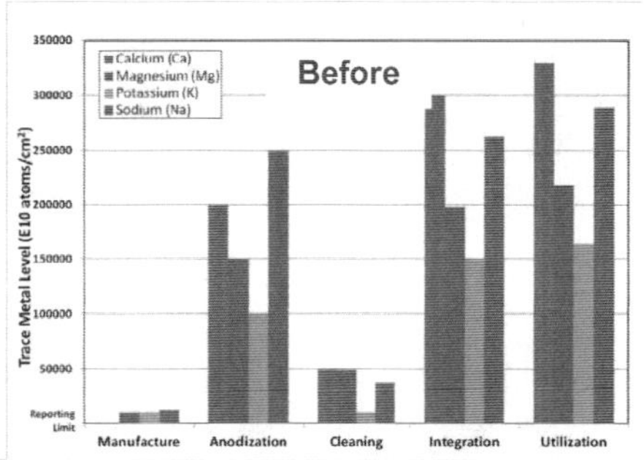

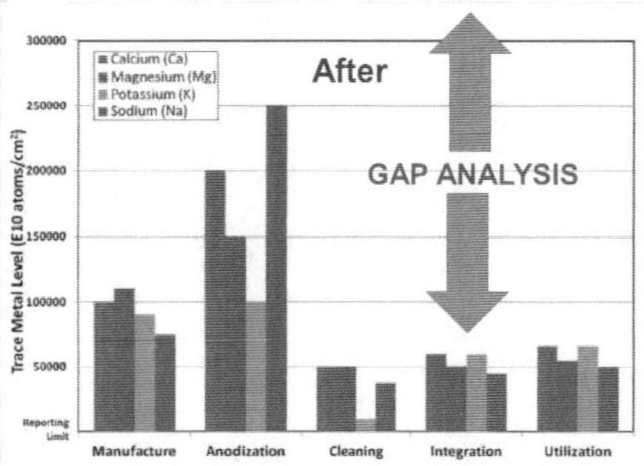

Contamination Mapping of Facility and Operation

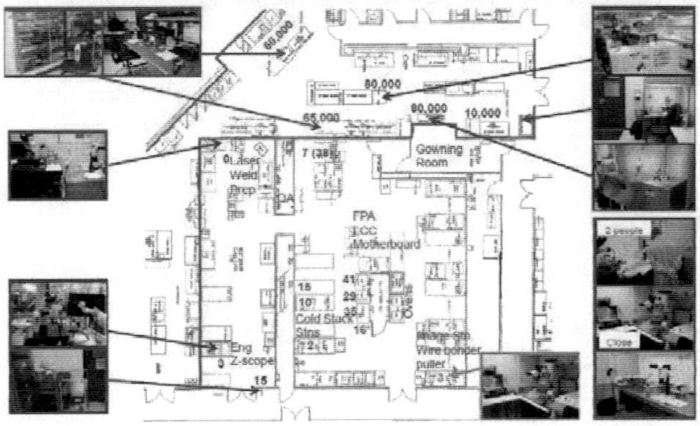

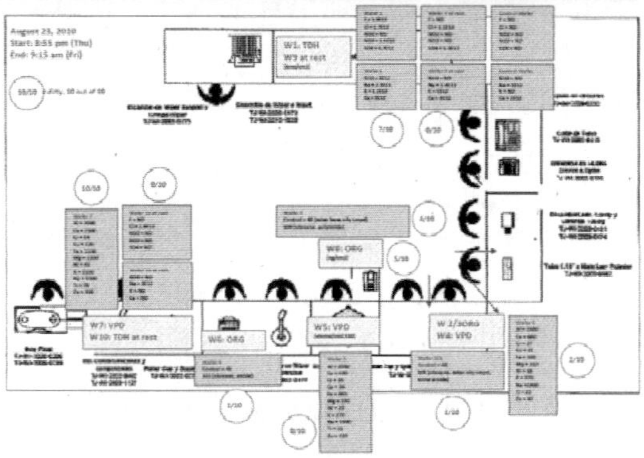

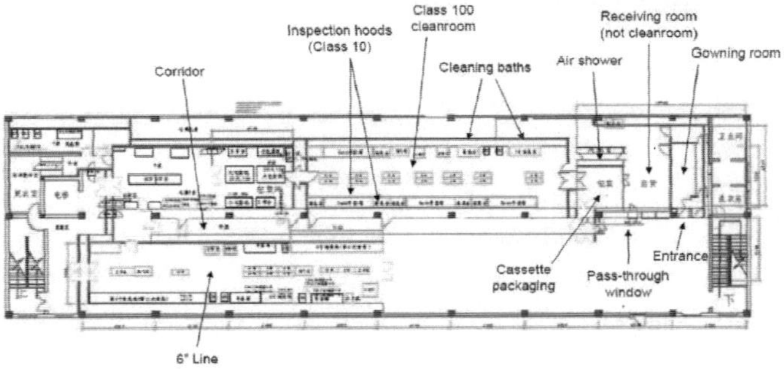

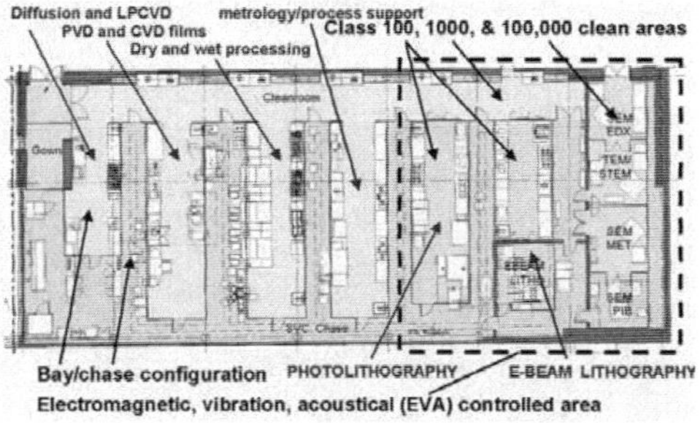

Contamination Risk Assessment

- Gap Analysis recommendations and action items concluded from Critical Wipe tests should be reviewed with the intent to calculate their risk posed to the product
- The "**Risk**" is likelihood of the occurrence multiplied by the "**Effect**" of the contamination hazard on the assembly or process step

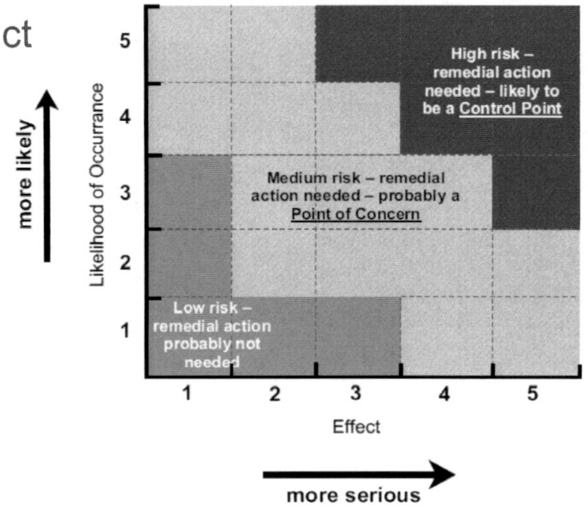

Summary

- With knowledge of the cleanroom use and how it is maintained, **Critical Wipe Testing is capable of:**
  - Rapid on-site testing for metals on critical surfaces
  - Being performed in a wide range of environments
  - Detect changes in surface metals on a part before and after it undergoes a process, after handling and after cleaning
  - Determine the surface metals on multiple components in wafer processing equipment
  - Quantify metal contamination at various locations in a cleanroom
- Once a historical **Critical Wipe Test data base** has been established, a contamination event (source and cause) will reveal itself

SMIC

A Study of Removing Scan Damage on Advanced ArF PSM Mask by Dry Treatment before Cleaning

Irene Shi, Eric Guo, Max Lu, Sandy Qian, Rivan Li

Semiconductor Manufacturing International Corp.
18 Zhangjiang Road, Pudong New Area, Shanghai, China
201203

Surface Preparation and Cleaning Conference
April 19~20, 2016 Santa Clara, California

SMIC

Outline

- Introduction
- Experimental
- Date & Analysis
- Summary & Acknowledgement

Surface Preparation and Cleaning Conference
April 19~20, 2016 Santa Clara, California

Introduction

EB (Electron-beam) repair is widely applied to deal with critical defects on advanced ArF PSM mask, and challenging defects may require several repair cycles. However, the cumulative scanning during repair processes would cause scan damage issues, shown as dark on CDSEM image, and transmission rate abnormal on Aims image.

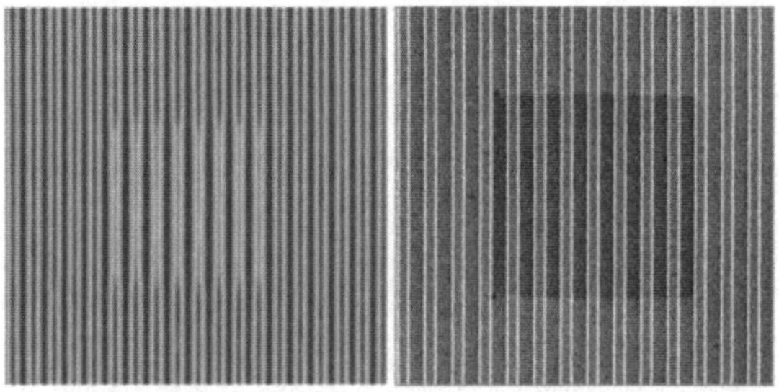

a. Aims image with scan damage   b. CDSEM image with scan damage

Figure 1 Repair Scan Damage Issue

Outline

- Introduction
- Experimental
- Date & Analysis
- Summary & Acknowledgement

Surface Preparation and Cleaning Conference
April 19~20, 2016 Santa Clara, California

Experimental

A novel approach using dry treatment based on reactive plasma Asher to deal with scan damage would be introduced. The process is to supply O2 mixed with innocuous H2O, H2 or other gases in the plasma Asher chamber, then gases are ionized into plasma and strike the mask surface.

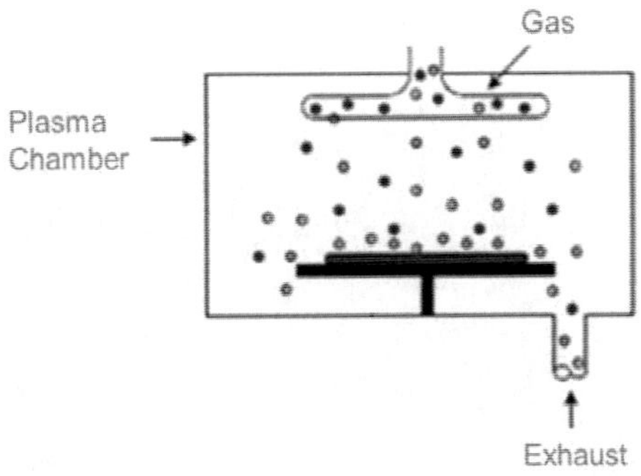

Figure 2 Dry Treatment with O2 mixed with innocuous H2O, H2 or other gases

Surface Preparation and Cleaning Conference
April 19~20, 2016 Santa Clara, California

Outline

- Introduction
- Experimental
- Date & Analysis
- Summary & Acknowledgement

Surface Preparation and Cleaning Conference
April 19~20, 2016 Santa Clara, California

Data & Analysis

Dry Treatment Effect on Scan Damage

SPM (a Sulfuric – Peroxide Mixture) cleaning method tends to relieve or remove scan damage partially, but O3 cleaning (an alternative of SPM to prevent surface haze formation), seems no effect on it. But it's found that O3 cleaning with dry pretreatment could remove the scan damage completely.

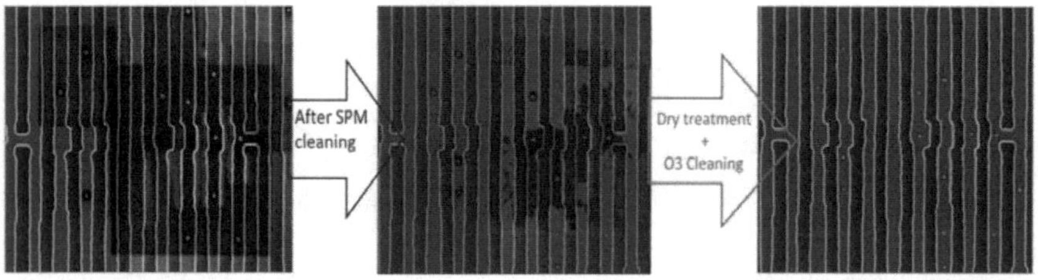

a. after O3 cleaning without dry treatment b. after SPM cleaning c. after dry treatment and O3 cleaning

Figure 3 Dry Treatment Removed Scan Damage induced by Etch Repair

Surface Preparation and Cleaning Conference

April 19~20, 2016 Santa Clara, California

Data & Analysis

Dry Treatment Effect on Scan Damage

Darken in CDSEM images may be not the only way of showing scan damage issues, Since we have once encountered image vibration phenomenon, highly suspicious as another behavior of scan damage. Conventional O3 clean doesn't work on it, but it has been removed by O3 clean with dry pretreatment assistance.

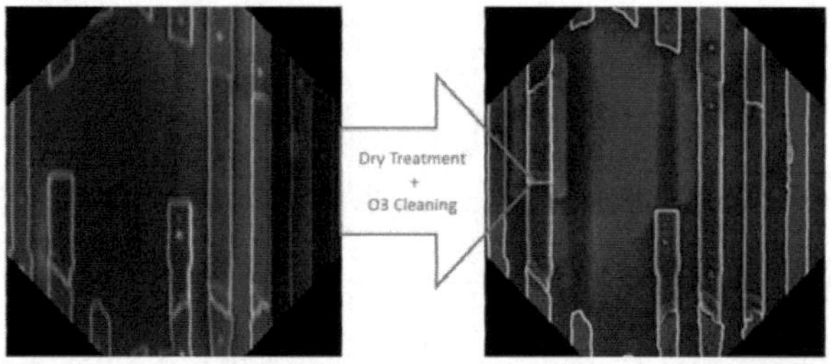

a. Vibrational Image due to scan damage b. Solved by O3 cleaning with Dry Pretreatment

Figure 4 Dry Treatment Solved Image Vibration induced by Deposition Repair

Surface Preparation and Cleaning Conference

April 19~20, 2016 Santa Clara, California

Data & Analysis

Dry Treatment Effect on Charge Effect

There are similarities between scan damage and charge effect induced by CDSEM scanning, and it's found that charge effect could be cleared by dry treatment only. On the other hand, Conventional O3 clean can also remove the charge effect easily.

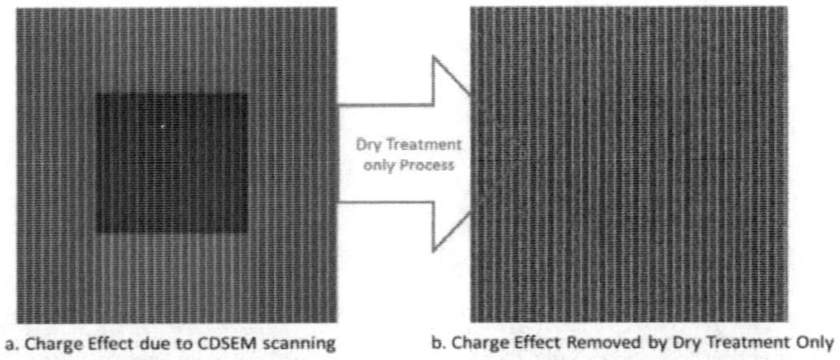

Figure 5 Dry Treatment could Remove Charge Effect

Data & Analysis

X-ray Composition Analysis of Scan Damage and Charge Effect

X-ray Composition Analysis results show that Carbon and Nitrogen elements increased for both scan damage and charge effect.

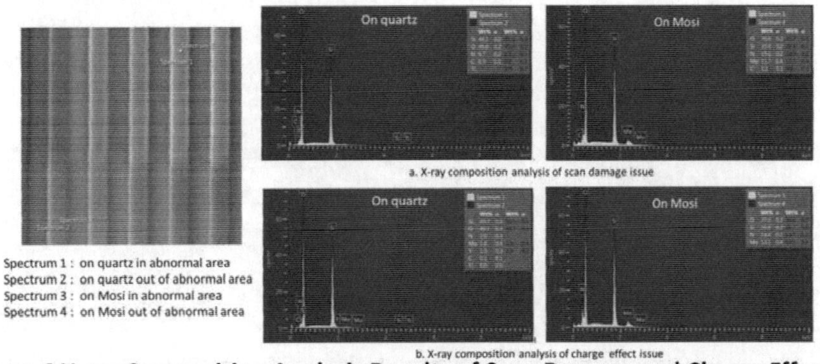

Figure 6 X-ray Composition Analysis Results of Scan Damage and Charge Effect

Surface Preparation and Cleaning Conference
April 19~20, 2016 Santa Clara, California

Data & Analysis

Mechanism for Dry Treatment

paper poses emitted secondary electrons absorbs hydrocarbon molecular, which has a hydrophobic property that is hard to clean. Dry treatment may change its hydrophobicity to hydrophilic.

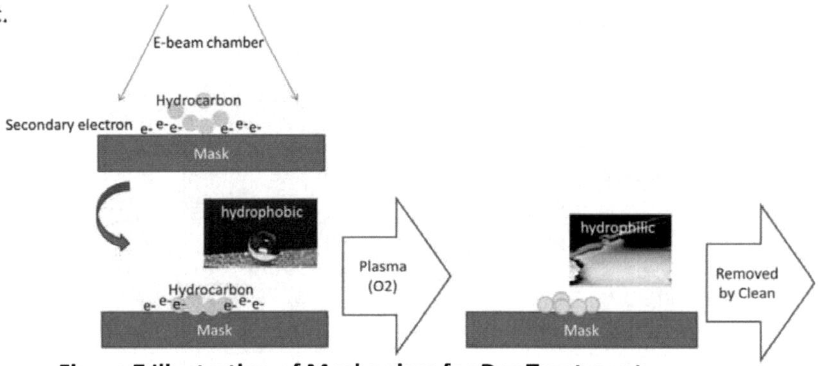

Figure 7 Illustration of Mechanism for Dry Treatment

Data & Analysis

Study on Dry Pretreatment Effect on Cumulative Doses of Scan Damage

Scan damage and charge effect is assumed to come from the similar mechanism, related to electron beam. We studied dry treatment and other wet clean methods effect on cumulative doses of scan damage. Results prove dry pretreatment is actually a dry clean method, which would gain double clean effect together with wet clean.

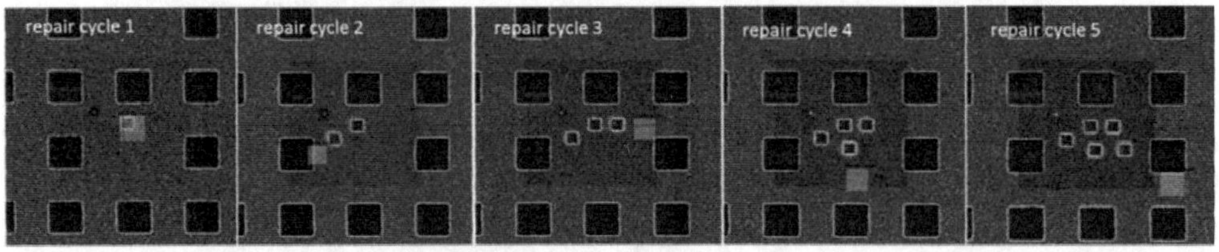

Figure 8 Scan Damage Accumulations in Repair Cycles

Surface Preparation and Cleaning Conference
April 19~20, 2016 Santa Clara, California

Data & Analysis

Study on Dry Pretreatment Effect on Cumulative Doses of Scan Damage

Clean with dry pretreatment provides stronger clean effect than SPM clean and O3 clean, but it still has an impassable limit for scan damage accumulation.

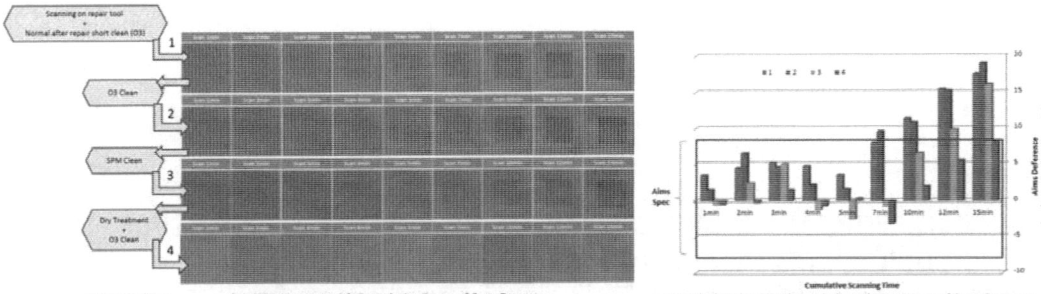

a. Process Flow b. CDSEM images with Cumulative Doses of Scan Damage c. Aims Deference Results with Cumulative Doses of Scan Damage

Figure 9 Dry Treatment Effect on Cumulative Doses of Scan Damage

Surface Preparation and Cleaning Conference

April 19~20, 2016 Santa Clara, California

Outline

- Introduction
- Experimental
- Date & Analysis
- **Summary & Acknowledgement**

Surface Preparation and Cleaning Conference

April 19~20, 2016 Santa Clara, California

Summary

Dry treatment method helps to remove scan damage and charge effect induced by electron beam source. It provides a substantial increase in cleaning capability on ArF PSM mask.

Acknowledgement

Sincerely thank the sponsor of the Program of Shanghai Subject Chief Scientist.

Thank You!

Post Tungsten CMP Cleaner Development with Improved Organic and Particle Residue Removal on Silicon Nitride and Excellent Tungsten Compatibility

Ching-Hsun Chao, Chi Yen, Ping Hsu, Eugene Lee, Paul Bernatis

EKC Technology, DuPont Electronics and Communications

Ken.Chao@dupont.com; Paul.R.Bernatis@dupont.com

Content

Introduction

- Problem statement & hypothesis

Experimental section

- Zeta potential analysis / Atomic-force microscopy / Cleaning efficiency estimation method / Tungsten compatibility analysis

Results and discussions

- pH & zeta potential relationship on Si_3N_4 & SiO_2
- Adhesion force & cleaning relationship on Si_3N_4
- pH & zeta potential relationship with cleaning efficiency on Si_3N_4
- Tungsten compatibility improvement with additive A
- Proposed clean mechanism on Si_3N_4

Summary

2

Problem statement

Clean efficiency on SiNx | **Tungsten compatibility**

RESIDUAL SLURRY | **SURFACE PARTICLE** | **W PLUG CORROSION** | **RECESSED W PLUG**

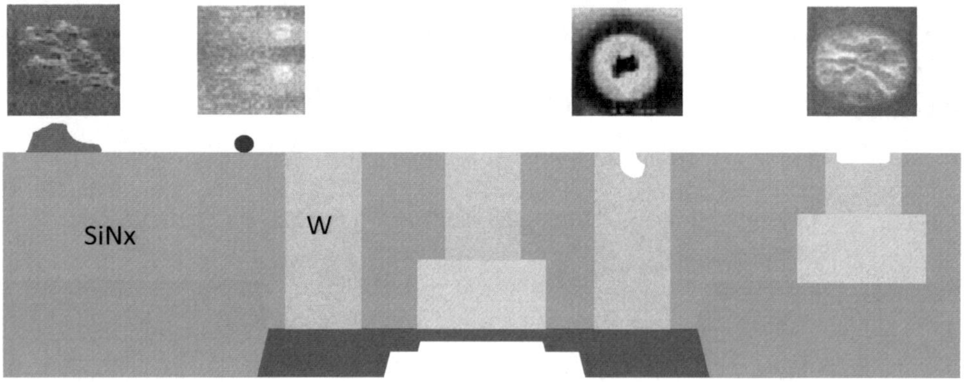

- There are four types of defectivity after post tungsten cleaning.
- Improve tungsten compatibility and cleaning efficiency to improve defectivity.

3

Zeta Potential Measurements on Wafer Substrates and Defect Sources

Zeta Potential vs. pH

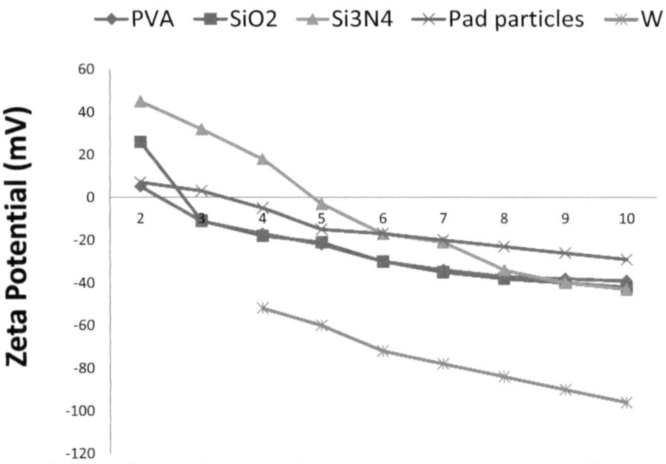

- Adjust formulation pH to measure zeta potential
- High pH value provides higher electronic repulsive force between contaminant particles and SiNx and tungsten wafers

4

Method to Estimate Cleaning Efficiency

- W & SiN wafer pre-treatment method:
  - Cut 3 cm square cubic area wafer coupons
  - Apply the slurry (with or with out centrifuge) on the W & SiN wafer coupon surface and allow it to dry it overnight.
  - Put the wafer coupon in 100 mL of post-CMP formulation for 1, 5 or 10 minutes. (500 RPM @ 50°C). Dry the coupon and analyze by SEM

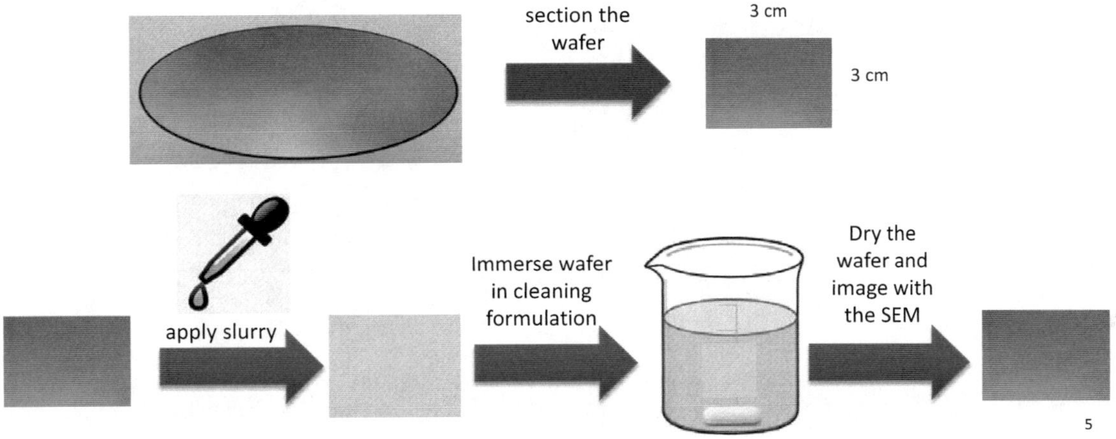

5

Atomic-force microscopy

Image of the cantilever mounted a bead

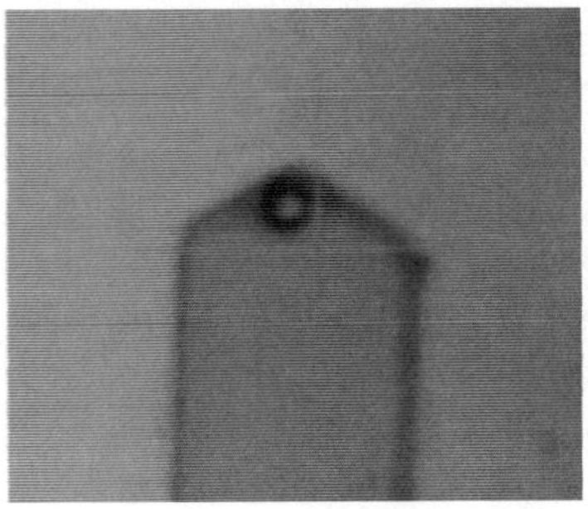

A schematic of adhesion measurement

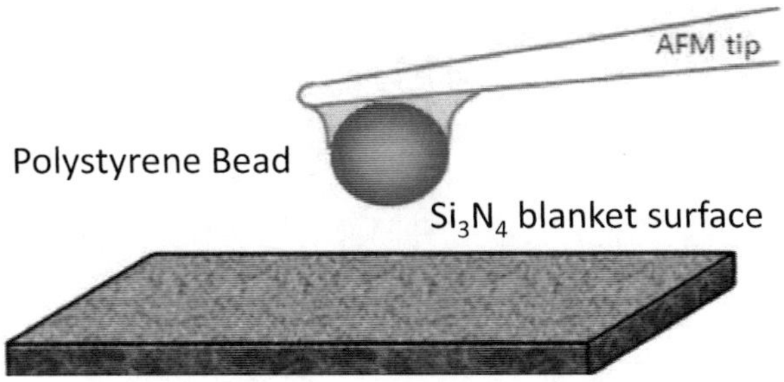

A technique using AFM (atomic force microscopy) was developed to measure the adhesion forces between colloidal silica particles (50 – 80 nm) and wafer surfaces in the cleaning solution. Additionally, by attaching a polystyrene bead (~5µm) to the AFM cantilever and measuring the force-distance curve to Si_3N_4, Cu and W wafers the adhesion force between polystyrene and the wafer substrate was obtained.

6

Tungsten compatibility analysis

- Pretreatment tungsten wafers
  - Cut 3 cm square cubic area tungsten wafer coupons
  - Immerse the coupon in 100 mL formulation for 10 minutes.
  - Measure the amount of tungsten dissolved in the formulation with ICP-MS

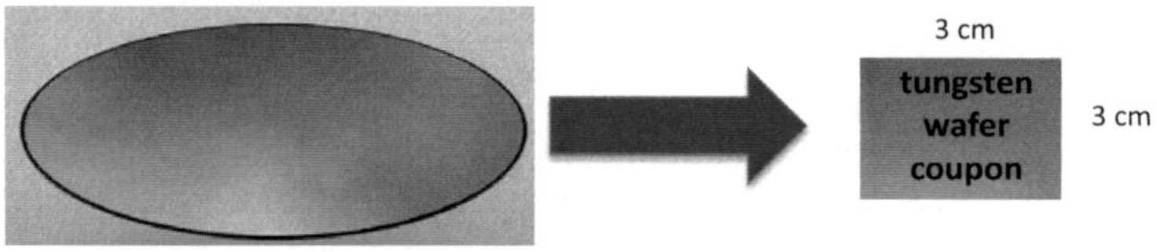

Immerse the tungsten coupon for 10 minutes in the cleaning solution

Remove the coupon and analyze the tungsten ion concentration by ICP-MS

7

pH & zeta potential of Si_3N_4 relationship

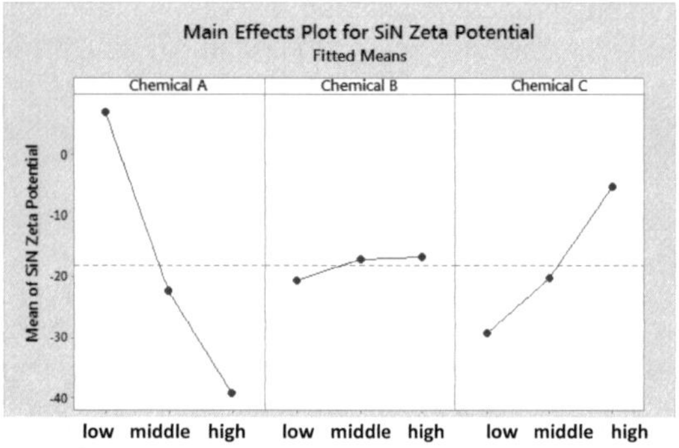

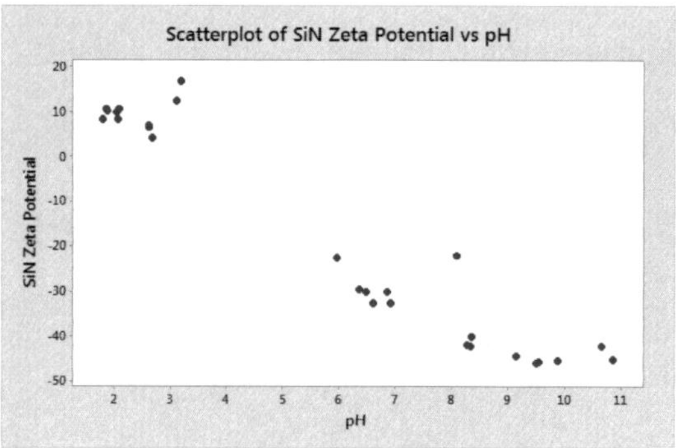

- **Chemical A** and **Chemical C** shifted the zeta potential of SiN_x. Increased levels of **Chemical A** raised the pH of the formulation and the zeta potential of SiN_x became more negative. **Chemical C** decreased the pH of the formulation as its concentration was increased and the zeta potential of SiN_x was increased.

8

pH & zeta potential of SiO_2 relationship

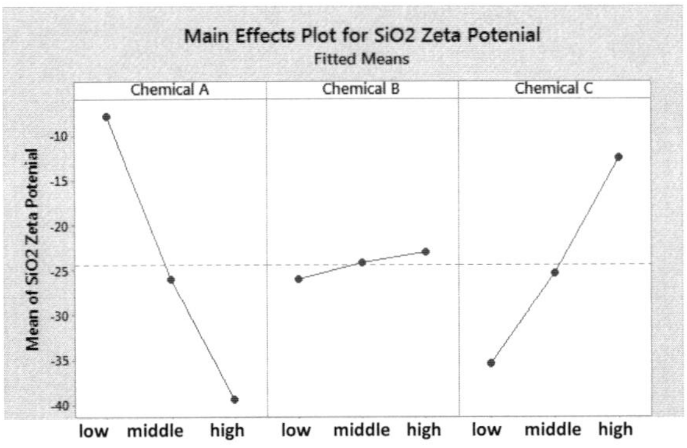

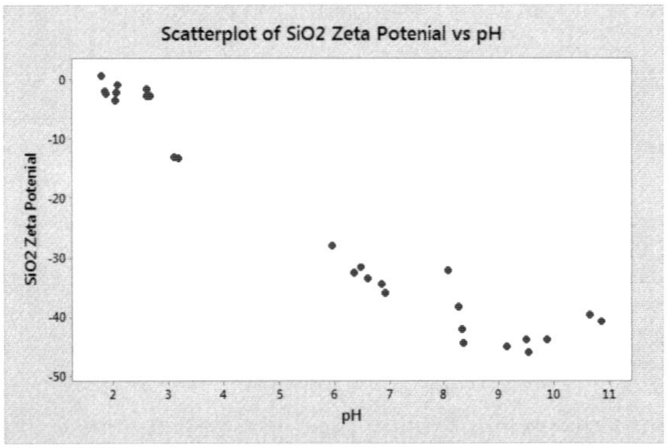

- **Chemical A** and **Chemical C** concentrations affected the zeta potential of SiO_2. Higher concentrations of **Chemical A** raised the pH and the zeta potential of SiO_2 became more negative. **Chemical C** had the reverse behavior and decreased the pH and increased the zeta potential of SiO_2.

9

Adhesion Force & Clean Relationship on Si_3N_4

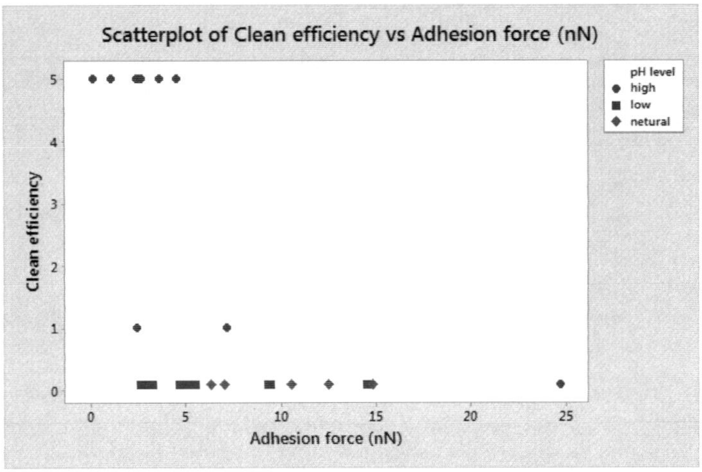

- The best cleaning efficiency was observed at high pH where the adhesion force between polystyrene and Si_3N_4 was reduced. Slurry residues would be more dissolved in an alkaline environment and the lower adhesion force will facilitate particle and organic residue removal.

10

pH & zeta potential relationship with clean efficiency on Si_3N_4

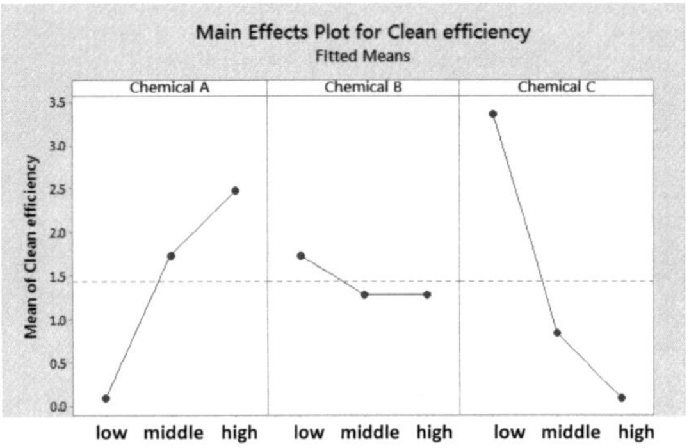

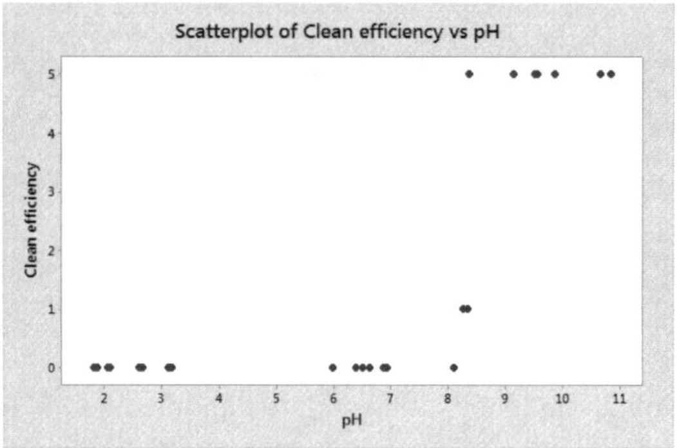

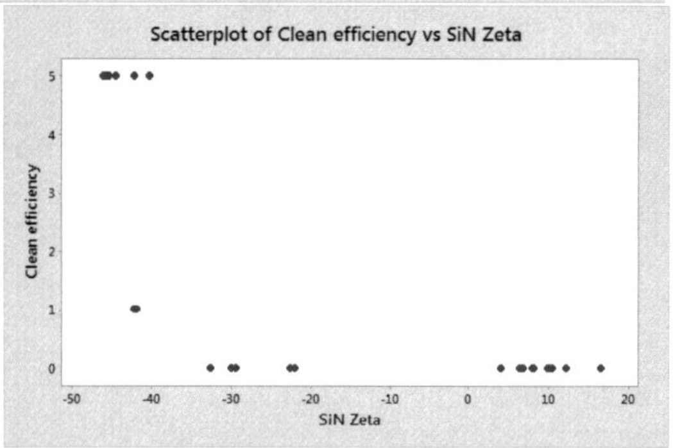

- From main effects plot, we found that increased amount of **Chemical A** would enhance clean efficiency but increased amount of **Chemical C** would decreased clean efficiency. The clean efficiency improvement came from pH effect or zeta potential effect because when we increased the amount of **Chemical A** the pH would increase at the same time.

Tungsten compatibility improvement by added chemical A

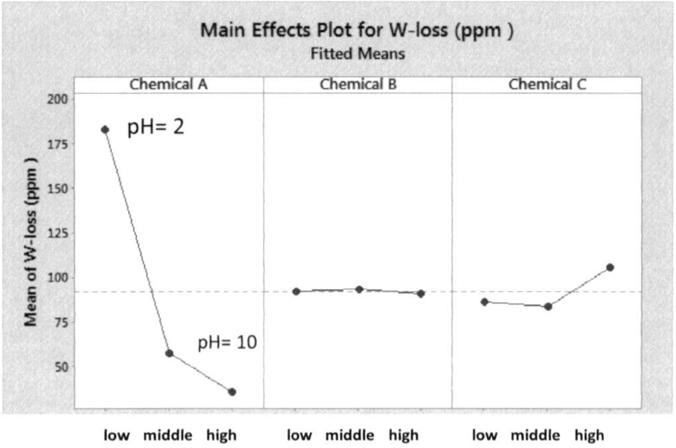

- Added **chemical A** could improve tungsten compatibility and reduce tungsten loss in the formulation. It is well known tungsten is easy to corrode in high pH region. However, **chemical A** showed good tungsten compatibility in the high pH region. When we added **chemical C** amount higher than middle level, it will increase tungsten loss.

12

Proposed clean mechanism on Si_3N_4

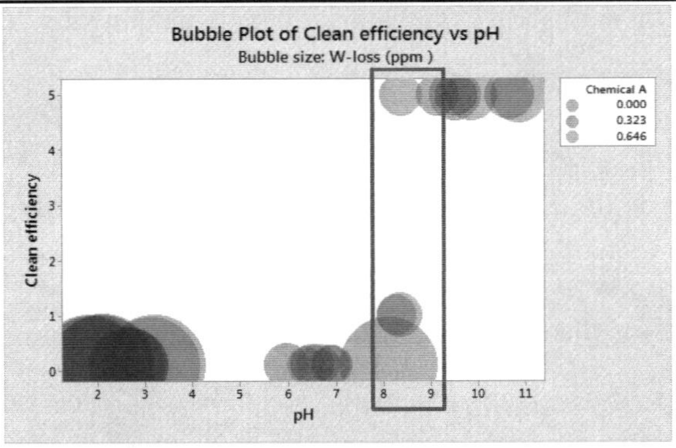

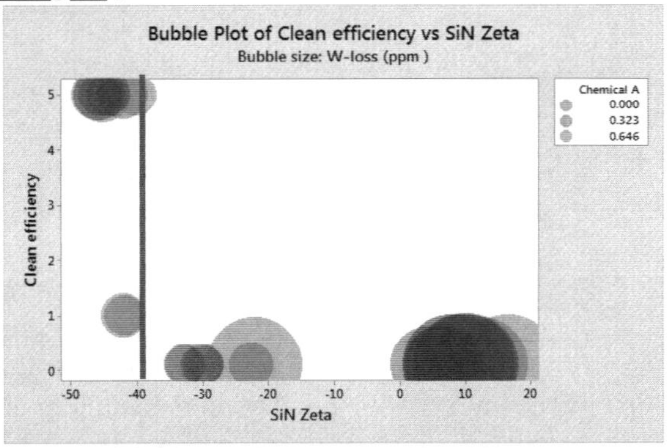

- We could find lower W-loss and better clean efficiency on pH 8 to 9. If zeta potential of SiN is less than -40 mV, it will showed better clean efficiency in the formulations. We also observe same trend on SiO_2 when zeta potential is less than -38 mV. Zeta potential effect maybe come from pH effect.

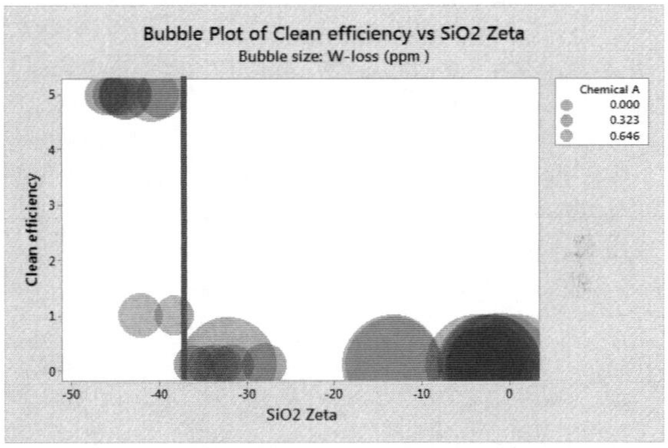

Summary

- ➢In the high pH region, lower adhesion force between polystyrene and Si_3N_4 will result in better clean efficiency.
- ➢ **Chemical A** could show good tungsten compatibility in the high pH region.
- ➢ If the zeta potential of Si_3N_4 & SiO_2 is less than -40 mV, the formulations yielded better cleaning efficiency

14

TEL
TOKYO ELECTRON

Inpria

Potential Challenges of Metal-Oxide Based Photoresist and Subsequent Rework Removal

Benjamin L. Clark[1], Michael Kocsis[1], Michael Greer[1], Andrew Grenville[1], Takashi Saito[2], Koichi Hontake[2], Lior Huli[2], Richard Farrell[2], David Hetzer[2], Hiroie Matsumoto[2], Andrew Metz[2], Shan Hu[2], Fitrianto[2], Jeffrey Smith[2], Shinichiro Kawakami[3], Koichi Matsunaga[3], Masashi Enomoto[3], Jeffrey Lauerhaas[4], Anthony Ratkovich[4], David DeKraker[4]

Inpria Corporation[1]
TEL Technology Center, America, LLC[2]
Tokyo Electron Kyushu Limited [3] / TEL FSI, Inc.[4]
April 19th 2016

EUV Photo Resist Design Principles

Small Molecular Building Blocks

Photocondensed Molecular Metal Oxides

High EUV Absorbance

Robust Etch & Mechanical Properties

Inpria

TEL

EUV Photo Resist Development Strategy

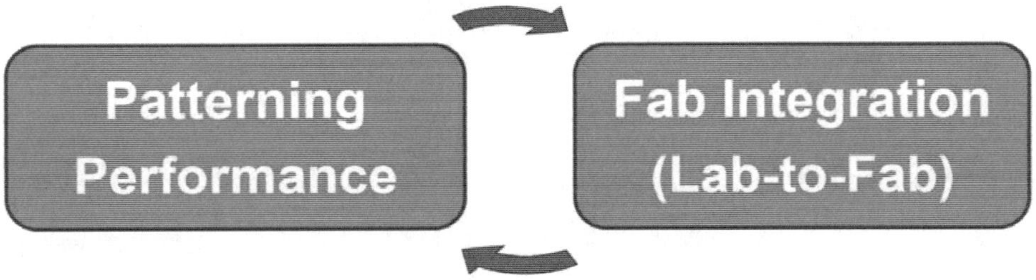

Working with equipment, materials, consortia, university, and device manufacturer partners

Transition From Lab-to-Fab

- Inpria Y-series photoresists contain organometallic complexes, which are soluble in commonly used fab solvents
- Demonstrating compatibility with fab equipment and processes is critical for integration in the fab
- Demonstrate:
  - Zero cross-contamination
  - Film coating and uniformity
  - Film defectivity analysis
  - Metal residue detection on Si backside and EBR region
  - Etch selectivity relative to conventional CAR
  - **Ability to rework without surface degradation**
    - **Particles**
    - **Metals**

Cross-contamination check

Experiment schematic:

Monitor wafers, processed through track, **PRE**-resist install

Coated 50 wafers wafers with Inpria Y-series resist

Monitor wafers, processed through track, **POST**-resist install

Both sides of all monitor wafers analyzed with two techniques:

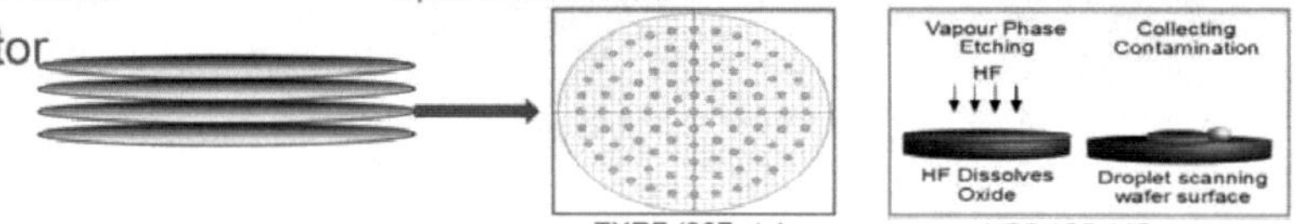

TXRF (397 pts)

VPD-ICP-MS

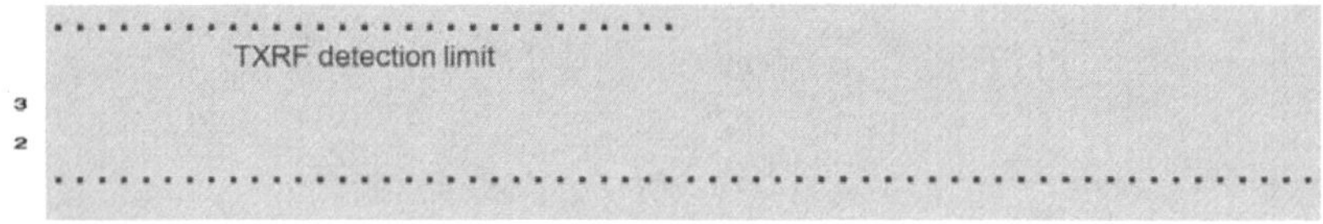

Typical fab spec

Inpria

TEL

Coating Metrics – Defectivity

- Defectivity improvement in resist manufacturing
  - Comparison between two resists synthesized with standard and improved filtration <u>without POU filter</u>

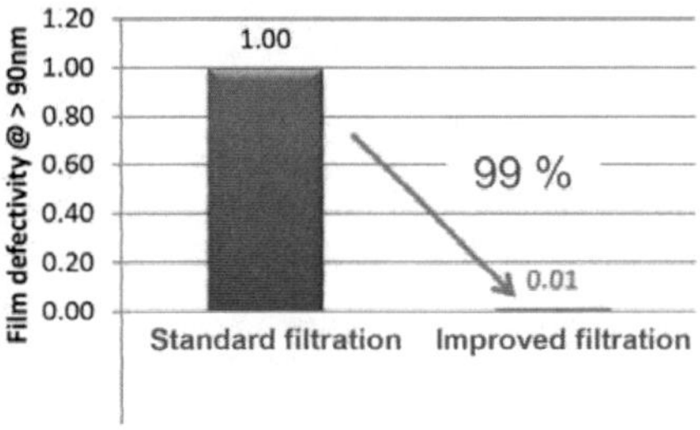

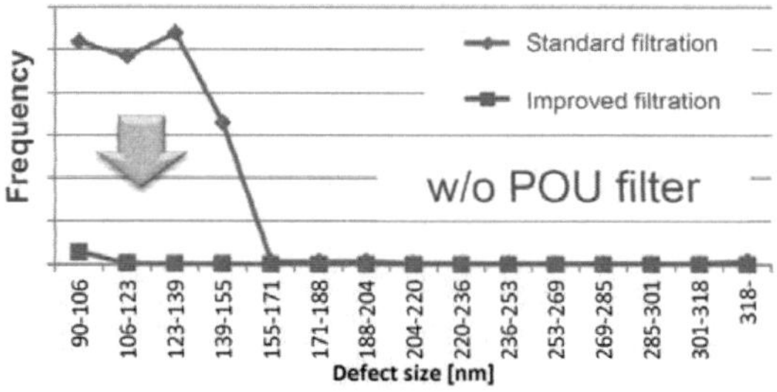

Improved filtration methods during resist manufacturing helps reduce film defectivity

Inpria

TEL

Coating Metrics – Defectivity

- Defectivity improvement by Coater / Developer
  - Comparison between standard dispense system and FEF (Filtration Enhanced Function) with POU filter

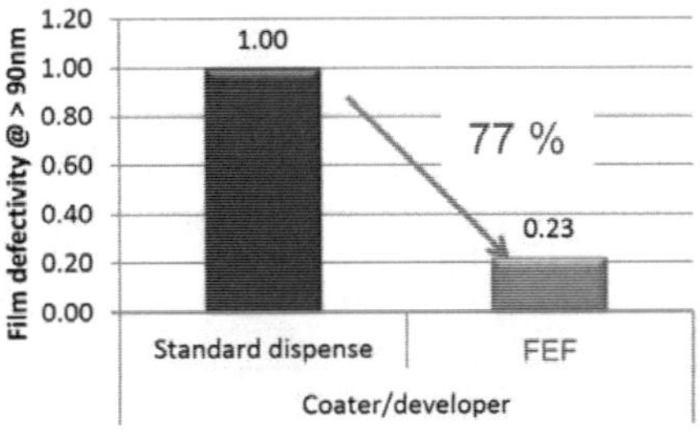

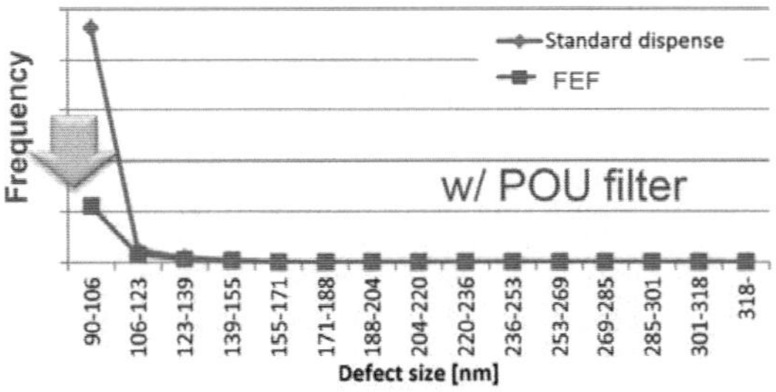

FEF on Coater / Developer system also helps to reduce film defectivity

Inpria TEL

Etch Rate Test

Etch rate comparison under standard Si-ARC/OPL etch condition

	Conventional CAR resist		Inpria YA series		Inpria YA series with hard bake	
	Initial	After etch	Initial	After etch	Initial	After etch
X-SEM images X300k	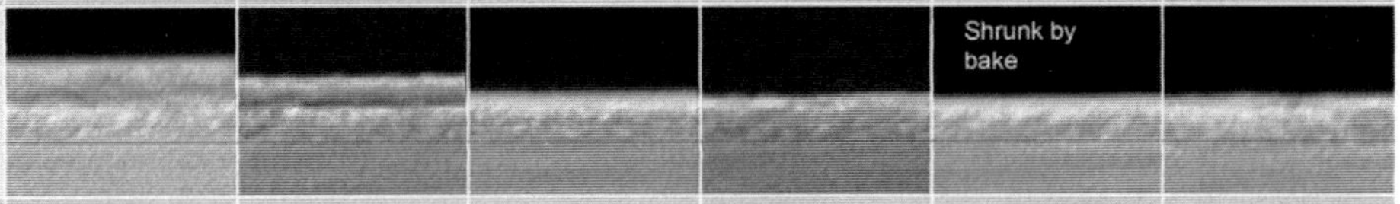				Shrunk by bake	
PR thickness	49 nm	30 nm	15 nm	9 nm	9 nm	7 nm
Initial – after	⊿19 nm		⊿6 nm	> 3 X	⊿2 nm	> 9 X

Inpria resist obtained more than 9X higher etch resistance compared to conventional EUV CAR resist under typical Si-ARC / OPL etch condition, same processing time.

Inpria TEL

TEL ORION™ –hp for Rework

- Unique closed chamber with novel ViPR™ process using SPM + Steam
- Aggressive strip process to remove resist and other organic containing layers with minimum process time & chemical use

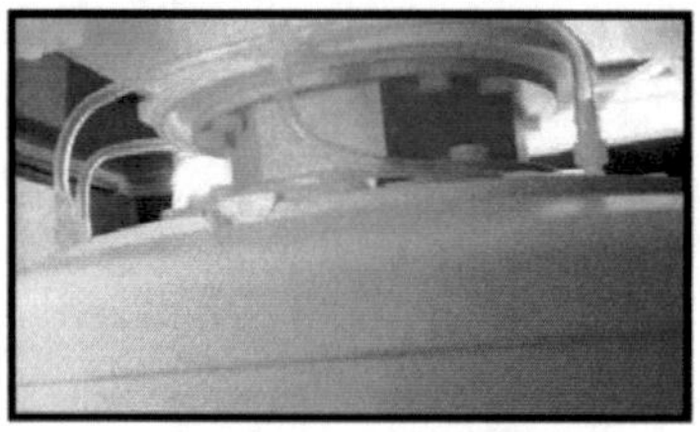

Challenge	Removal
High dose (>E15) implant DUV	✓
Tri-layer (Si-ARC up to 43% Si)	✓
Amorphous C	✓
Plasma Doped (PLAD)	✓

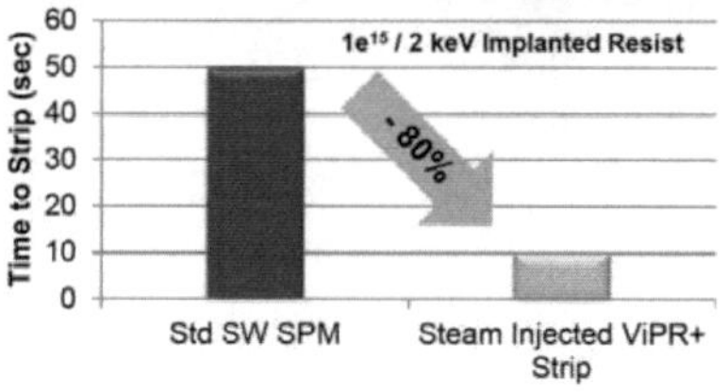

Inpria

TEL

Rework Test

ORION™-hp

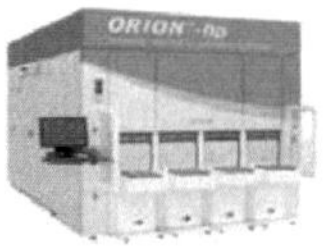

- **Resist strip test with steam injected SPM + SC1**
  - **Steam Injected SPM time varied; SC1 time fixed**

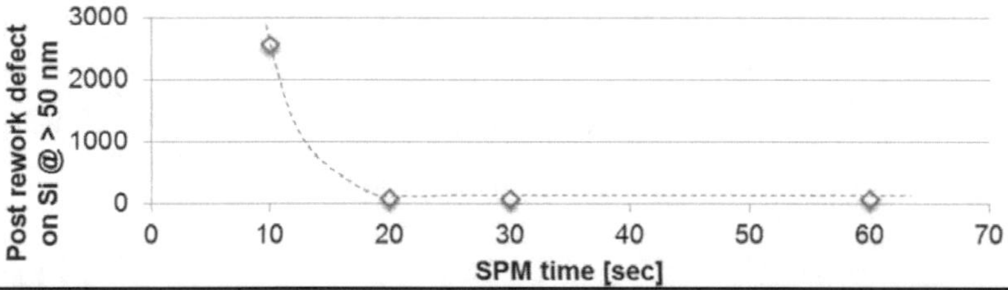

Resist component	Method	Detection Limit (E10 atoms/cm2)	Post rework			
			10 sec	20 sec	30 sec	60 sec
Metal	TXRF	5	< 5	< 5	< 5	< 5

- Inpria resist stripped successfully with steam injected SPM
  - Metal level below detection limit with ≥ 10sec process time
  - Defects at baseline levels with ≥ 20sec process time

Inpria

TEL

Conclusions

- Inpria resist compatibility with CLEAN TRACK™ LITHIUS Pro™ - EUV coater/developer has been demonstrated
- Inpria resist stripped successfully with steam injected SPM
- Metal level below detection limit with ≥ 10sec process time
- Defects at baseline levels with ≥ 20sec process time

Inpria TEL

Acknowledgements

- Special thanks to ASML for NXE3300 exposures
- TEL™ personnel at ASML and IMEC
- TEL ES / SPS personnel at TEL Technology Center, America, LLC / TEL FSI, Inc.

Inpria TEL

Thank you

TEL™

TOKYO ELECTRON

Effective Clean in 450mm Scrubber Tool for Throughput and Defect Removal Improvement

Chung Ju Yang[1], Jamie Prudhomme[1], Chih-How Huang[1], Takahiro Yamaguchi[2], Akira Morita[2], Eiji Kawabe[2]

[1]:Global 450mm Consortium, 141 Fuller Road, Albany, NY, 12203, USA

[2]: SCREEN Semiconductor Solutions Co., Ltd., Tenjinkita-machi 1-1,Teranouchi-agaru 4-chome, Horikawa-dori, Kamigyo-ku, Kyoto, 602-0087, Japan

ChungJu.Yang@g450c.org
TEL : +1-(518) 649-7568

Abstract

- **An effective brush clean method was demonstrated on a 450mm scrubber tool.**
- **450mm equipment development is a challenge in both technology and throughput to fulfill sub 10nm process requirements and cost reductions compared with a 300mm tool.**
- **The new brush clean method can significantly improve the particle removal efficiency and achieve higher process throughput than 300mm tools.**

New Design Concept

– **450mm scrubber tool implemented "Dual brush" as standard configuration.**

– **Both brushes move simultaneously for wafer cleaning during process.**

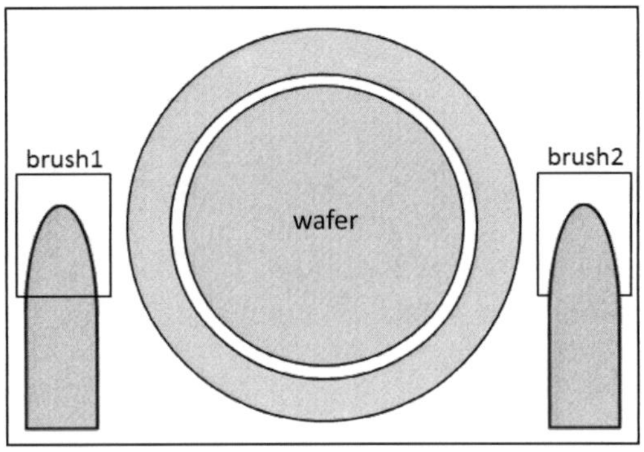

Process tuning factors :

- *Cleaning time*
- *Wafer rotation speed*
- *Flow rate*
- *Brush start/stop position*
- *Brush speed*
- *Brushing method*

Experiment

– **Particle removal efficiency (PRE) has been evaluated to identify wafer cleaning capability.**

  - **3 tests were performed with variable defect deposit conditions from easy to tough removal.**

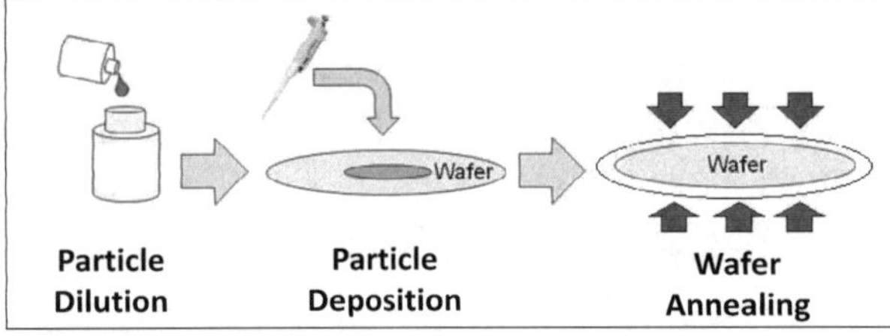

Control factors :

- *Defect type*
- *Defect size*
- *Dilution ratio*
- *Anneal temperature and time*

– **Long term particle performance has been collected to verify process stability.**

Experiment Result

G450C

– **Particle removal efficiency evaluation :**

- **Dual brush process shows +20% PRE improvement in tough defect samples compared with 450mm POR.**
- **450mm POR has benchmarked with 300mm in process performance, which indicates dual brush is potentially +20% better than a 300mm standard process.**

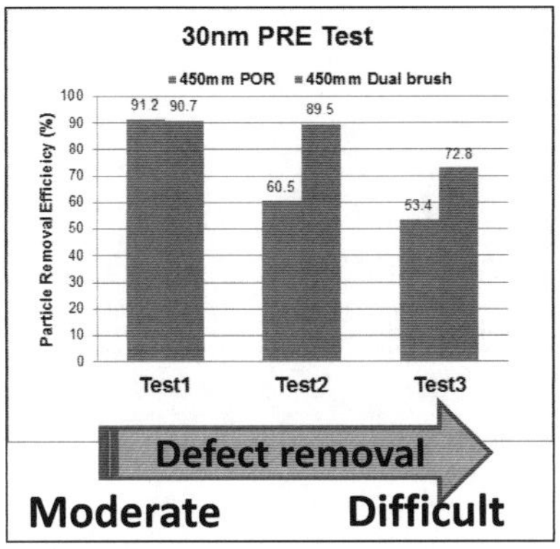

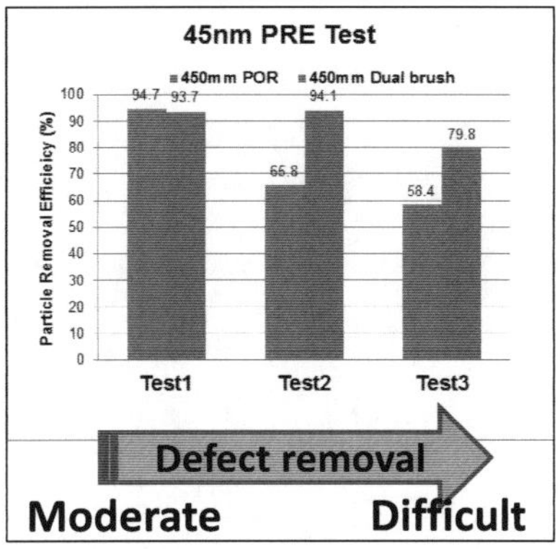

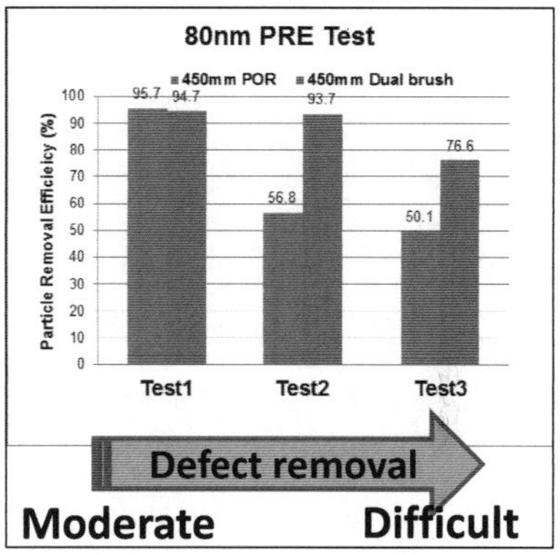

Experiment Result

G450C

– **Defect map for analysis:**

- **Dual brush process performed better particle removal on wafer edge area compared with POR.**
- **Overall PRE for the dual brush process are higher than POR, but dual brush process also found less particle removal in wafer center compared with the wafer edge area. Some process parameters need to be optimized for further improvement.**

	Defect map after deposition (Before clean)	Defect map after clean
450mm POR	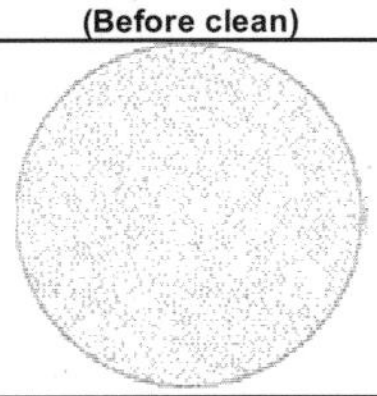	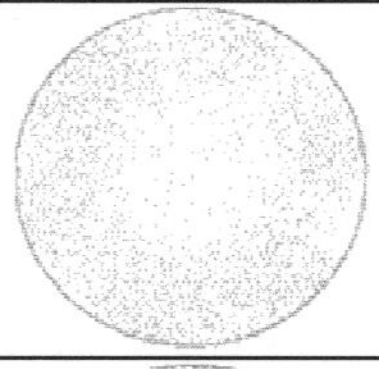
450mm Dual Brush	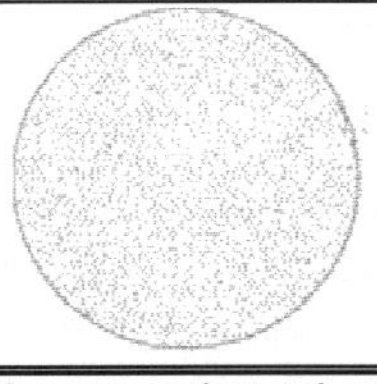	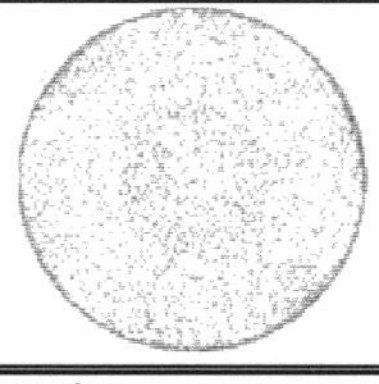

※ : Defect map on the tough removal sample

Experiment Result

– **Long term particle performance show comparable results between POR and dual brush process in 45nm particle adders.**

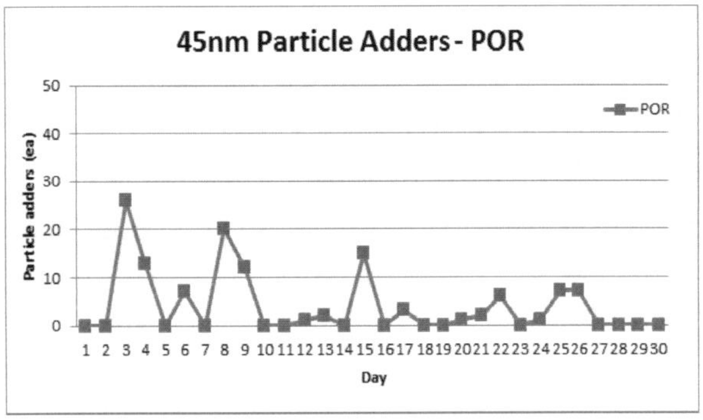

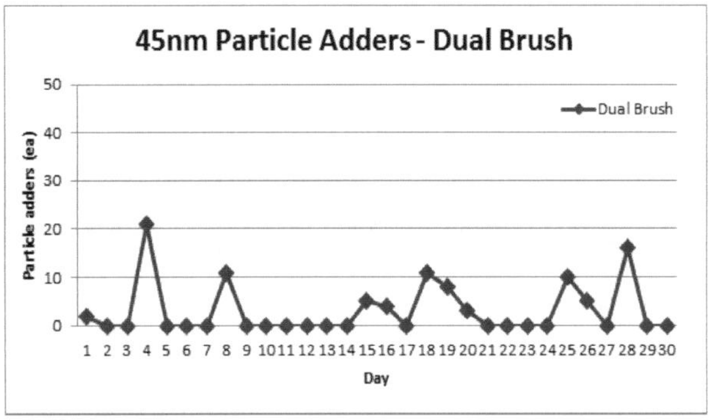

Throughput Assessment

– **Wafer Throughput Evaluation:**

- **Dual brush can significantly shorten process time with better process performance due to simultaneously brushing in both arms.**
- **WPH simulation on a 450mm HVM production tool (multi chambers) was performed based on real process time.**
  - **Dual brush can be 10% higher than 300mm POR and 50% better than 450mm POR.**

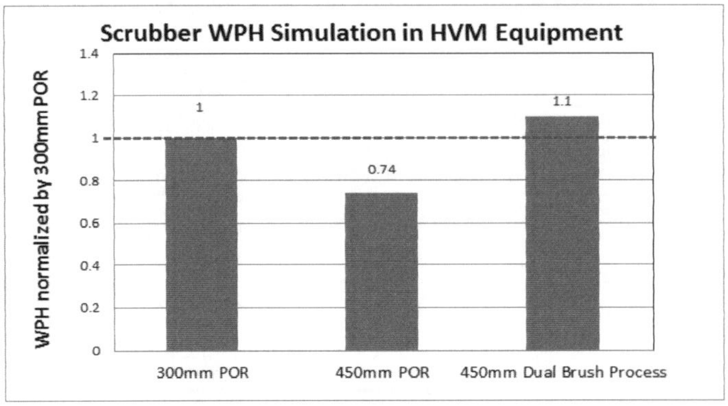

Conclusion

G450C

– Dual brush scrubber clean has demonstrated great potential for defect removal and wafer throughput improvement on 450mm tool.

– Further process tuning will be needed to optimize the particle removal in wafer center.

AUTHOR INDEX

AUTHOR INDEX

Surface Preparation and Cleaning Conference (SPCC)
c/o Mark Thirsk
Linx Consulting-PO Box 384
Mendon, MA 01756-0384
USA

CURRAN ASSOCIATES INC.
proceedings.com

ISBN 978-1-5108-2572-7
9 781510 825727